Information Sources in
Physics

A series under the General Editorship of
I. C. McIlwaine, BA, PhD, FLA
M. W. Hill, MA, MSc, MRSC, FIInfSc

*This series was known previously as 'Butterworths Guides to
    Information Sources'.*

Other titles available include:

Information Sources in Chemistry (Fourth edition)
    edited by R. T. Bottle and J. F. B. Rowland
Information Sources in Grey Literature (Third edition)
    by C. P. Auger
Information Sources in the Environment
    edited by Selwyn Eagle and Judith Deschamps
Information Sources in Finance and Banking
    by Ray Lester
Information Sources in Music
    edited by Lewis Foreman
Information Sources in the Life Sciences (Fourth edition)
    edited by H. V. Wyatt
Information Sources in Engineering (Third edition)
    edited by Ken Mildren and Peter Hicks
Information Sources in Sport and Leisure
    edited by Michele Shoebridge
Information Sources in Patents
    edited by C. P. Auger
Information Sources for the Press and Broadcast Media
    edited by Selwyn Eagle
Information Sources in the Medical Sciences (Fourth edition)
    edited by L. T. Morton and S. Godbolt
Information Sources in Information Technology
    edited by David Haynes
Information Sources in Pharmaceuticals
    edited by W. R. Pickering
Information Sources in Metallic Materials
    edited by M. N. Patten
Information Sources in the Earth Sciences (Second edition)
    edited by David N. Wood, Joan E. Hardy and Anthony P. Harvey
Information Sources in Cartography
    edited by C. R. Perkins and R. B. Parry
Information Sources in Polymers and Plastics
    edited by R. T. Adkins
Information Sources in Economics (Second edition)
    edited by John Fletcher

Information Sources in
Physics

Third Edition

Editor

Dennis F. Shaw

London · Melbourne · Munich · New Jersey

Series editor's foreword

As is obvious, any human being, faced with solving a problem or with understanding a task, reacts by thinking, by applying judgement and by seeking information. The first two involve using information from the most readily available source, namely one's memory, though searching this and retrieving what is wanted may be conducted below the level of consciousness. If this source does not provide all that is needed then the information searcher may turn to all or any of three external sources: observation (which can consist simply of 'going and looking' or can involve understanding sophisticated research); other people, who may be close colleagues or distant experts; and stores of recorded information, for example a local filing system or an electronic databank held on a computer network or even a book or journal in a national library.

The order (observation; other people; recorded information), is not significant though it is often a common sequence. Certainly it is not intended to impute an order of importance. Which of the three or what combination of them one uses depends on a number of factors including the nature of the problem and one's personal circumstances. Suffice to say that all three have their place and all three are used by every literate person.

Nowadays the amount of information in any field, even if one can exclude that which has been superseded, is so large that no human being or small group of people can hope to know it all. Thus, company information systems, for example, get bigger and bigger even when there are efficient means of discarding unwanted and out-of-date information. Managing these systems is a complex and full time task.

Many factors contribute to the huge information growth and overload. Throughout the world, large amounts of research continue to be undertaken and their results published for others to use or follow up. New data pours out of the financial markets. Governments keep passing new legislation. The law courts keep generating new rulings. Each organization and everyone in it of any significance, it seems, is in the business of generating new information. Most of it is recorded and much of it is published.

Although there is a growing tendency to record information in electronic media and to leave it there for distribution via electronic networks of one sort or another, the traditional media are still in use. Even tablets of stone are still used in appropriate circumstances but, of course, it is paper that predominates. The electronics age has not yet led to any reduction in the amount of printed material being published.

The range of types of published or publicly available information sources is considerable. It includes collections of letters, monographs, reports, pamphlets, newspapers and other periodicals, patent specifications, standards, trade literature including both manufacturers' product specifications and service companies' descriptions of their services, user manuals, laws, bye-laws, regulations and all the great wealth of leaflets poured out by official, semi-official and private organizations to guide the information in other than verbal form; maps, graphs; music scores; photographs; moving pictures; sound recordings; videos. Nor, although their main content is not published information, should one forget as sources of information collections or artefacts.

In an attempt to make some of the more frequently needed information more easily accessible, these sources of primary information are supplemented with the well-known range of tertiary publications, and text books, data books, reviews and encyclopedias.

To find the information source one needs, another range of publications has come into being. To find experts or organizations or products there are directories, masses of them, so many that directories of directories are published. To find a required publication there are library catalogues, publishers' list indexes and abstracting services, again a great many of them.

For librarians and information specialists in the industrialized countries, access to abstracting services is now normally achieved online, i.e. from a computer terminal over the telephone lines to a remote computer-based database host. Since 1960 the use of libraries by information and even document seekers has changed considerably and can be expected to change further as the British Library study Information 2000 indicates. More and more primary information is being stored electronically and more and more copies of printed documents are supplied via telephone or data networks. Sets of newspapers or other major publications can be acquired on optical discs for use in-house. Scientists in different universities, perhaps working on a common project, are sharing their results via the medium of electronic bulletin boards. The use of electronic messaging systems for disseminating information is now commonplace. Thus the combination of computer technology with telecommunications engineering is offering new ways of accessing and communicating information.

Nevertheless, the old ways continue to be important and will remain so for many years yet.

The huge wealth of sources of information, the great range of resources, of means of identifying them and of accessing what is wanted increase the need for well aimed guides. Not all sources are of equal value even when only those well focused on the required topic are considered. The way new journals proliferate whenever a new major topic is established, many it seems just trying to 'climb on the bandwagon' and as a consequence substantially duplicating each other, illustrates this. Even in an established field the tendency of scientists, for example, to have a definite 'pecking order' for selecting journals in which to publish their research is well known. Some journals submit offered articles to referees; some others publish anything they can get. Similar considerations apply to other publications. The degree of reliance that can be placed on reports in different newspapers is an illustration. Nor is accuracy the only measure of quality. Another is the depth to which an account of a given topic goes.

The aim of this Guides to Information Sources series is to give within each broad subject field (chemistry, architecture, politics, cartography etc.) an account of the types of external information source that exist and of the more important individual sources set in the context of this subject itself. Means of accessing the information are also given though only at the level of the publication, journal article or database, individual chapters are written by experts, each of whom specializes in the field he/she is describing, and give a view based on experience of finding and using the most appropriate sources. The volumes are intended to be readable by other experts and information seekers working outside their normal field. They are intended to help librarians concerned with problems of relevance and quality in stock selection.

Since not only the sources but also the needs and interests of users vary from one subject to another, each editor is given a free hand to produce the guide which is appropriate for his/her subject. We, the series editors, believe that this volume does just that.

Michael Hill
Ia McIlwaine

About the contributors

Heinrich Behrens
Graduated in Physics at the Technische Hochschule, Munich 1962 (Dipl.-Phys) and took his doctor's degree in Physics at the Technische Hochschule, Karlsruhe 1966. From 1962 to 1974 he worked at the Institute for Experimental Nuclear Physics of the Nuclear Research Center, Karlsruhe, first as a scientist, later on as group leader. His main area of research was the field of beta-decay. In 1975 he changed to the FIZ Karlsruhe. Here he was involved in the establishment of data compilations and of computerized databanks, mainly in the field of physics. Dr. Behrens is now Head of the Department of Physics and Astronomy at FIZ Karlsruhe, and in this function is responsible for the production of databases in physics.

Robert Berman
A Cambridge graduate, he worked in Oxford on the war-time project for separating uranium isotopes by gaseous diffusion, being the first person in the country to separate the isotopes in this way. He later worked in Oxford on thermal properties of solids. He edited *Physical properties of diamond* (1965) and published *Thermal conduction in solids* (1976), as well as many articles in scientific journals. He was University Lecturer in Physics and is now Emeritus Fellow of University College, Oxford.

John Chillag
Is currently Information Officer at the Mid Yorkshire European Information Centre, Leeds Metropolitan University. Until his retirement in 1990 from the British Library, he spent 27 years at the present Document Supply Centre in acquisition and handling of grey literature. He was involved with *SIGLE* during its formative years. He has published extensively on grey literature topics.

Stephen van Dulken
Is a chartered librarian and an Open University graduate. He has been working at the British Library since 1974, since 1987 specializing in patent documentation. He edited *Introduction to patents information* and the last two editions of *International guide to official industrial property publications*.

Eric Finch
Is a Senior Lecturer in Physics at Trinity College, University of Dublin. He obtained a DPhil from the University of Oxford, and has worked in several areas including the physics of nuclear radiation detectors and radiological protection. His book *Radiation detectors: physical principles and applications*, co-authored with C.F.G. Delaney, was published by the Clarendon Press, Oxford, in 1992. His teaching duties include giving lectures on waves and sound and (to music undergraduates) musical acoustics.

Nancy Fjällbrant
Graduated from the University of Edinburgh with a BSc in physiology. In Sweden, she turned to work in the information sector, working as a lecturer at the Swedish College of Librarianship and at Chalmers University of Technology, Gothenburg. After taking a PhD at the University of Surrey, she became Head of the Library User Education Department at Chalmers University. Dr Fjällbrant is a well-known author on user education and the use of information technology in libraries.

Tore Fjällbrant
Studied at Chalmers University of Technology, Gothenburg where he received his MEng and DrEng degrees. He is at present Professor of Applied Electronics at the University of Linköping in Sweden. Professor Fjällbrant has specialized in signal processing and in particular speech transmission and recognition. He is currently working on models of the auditory system and biomedical applications.

Willem Hackmann
Is Assistant Curator at the Museum of the History of Science, University of Oxford and a Fellow of Linacre College. His interests range widely, from the history of scientific instruments, and the interaction between art and science, to twentieth-century warfare technology, in particular sonar and radar. His publications include *Electricity from glass* (1978), *Seek and strike. Sonar, anti-submarine warfare and the Royal Navy 1914–54* (1984), and *Scientific amusement arcade* (1988), the catalogue of an interactive science exhibi-

tion for children commemorating the 150th Meeting of the British Association for the Advancement of Science which took place at Oxford.

Julie Hurd
Earned the MS in Physical Chemistry from Michigan State University and the PhD in Theoretical Chemistry from the University of Chicago. She has done interdisciplinary research in theoretical physical chemistry and in geochemistry before earning the MA in Library Science from the University of Chicago. She has taught courses in science information and online searching in library schools and has worked in science libraries at the University of Chicago, Michigan State University, and the University of Illinois at Chicago, where she is currently Science Librarian. She is an active member of the American Society for Information Science and the American Library Association, and has publications on topics related to scientific information and use of information technology.

Manfred Kremer
Graduated from the University of Heidelberg in the field of Biophysics. He obtained a PhD in Theoretical High Energy Physics from the University of Heidelberg and then worked at the Universities of Heidelberg and Mainz before joining the German Supercomputer Centre HLRZ in 1988.

Elizabeth Marsh
Founded her own information consultancy in 1988 specializing in the acquisition, analysis and distribution of information including market research, user requirements and the technology base. She was previously Library Manager with the *Financial Times* and Librarian with the Rutherford Appleton Laboratory of the Science and Engineering Research Council.

Robert Michaelson
Is Head Librarian of the Seeley G. Mudd Library for Science and Engineering at Northwestern University, Evanston, Illinois, USA. He completed a PhD in Physical Chemistry at Yale University and an MA in Library Science at the University of Chicago. He is the contributor for the chemistry and physics sections of the *Guide to reference books* (American Library Association).

Robin Nicholas
Is a Reader in Physics at Oxford University. He runs a research group looking at the magneto-optical and magnetotransport properties of a wide range of semiconductor heterostructures. He

graduated and completed a DPhil at Oxford University, and has spent periods working in Grenoble, Stuttgart, Leoben (Austria) and Tokyo.

Steve O'Neale
Is responsible for the offline computing in the data acquisition and analysis section of the OPAL experiment at LEP. He has promoted the use of heterogeneous workstation systems and third party disks and tapes for economical computation and data access. He completed his PhD in particle physics at Birmingham University in 1978 and has worked on quality control and data management for bubble chamber and counter experiments at CERN, FERMILAB and SLAC.

Philip Ponting
Is a corporate member of the Institution of Electrical Engineers and a Chartered Engineer who has been a staff member of the European Organization for Nuclear Research (CERN) in Geneva since 1963. He leads a group developing electronic systems for high energy physics experiments and is a member of the Executive Electronics Board of the CERN Research Divisions. He has been involved in Standards activities for many years and is currently a member of the Executive Group and Secretary of the Committee for European Studies On Norms for Electronics (ESONE), a body linking many of the major European research institutions in an effort to encourage the adoption of common practices in electronics and related domains. He is the author of a series of papers describing the origins, status and prospects of systems in current use in his field of technology.

Francis Pratt is currently a Royal Society Research Fellow at the Institute for Materials Research, Tohoku University, Japan. Since completing his DPhil in far infrared spectroscopy at Oxford University in 1985, Dr. Pratt has been working at the Clarendon Laboratory, Oxford, where he has carried out research on a wide range of materials. His main research interests are in electronic properties of organic metals.

Alan Rodgers
BSc PhD CPhys FInstP has retired from RMCS (Cranfield) where he held a Personal Chair in Directed Energy Applications. His main research interest was in laser beam interaction physics, and currently he is a consultant on laser safety.

Dennis Frederick Shaw
CBE MA DPhil FInstP is a Fellow of Keble College, University of Oxford, a Chartered Physicist and former Keeper of Scientific Books in the Radcliffe Science Library. For twenty-five years he lectured to undergraduates on electronics and has published research papers on experimental nuclear and particle physics and instrumentation. During the years 1986–90 he was President of the International Association of Technological University Libraries.

Robert Wyatt
Gained a postgraduate diploma in Librarianship from the University of Wales in 1980. He has subsequently pursued his career in librarianship at the Radcliffe Science Library, Oxford. After working in readers' services and as a cataloguer he became Serials Librarian in 1989.

John Ziman
Is Emeritus Professor of Physics of the University of Bristol. He was brought up in New Zealand, studied at Oxford, and lectured at Cambridge, before becoming Professor of Theoretical Physics at Bristol in 1964. His researches on the theory of the electrical and magnetic properties of solid and liquid metals earned his election to the Royal Society in 1967. Voluntary early retirement from Bristol in 1982 was followed by a period as Visiting Professor at Imperial College, London, and from 1986 to 1991 as founding Director of the Science Policy Support Group. He was Chairman of the Council and Society from 1976 to 1990, and has written extensively on various aspects of the social relations of science and technology.

Contents

Preface

Eight years have elapsed since the publication in 1985 of the second edition of *Information sources in physics*. The authors of the chapters in this new edition (and the publisher) have thus satisfied one of our reviewers who expressed the hope 'that Butterworths will produce a third edition of this worthwhile work in less than the ten years which were allowed to elapse between first and second editions'. We hope we have satisfied the many other reviewers who welcomed the second edition as a definitive guide to the literature of physics and 'the standard work ... for both the librarian and the physicist'.

The format of the third edition follows closely that chosen for the second although there have been several significant changes to encompass recent developments in physics and related subjects. There are two new chapters on quantum optics and physics of materials which recognize the rapid growth in importance of laser physics and materials science; and eight of the remainder have been completely rewritten by new authors. The chapter on patents has been rewritten and now covers the important areas of European and international patents. With two exceptions (see below) the other chapters have been fully revised and updated to include the most important works published since 1984 and to eliminate obsolete or outdated editions.

The two exceptions reflect major changes on the one hand in the development of the subject and on the other in the development of new media for the supply of information in physics. I am grateful to Professor Leonardo Castillejo and Dr Ian Aitchison for their valuable comments by correspondence and in discussion which led to the decision to exclude the chapter on theoretical physics preferring to transfer the relevant material to the other subject chapters where they fit much better. To quote from Professor Castillejo 'on one hand the enormous growth of physics has compartmentalised the subject so that each area has its own literature and favoured publications, while on the other hand recent development of many of the fundamental new theoretical ideas have spanned several fields. This makes the separation into *subject specific* and *theoretical* particularly hard'. For these reasons, authors of the individual chapters have included the more important theoretical aspects in their selection of source material. This has resulted in a significant expansion of the literature coverage in the chapters on condensed matter physics (experimental heat and low temperature physics, materials physics and semiconductor physics) as well as those on mechanics and acoustics, nuclear and particle physics, quantum optics and spectroscopy.

The other factor which has influenced the presentation has been the growth in importance of digitally-stored information sources both online and on disc. This has resulted in the inclusion of references to these sources in all the subject chapters as well as in the chapter on grey literature. The separate treatment of online services was thus deleted, and the material on abstracting and indexing services was adequately covered elsewhere.

A further change is the omission of the name index. The version provided in the second edition served a useful function in scientometric analysis, but did little to assist the reader seeking an information source. Thus, the value of this item did not justify the effort required and the expense of compiling and printing. In compensation the contents section and subject and title index have been greatly expanded. The reader is recommended to use these two sections of the book to become familiar with the contents.

The problem of limiting the size of this book to manageable proportions has been aided by the existence of related works in the series, particularly *Information sources in energy technology, Information sources in metallic materials, Information sources in the earth sciences, Information sources in the life sciences* and *Information sources in engineering*. For this reason there is only passing reference in this work to the following subjects: biomedical engineering, biophysics, energy research, physics of the environment, medical physics, nuclear engineering, physics instrumentation, and rheology.

A work which is the combined effort of twenty authors inevitably will contain some duplicated references. In editing the chapters the attempt has been made to eliminate unnecessary duplication. However, it is distracting for the reader to be directed frequently by cross-reference to another chapter for an essential item, and some standard works will therefore be found listed in several parts of the book. Each item is described briefly but in sufficient detail to provide unique identification, thus enabling the reader to trace it through any good library catalogue. Accuracy and completeness of the bibliographic elements have been checked by reference to standard bibliographies, particularly R.R. Bowker's *Books in Print* and the INSPEC *List of Journals and other Serial Sources 1992/93*. The international coverage of the work can be confirmed by reference to the addresses of publishers listed in Appendix C. This Appendix is important in providing the full name in cases where an abbreviation is used in the text. Some of the works cited are now known to be out of print and therefore may be found only in national and other major libraries.

Besides the collective wisdom of the authors, whose ready co-operation and generous gift of time is gratefully acknowledged, there are several persons who have contributed valuable expertise and time to assist in the production of this book. The staff of INSPEC have supplied invaluable information from their computer database and have responded to several enquiries. The British Library at SRIS has helped to fill gaps in reference sources not now available in Oxford and I am particularly grateful to Mr Alan Gomersall. The staff of the Radcliffe Science Library have given much valuable advice on bibliographic searches and have traced several new publications which were not listed at the time of drafting the text and this service is gratefully acknowledged. Miss Barbara Challoner has made a valuable contribution in keyboarding the contents of two major chapters as well as assisting with the production of the Appendices. My wife was spared the effort of compiling the indexes for this edition due to the sophistication of the software on my personal computer but she has contributed much valuable advice as well as showing great tolerance in enabling me to complete my task within the publishers deadlines.

Finally, the cooperation of the publishers in the planning and production of this edition is much appreciated. The series editors have given careful consideration to all the proposed changes and made valuable suggestions both for inclusion and exclusion of material. The sub-editor has provided an important service in standardizing the presentation and correcting omissions overlooked by the Editor.

Oxford, August 1993

CHAPTER ONE

Introduction

J. M. ZIMAN

Where might one expect to find 'Physics' but in its literature? The knowledge of any one physicist is the minutest fraction of all that is written down, somewhere, recording the observations, opinions, conjectures and measurements of hundreds of thousands of people. The apparatus by which this information was obtained is merely an assembly of mechanical devices, of no significance out of the context of the knowledge to which they can be made to contribute.

What indeed, is any science but a body of public, organized *knowledge*. Apart from the actual use of this knowledge in the affairs of men – which may be the prime employment of many readers of this book – the goal of scientific research is to collect information about the natural world, and to structure this information into a coherent, comprehensible form. The classical philosophical analysis of scientific method stops short, however, at the moment when the scientist, having arrived at some brilliant discovery, 'writes it up', for the benefit of an expectant world, and sits back waiting for the applause and bouquets. The truth is that new scientific information, however original and earthshaking, does not become scientific *knowledge* until it has been communicated, criticized, verified and remains unfalsified, by those sceptical vultures – other scientists in the same field. We have come now to realize the very great importance of the *communication system* of science as the instrument by which the results of individual private research projects are transformed into the *social* product of scientific knowledge.

The nature of this communication system is thus of the highest significance, especially for those who are professionally committed to basic research, or who have great need of particular types of scientific knowledge for technical application. The government bureaucrat without experience of research may imagine that a large, expensive, but essentially automatic computerized index will do all that is needed: every graduate student, grappling with the background bibliography of his PhD thesis topic, quickly learns differently. But not every high and mighty professor, with a hundred papers to his/her name, grasps the full subtlety and complexity of the system as it has evolved over 300 years.

The title of this book, which specifies *information* sources (whereas the first edition specified *literature*), recognizes the development of non-printed sources which were acknowledged for the second edition but have increased even more in importance during the last nine years since the second edition was published (as we shall see below). However, many things that 'every physicist knows' are never written down but have been passed on, by word of mouth, from generation to generation. It is almost impossible to describe the precise manual procedures employed by a skilled craftsman: we learn them only by attempting to copy them under his personal supervision. Every good laboratory has its wise person, who can tell you the answer to a practical question that may never have been discussed at all in the official literature. In a complete account of the communication system we should also take into account all those little chats, private letters, seminars, conference talks, etc., which occur in the normal social contacts between scientists in the same discipline and which play such a very great part in the early stages of the formulation of a new scientific idea. The substance of some of these informal verbal communications does often get into print eventually, but long after the moment when they were truly active upon their recipients. What Einstein wrote to Bohr in 1927 is of great historical, biographical and philosophical interest, but it is not part of the current literature of physics in the narrow technical sense.

Nevertheless, the 'official' literature of physics is of perplexing diversity. A familiar error is to suppose that only the 'primary' literature counts – those innumerable, unintelligible, highly specialized accounts of particular little researches that are thrown together, quite without order, between the covers of the *'primary' journals*. To compound this error, it is widely held that only a very few leading journals really count – that if it has not appeared in the columns of the *Physical Review* it is not genuine physics, or can be disregarded for all practical purposes.

The fact is, however, that primary physics research is published in a large number of different journals, in many languages. It is quite true that various social processes tend to concentrate the work of the most active scientists in any given field into quite a small number – a few dozen, say – of internationally recognized journals, but this does not mean that a search for particular knowledge in the other journals is invariably useless. If we are not to go over to a monolithic, totally inflexible system of scientific publishing in which there is one and only one worldwide journal for each separate speciality, we must continue to insist that it is the *responsibility of every physicist to inform himself of the full literature of his subject, and to cite all relevant, reputable sources, at home and abroad.*

Notice, on the other hand, that there is very considerable specialization of topics, even in primary journals claiming all of physics for their province. This *Hatton Garden effect* (the district in London where the wholesale jewellers congregate) is a natural evolutionary process, which can be artificially accelerated by deliberate planning (for example, by the European Physical Society in its Europhysics journal scheme) but cannot easily be halted. The subdivision of some of the major journals of the big physical societies follows a similar trend. To 'know the literature' of a particular branch of physics is to know where one might expect to find the best new papers in that field, with perhaps an intuitive feeling for the scholarly style that is likely to be prevalent there, and a subtle grasp of the appropriate vehicle for some new paper that one would like to present to a particular circle of readers.

Prior to the primary journal in the public relations schedule of a new scientific discovery may come that bastard progeny of priority neurosis and reproductive technology – the *preprint.* Well, of course, it is very useful, occasionally, to know about some really good scientific idea a few months before it can actually be printed in hard type and distributed across the oceans by slow old mail boats, but this indiscriminate, *ad hoc* private distribution of badly printed versions of uncriticized scientific work evidently wastes a great deal of money, and dissipates the pressure that would otherwise build up to keep the official *public* journals running more efficiently. For my part, I do not count preprints as belonging to the literature of physics, and value 90 per cent of those I receive chiefly as good scribbling paper!

The *letters journal* is more legitimate, but is often weakened by the attempt to perform two functions simultaneously. There is a need, indeed, for very rapid publication of preliminary brief reports of outstanding and essentially revolutionary new discoveries. There is also some need for a medium for the publication of quite short communications – not necessarily of great significance, but sufficiently self-contained and relevant to be useful to specialists in that field. Unfortunately, these functions are often confounded; most 'letters' in this type of journal are not complete in themselves, and yet do not promise such marvellous consequences that they deserve publication in advance of an ordinary primary paper in which all details would be made clear. But here, perhaps, we see most clearly the effects of the human pressure to get into print with original work as early as possible, without regard to personal reputation for clarity, accuracy or completeness of proof, and for our own needs as readers of all this stuff.

Even if we cannot expect all the scientific virtues from every scientific paper we are forced to read, we do insist on a certain level of scholarly accuracy and credibility. Indeed, a most extraordinary feature of all scientific research is the immense trust that we must put in the honesty and good faith of other scientists. We must simply assume, in the first instance, that when they say they have measured some property, or evaluated some formula, then this is literally true. Of course, after a little while, by critical comparison with our own results, or by contradiction with other equally sincere results quoted by other authors, we may come to doubt the accuracy of their work, and curse them for careless fools – but we know very well that reliable scientific results are very hard to achieve, and we take it for granted that they did their best.

This child-like trust in fallible, corruptible humanity is preserved by the *referee system*. When primary scientific journals were first invented, at the end of the seventeenth century, they carried a splendid miscellany of strange observations, factual or fictional, controlled only by the acumen or prejudiced judgement of the editor. But since these journals were usually published by learned societies, the privilege of publishing pretty much what they liked in them was restricted to the members of each society, or to their friends and clients who would 'communicate' through them. There still remain one or two journals of this kind – and there have been a few cases where a very distinguished scholar, who has gone a bit gaga in his old age, has given a very idiosyncratic twist to the publications in journals under his thumb. I am told, moreover, that the system of quality control in the Russian scientific literature has depended more on the authority of the director of each author's institution than on the judgement of independent referees. This may change now as a result of *perestroyika*.

Generally speaking, however, a primary paper in a reputable physics journal has been subjected to scrutiny by one or more experts – almost always anonymous to the author – before being accepted for publication. The conflicts that can arise between authors and referees are legendary and searing to the spirit, but the experience of sitting on the editorial board of a journal and reading both sides, will soon convince one that this system performs an essential service. This is the main bulwark against the publication of all manner of uncritical, irrational, essentially misleading and unsound work purporting to be science. Without some such explicit criticism, there would be no standards at all in research. The idea that good physics is, so to speak, guaranteed by the use of expensive equipment and complicated equations by people with PhDs from accredited institutions is all nonsense. Research is an art that can be done very badly indeed, and the literature would be useless as a repository of scientific knowledge if it were not accepted that every paper that appears in a scholarly journal had been judged at least superficially sound by competent experts. The price to be paid when the experts mistakenly reject or delay an occasional good paper is negligible by comparison with the confusion, lack of confidence and general decline of scientific standards that would ensue if the anonymous referee system were simply dropped.

As I have indicated, the difficulty of doing good physics is so great that a high proportion (say 90 per cent) of the papers that get published do not prove of permanent value. How does this process of purification and recrystallization of the scientific literature actually occur? By what means are students, and others unfamiliar with a field over many years, guided to the choice of the truly significant works, containing the currently agreed knowledge of the topic in question? This is the function of the *review article*, the *monograph* and the *textbook*. The task of bringing together the results of primary research papers and presenting some unified account of progress in the solution of some particular scientific problem is of much greater importance and responsibility than is given credit for in the reward system of the scientific profession. It is recognized to be something of an art to do well, but by a shallow psychological assessment is accorded little originality by comparison with any petty application of conventional techniques to a 'new' primary research problem.

Of course, if a review is merely a descriptive bibliography of all papers on the subject, ordered by crude categories such as 'Experimental', 'Theoretical', 'Low Energy Experiments', 'High Energy Experiments', etc., then it demands little more than assiduity and a well-kept card index. This is the intellectual level at which such mechanical systems as abstract journals, information-retrieval systems and current-awareness and alerting services can work with maximum efficiency. It must be emphasized, however, that despite their great utility as tools for *cataloguing* the primary literature of physics, and thus saving the immense labour of searching for potentially relevant material, these devices do not, and cannot by any stretch of the imagination, act as a substitute for the intellectual work of making sense of what has been published, accentuating what is acceptable and rejecting what can no longer be considered worth attention.

There is a tendency to treat 'review articles' as a single category in the literature, and yet they are a genus with many species. Quite apart from specialization by field – nuclear physics, solid state physics, etc. – they must be considered both as to audience and to aim. One journal, for example, may be an excellent medium for rather speculative reviews, very close to the primary literature and to current controversies, flying the kite of some particular theory over a somewhat wider range of potentially relevant evidence than would normally be allowed in an ordinary primary paper. Another type of review that is similarly addressed to the specialist research worker in a particular field might be a very careful compilation of experimental data, critically assessed and reduced to standard form for reference. Or it might be a formal account of a mathematical technique, bringing together and expounding implicit axioms, theorems and potentially useful formulae.

The next stage would be the specialist 'critical review', which would attempt to cover all the published literature on some particular problem over some particular period, indicating those points on which there is agreement, attempting to resolve contradictions of observation or theory, and setting out the elements of conflicting, controversial interpretations or explanations. Such a review is again usually intended for experts – and calls, of course, for considerable tact, generosity of spirit and intellectual firmness. If there is any meaning to the title of being an 'authority' on a subject, it must signify the ability to write such a review without losing either the respect or the friendship of one's scientific colleagues!

From the critical review, we move to the 'didactic' article, intended more for beginners in the subject, such as graduate students. Here the emphasis is on the clear exposition of what is well understood, in language intelligible to the relative outsider. This calls for special skills of style and thought, and the will to resist the blandishments of current controversies and speculations. It is particularly important to cultivate a tone of scepticism concerning many fashionable doctrines: this is the point where the unwary student may pick up paradigms that have not really been fully tested, and yet fail to grasp various simple basic points that every experienced worker in the field knows to be sound.

This is the level at which the coherent *monograph*, covering a much wider field than a single review article, can be extremely valuable. The greatest weakness of the communication system of physics is the absence of any mechanism for the deliberate production of syntheses of knowledge over any field broader than the conventional specializations of individual research workers. Everybody will agree that such works are of the greatest value when they happen to get written, but there is no machinery for ensuring that appropriately qualified physicists actually embark on such projects and are appropriately rewarded at the end of them. Yet such a task is worthy of the very best efforts of any first-class scientist, over a period of years, and if carried out satisfactorily will make her a real authority over the whole field. It is true that symposia of separate reviews by independent authors are organized, collected together and printed as single volumes, under the guidance of energetic and learned editors, but these are usually lacking in intellectual coherence, and miss many valuable opportunities to make comparisons, draw parallels, carry over techniques from one problem to another, and generally to show that the world is not quite so complicated as the little bits of the jigsaw puzzle into which it is arbitrarily divided by scientific specialization.

Presumably, this is the point where one should mention a class of literature that occupies quite a proportion of the shelf space in any physics library – the miscellaneous volumes of *conference proceedings* printed and sold to the great profit of many commercial publishing houses but reflecting no great credit on those who conceive them scientifically. Superficially, it is an attractive idea to collect together and publish the papers presented by the galaxy of scientific stars that have gathered together for a week, from the four corners of the globe, on some delightful Aegean island, to scintillate to one another on, say, 'The Medium Low High Energy Electron-Proton-Meson Interactions in Galactic Clusters of Type 57b'. In practice, however, the 'contributed' papers are mostly shortened and inferior versions of work that is about to be, or has been published elsewhere, together with a few agreeable, racy, 'invited' papers which pretend to review the current situation but mainly convey the personal opinion of the speaker concerning the relative merits of his own and other people's contributions to the subject. Scientifically speaking, most volumes of this kind are almost worthless, and date rapidly, but they can, on occasion, be useful as the starting point for a proper literature search. If I really wanted to know about Electron-Proton-Meson Interactions . . . then I might glance through this conference report and note the existence of certain lines of argument, and go to the references cited in the various articles for a more accurate account. In other words, these are an expensive form of abstracting service, suitable for the lazy person who has not kept up with all the literature.

The typical undergraduate or graduate *textbook* is a different animal again, organized around the peculiar needs of a course of study of so many weeks at such and such a level. It is characteristic, again, of the chaotic incompleteness of the literature of physics in the book format that there are innumerable excellent textbooks on certain standard courses – for example, undergraduate quantum mechanics or statistical mechanics – but almost nothing on new subjects that are only just entering the university curriculum. The commercial textbook publishers, by ruthless competition, are partly responsible for this unsatisfactory situation. Each publisher insists that he must offer a complete list of books on the conventional topics (not realizing that academic physicists care nothing for an imprint, and only seek works by the best available author) instead of encouraging the publication of quite new books that will create their own fashion. And for any physicist who regards the writing of textbooks as mere drudgery, let me remind you of the historical fact unearthed by Holton (1969; 1973) – that Einstein got some of the essential elements of relativity theory from reading the perceptive textbook on electromagnetism by Föppl (1894), who explicitly pointed out the anomalies in Maxwell's treatment of a conductor moving in a magnetic field. Textbooks eventually determine the scientific viewpoint of the generation who read them as students: there is as much opportunity for influence over the progress of science in this task as in many a great 'discovery'.

Scientific snobbery conventionally excludes from the 'literature' all works of popularization. Yet journals such as *Scientific American* and *La Recherche* are immensely valuable, not only to the 'lay person' but also to the professional scientist who wants a general picture of the present state of some distant branch of their own subject, or who would like to know, in simple terms, what all the fuss is about some fashionably famous discovery. The habit of reading such journals regularly is worth acquiring: it gives the physicist that continuing education which she cannot hope to get by more formal procedures. Every school and university library should be well furnished with popular scientific books and magazines, to provide a background of general scientific culture, and a wider context for specialized courses of study; and every professional scientist should regard it as a duty to contribute to works of this kind, either directly, if one has the literary gifts, or through the intermediation of a skilled science journalist.

The 'information sources' detailed in this book cover what is known as the 'literature of physics' but, the more one thinks about it, the more difficult it becomes to delimit the subject matter. In one direction we go into chemistry, in another into mathematics. Philosophy is not irrelevant to many of our deeper problems, and the practical connection with engineering is a broad artery along which valuable cargoes of information travel in both directions. Only a technical philistine would cut physics off from its own history; its place in modern civilization; its cultural links; its religious, spiritual, aesthetic significance. If 'the literature of physics' means any book or journal from which a physicist might gain information that would profit him in his professional scientific activities, then it would surely not exclude the writings of Plato or the drawings of Leonardo!

But a literature is not merely the books on the shelves in some appropriately labelled bay of a library stack: it is a language, a style, a manner of talking. In some respects, the literature of physics is very homogeneous, with only a limited vocabulary and a restricted subject matter. It is well known that a large proportion of the technical literature of physics is written in a mixture of two international languages – bad English and algebra. Physicists of all nationalities are more concerned to put their ideas, however ill-expressed, to their colleagues in other nations than they are to use, with full elegance and subtlety, their mother tongues. On the face of it, one would suppose that this would give a decisive advantage to those who have spoken some dialect of English from early childhood, but this is not evident from the literature itself. Perhaps the effort of expressing oneself in a foreign language clarifies thought; the worst scientific writing often comes from those who are so familiar with a language that they dash down the phrases as they come into their heads, stringing together cliché after cliché in the hope of muddling through to some final conclusion. There have been periods when scientific papers have been translated very clumsily into English by authors who have been cut off by war from close contact with that language, but this is not the present situation. There is no doubt, however, that insistence upon English as the *sole* language of physics, whether written or spoken, would impoverish our linguistic resources. There is more to be learnt from a sound professional translation of a paper conceived by its author in his own good Russian, Chinese or Spanish than from the conventional sentences in a standardized technical vocabulary in which he would be constrained to write if he were not really fluent in English. And it is for those of us who speak the various versions of English to respect this scientifically irrelevant factor, and to open our ears to foreign tongues and our minds to foreign thoughts.

It is true, nevertheless, that the subject matter of physics is so universal, so divorced from the varieties of individual human experiences, that it can best be expressed in a very narrow formal language where the words are deliberately stripped of their emotive significance. It is interesting to observe the evolution of this language. A vivid word used in a metaphorical sense to express a new concept – 'strangeness', 'beauty', 'spin-flip', 'tunnelling', 'saturation' – quickly takes its place amongst various more or less barbarous neologisms or acronyms – 'superparamagnetism', 'laser' – in the jargon of our subject. In fact, the literature of physics is not so full of jargon as many other scientific disciplines – think of biochemistry, or geology – but a conscious effort is needed not to fall too deeply into an elaborate, stereotyped mode of expression. On the one hand, there is much need for a fairly precise vocabulary of words whose meaning can be clearly defined in a strict technical context; on the other hand, one should use a simple, clear, direct style, in words drawn from basic English or from more charmingly poetic sources, to impress the argument into the head of the reader. Such a style is not achieved without a real effort by the writer. It is often necessary to draft and redraft a paragraph a dozen times, listening to the words as if one had never heard them before, setting them in the most efficient logical order for the expression of the argument. The goal of scientific writing is not to initiate some outburst of passionate feeling in the readers but to persuade them that what you have to say could scarcely be doubted by any reasonable person. If you are confused, if you make grandiose claims, if you niggle away at irrelevancies while leaving great gaps in the argument, if you unconsciously contradict yourself – in other words, if the reader has any cause to suspect that you don't really know what you are talking about – then all your efforts are wasted. Your role is that of the friendly guide, taking readers by the arm and gently leading them on, step by step, letting them see for themselves the connections between the various points to which you will draw their attention, so that they feel, in the end, as much at home inside your argument as if they had lived there all their lives. Physics is difficult enough to understand without the obfuscations of pretentious language, artificial elaborations of syntactical structure and similar devices used by vain simpletons to parade their superior learning.

The other major component of the literature of physics is *mathematical symbolism*. Now, of course, physics is deeply committed to the algebraic method of thought. I would, myself, define it as 'the science that attempts to describe the natural world in mathematical terms', emphasizing not merely quantitative measurement but also the creation of strictly defined models whose theoretical properties mirror those observed in reality. The other major scientific disciplines define themselves by a natural realm of subject matter – the rocks of the earth's crust, say, or the human mind – and exercise all forms of rational observation and discourse on that subject matter. In physics we have a special, immensely powerful intellectual technique, which we turn on nature in all its aspects but which only yields comprehensible results in particular circumstances. Physicists study atoms, electrons, nuclei, pure crystals, gases, etc., not because these are the topics allotted to us in the universal curriculum but because there are objects or systems that are actually simple enough to be described with fair accuracy by finite, soluble mathematical models. Mathematical symbolism is therefore an integral part of the literature of physics, and a physics paper without a single algebraic equation, integral or table of numerical data can scarcely be imagined.

On the other hand, physics is not 'merely' mathematics. Suppose one could express all the hypothetical properties of any of our theoretical models in the totally symbolic form of *Principia mathematica*; there would still be the task of interpreting the symbols into the descriptive words of the world that they were supposed to represent. The programme of axiomatization, being endless, turns out to be fruitless; at a certain stage we return to ordinary verbal propositions to express our discoveries and theories.

The choice between mathematical and verbal exposition is therefore, to some extent, a matter of taste. An equation may be immensely powerful and exact, but the symbols are cryptic, and do not speak so directly to the comprehension. It takes long practice and great concentration to 'read' a page of algebra and grasp its significance. To the extent, therefore, that the literature of physics consists of mathematical manipulations and formulae written out at full length, it becomes esoteric, and intelligible only after deliberate and laborious deciphering. The function of mathematical symbols is to allow one to carry out a sequence of complicated logical transformations and operations purely mechanically, without recourse to intuition. It is not very instructive, therefore, to set down in full all the successive stages of such a process; the important stages are at the beginning, when physical ideas are being represented by symbols according to some scheme of assumptions and approximations; and at the end, when the results of the calculation, numerical or graphical, are to be compared with experiment and interpreted. The experienced reader of physics does not try to follow through the mathematical argument, step by step, at a first reading, but jumps from verbal formulation of the first few equations to the discussion of the results, being capable of visualizing for himself the general flow of the argument in between. It is the responsibility of the author of such a paper to bring into the open precisely these points of uncertainty of technique and interpretation, and to keep for his own notebooks and computer printouts the mechanical 'workings' of his calculations.

What is not, perhaps, adequately appreciated is the extent to which we actually depend on visual images rather than algebraic manipulation in thinking about physics. A deplorable tradition in applied mathematics – perhaps an excess of zeal by the followers of Descartes, in revolt against the geometrical proofs of Newton's *Principia* – was the deliberate banishment of diagrams from the exposition of physical theory. The literature of physics has now recovered from this impoverishment of the imagination, but there are still some inhibitions, in the formal primary literature, against the pictorial representation of physical ideas. In this respect, the 'vulgarizers' have taught us a lot; faced with the difficult task of explaining very difficult ideas to the lay public, they have discovered, quite naturally, that a few lively diagrams speak far more eloquently than strings of words and symbols. There is no reason, except academic snobbery, why the same method should not be used to convey an intuitive notion of a new concept from the very moment of its birth.

The richness and variety of the literature of physics is confusing: the scale of it is daunting: the current journals on display in their hundreds, the stacks of bound volumes of journals, monographs, textbooks, data collections, annual reviews, etc. How can we find our way through such a mass of material to the few simple facts that we need to know? Who can read those 100 000 papers that are published each year? What does it all *mean*?

The great advantage that we have in physics is that the subject has a strong internal intellectual structure. Precisely because we seek a logical mathematical explanation of the phenomena that we study – indeed, confine ourselves to the study of phenomena that can be given explanations of this kind – we may discern in the physics literature a rational ordering of topics. It is evident, for example, that the study of the structure of the nucleus can be clearly distinguished from the theory of sound waves in liquids by the scale of the phenomena, in spatial extent, energy, etc. It is not too difficult, therefore, to set up a classification scheme, for the library shelves or for the index of a volume of abstracts, in which every topic seems to have its proper place.

Nevertheless, such schemes are often profoundly misleading. Nuclear structure is closely related to the dynamics of liquids by the 'liquid drop model'. The excitation spectra of atoms, of nuclei and of nucleons follow essentially the same quantum mechanical principles, although differing in energy scale by factors of 10^5. An alternative classification scheme, in terms of mathematical equivalence of models, would put all these phenomena on the same library shelf, at least within the same volume of an encyclopedia of theoretical physics. In other words, the store of physics information, unlike a biological 'family tree', is multiply connected and capable of classification in many orthogonal dimensions (as is explained by Robert Wyatt in Chapter 3). For the librarian this may not be very important – he merely wants to find the proper niche for each book along some quasilinear scale. But for the reader, the searcher and, especially, the browser it is essential to know something of these connections, and to imagine the possibility that relevant information will be found in some quite different corner of the library, through a theoretical analogy, an instrumental similarity or the intervention of some mechanism on a completely different scale from those considered active. It is an astonishing fact about modern physics that the whole of our knowledge of the physical universe is interrelated: it is quite conceivable that our present uncertainties concerning *quasars* – the largest, most energetic, astronomical objects – may not be resolved until we understand about *quarks* – supposedly the smallest constituents of matter. To make full use of the literature of physics, one must really have an adequate grasp of this general structure, and not be constrained within the conventional boundaries of a subject as taught, of books as written, of libraries as arranged.

I have written almost as if the literature of physics were something external, public, to be found in libraries and consulted when necessary. But, of course, it is so much an integral part of the active life of a practising physicist that he must come to feel that he owns a bit of it himself. Books and journals are expensive to produce and to buy – but, to the extent that they contain our own products and our own needs, we must surely never grudge the cost. It is easy enough to estimate that the transfer of scientific information costs a few per cent of the whole of the research with which it deals – essentially a negligible proportion by comparison with its value to the scientific community. No serious physicist should, therefore, think twice before spending what is necessary for this purpose, whether by subscriptions to learned journals, the purchase of his or her own collection of books and reviews, or in support of a convenient library. Unfortunately, the system of grants, etc., which provides the material resources of research, does not always provide funds for such purposes, so that the 'literature' is treated as a marginal expenditure, coming from the taxed income of the individual research worker or from the tail-end of the university budget.

This is particularly the case in just those countries where the scientists are cut off, by distance and general poverty, from close personal contact with current research; nothing is more depressing, in a visit to a university physics department in a developing country, than to see the few meagre journals, often months late, on the library shelves, and to realize how weakly their attempts to contribute to scientific knowledge are sustained by a grasp of what has been and is being done by other people. But there is a 'parochialism of the heart', as well, to be found in scientific groups who have all the resources they need for such contacts, and who are yet too lazy, timid or foolishly vain to make themselves familiar with all relevant publications on their subject. It is often asserted that the literature is too vast, too incoherent, too overwhelming to be comprehended and used by the individual scientist, except perhaps with the aid of some mechanical, computerized, retrieval device. This does not seem to be the real truth. In an active research field the interests may be so narrow at each particular moment that only a few papers really count, and these can easily be discovered.

Let me repeat: it is the duty of the serious physicist to make oneself aware of the sources of information for one's subject – on the narrow front of one's immediate research topic; more generally, on the broader front of one's field; and in the large, as a person of genuine learning, over all of physics and the other sciences. Besides, it is all so *interesting*: are we not to be envied, in being permitted to devote our lives to such matters?

References

Föppl, A. (1894) *Einführung in die Maxwellsche Theorie der Elektrizität.* Teubner.

Holton, G. (1969) *Isis*, **60**, 133–196.

Holton, G. (1973) *Thematic origins of scientific thought*, p. 219, Cambridge, MA: Harvard U.Pr.

CHAPTER TWO

The scope and control of physics, its literature and information sources*

DENNIS F. SHAW

Physics is the basic physical science. It deals with those fundamental questions on the structure of matter and the interactions of the elementary constituents of nature that are susceptible to experimental investigation and theoretical enquiry' (Weidner, 1974). This statement by Weidner gives a succinct image of the fundamental role of 'physics' in the natural sciences. However, it needs to be amplified in order to convey clearly to the reader the full scope of the subject. This is best achieved by stating the broad classes of natural phenomena with which physics is concerned. These are:

- Mechanics
- Thermodynamics and Heat
- Electricity and Magnetism
- Optics
- Solid State Physics
- Molecular Physics
- Atomic Physics
- Nuclear Physics
- Particle Physics
- Quantum Mechanics
- Relativistic Mechanics
- Conservation Laws
- Fundamental Forces and Fields.

The treatment of the individual subjects in this book is based on this analysis.

* based on the first edition chapter written by H. Coblans

Physics literature

The traditional source of information in any branch of learning is the written word. When Coblans produced the first edition of this book in 1975 he gave it the title *Use of physics literature* (Coblans, 1975). It was recognized then that other forms of publication and dissemination of information, particularly through the use of electronic storage, were likely to grow in importance. During the decade following, many scientists were predicting that electronic publishing would render hard copy publishing techniques obsolete. It is true that digitally stored texts are growing in importance in the field of scholarly publishing, nevertheless, as noted by other authors in this third edition, books and periodicals printed on paper still comprise the overwhelming majority of the literature of physics which provides the researcher with essential sources of information. The contents of electronic databases largely comprise data derived from conventionally printed publications; although the converse is also true, since the hard copy for conventional publications is often produced from a master version stored in a computer file. However, the time is now with us when some of these databases have no hard-copy equivalent version (see, for example, Chapter 7 by M.Kremer and S. O'Neale and Chapter 15 by H. Behrens). Of course, a printed version can be obtained by transcription through one of the several computer output media, but this is not the same as the publication of a definitive hard copy produced to a recognized standard and identifiable through a bibliographic description. For these reasons, the title *Information sources in physics* was adopted in 1985 and is retained for the third edition of this work.

The literature of physics comprises mainly journals, encyclopedias (or handbooks), textbooks, monographs in series, progress reports and annual reviews, conference proceedings, dissertations and databooks. To this list we may add certain official and commercial publications (e.g. patent specifications, computer documentation) and ephemera (mainly preprints and technical reports). This variety of forms has evolved, during the last 300 years mainly, in response to the need for physicists to communicate their ideas and discoveries to each other, to scientists working in other disciplines, to students and, more recently, to manufacturers applying the results of physical discoveries in industry. The pattern is common to the majority of scientific disciplines.

Besides being of general interest and utility to the physicist, a general understanding of the information-transfer processes used by physicists is important for those reponsible for documentation and also those responsible for formulating science policy both nationally and internationally. It is therefore not surprising to find that the physics information flow has a well-established pattern supported by governments in many countries and involving major international organizations. This book serves both to describe the main individual items where physics information may be found and also the underlying systems through which that information is made available to those outside the physics profession. The author of this chapter, who is also the editor of the whole work, is himself a professional physicist. He has learnt a lot from preparing the outline for this work and believes many of his colleagues in the profession will find much here which is new and interesting, perhaps even of importance!

Primary publications

The primary publications in any field of knowledge are those which contain the work of contemporary scholars who are contributing to the advancement of the subject. These publications, therefore, constitute a major source of information of primary importance. However, John Ziman in his introductory chapter has drawn attention to the wide diversity of the primary literature in physics. There have been two very important studies of the complex problems relating to physics information flow; one conducted by the Physics Survey Committee of the US National Research Council was published in 1972 by the National Academy of Sciences in Washington (NAS, 1972). More recently, a report of a study commissioned by the Institute of Physics in London which set up a Physics Information Review Committee in 1974 was published by the British Library (Singleton, 1978). Everyone concerned with physics information, whether in discovering, disseminating or using it, should be familiar with the contents of these reports. In the hope that some of the readers of this book may be sufficiently concerned to act, the conclusions are summarized here.

Summary of conclusions

1. There is a general lack of information about the system for primary publication and the interaction between its component parts. Financial flows are of prime importance and need to be more clearly identified and related to the performance of the system.
2. Conferences are expensive but there has been little systematic study of their nature, organization and control. Particular emphasis should be placed on the need to improve oral presentation and audio-visual aids. The reports of conferences constitute a particularly difficult set of bibliographic material; there

is a particular need to examine the extent to which republication elsewhere constitutes a duplication of effort and waste of resource. (This applies also to doctoral dissertations.)

3. A vast amount of applied physics literature appears in report form, some of which is well published and controlled but the majority of reports are 'semi-published or unpublished', even though they exist in hard copy. There is a need to provide education in communication skills for physicists and for studies on the content of conference reports and their source and availability. Attention is drawn to the enormous waste of resources due to commercial secrecy resulting in the unnecessary withholding of results from research conducted in industrial laboratories.

4. There is a need for clearer exposition of the practices employed for the control of standards and quality of the primary literature. Primary publication is an industry in its own right, but little quantitative information exists (or at least is readily available) concerning its economics. It is particularly important that publishers of journals should state their policy clearly both to authors and referees. Besides the changes expected to result from a rapid spread in electronic data-transmission systems there is potential for innovation in conventional publishing methods, and the system is closely coupled in some aspects but not in others.

5. The letters to editors, letters journals and short notes constitute a substantial proportion of the primary literature of physics and have achieved an unexpected importance. Studies are needed to establish more clearly their status and archival importance.

6. The virtual disappearance of the private subscriber to the majority of journals leaves editors and publishers with no clear picture of the readership for which they are catering.

There is still concern amongst professional scientists that insufficient attention is paid by governments and institutions to the problems of maintaining the conventional means for the dissemination of the results of scientific research. This led The Royal Society, the British Library and the Association of Learned and Professional Publishers to initiate a study of the increasing signs of stress and strain in the whole of the science, technology and medicine (STM) information system. The objectives of the study are outlined in an announcement in the *Royal Society News* (Royal Society, 1992).

Secondary services

The term 'secondary services' is used to describe those services essential to the easy flow of information which, although not original in their content, nevertheless constitute a vital element in the process. The main elements of the secondary services are abstracts journals and the associated computer databases, together with monographic series of reviews and advances, and encyclopedias (or handbooks).

There are essentially four fully comprehensive services providing abstracts of the literature in physics:

Physics Abstracts (INSPEC)
Physics Briefs/Physikalische Berichte (Physik-Verlag)
Bulletin Signalétique (CNRS)
Referativnyi Zhurnal: Fizika (VINITI)

Physics Abstracts: Science Abstract Series A (formerly *Science Abstracts, Section A – Physics)* started in 1898 and now has two issues per month. It is the most widely used of all the English-language abstracting journals in physics. Its coverage of 'published' literature from all countries and in all languages is excellent. To a lesser extent it provides coverage of 'report' literature. Entries are completely in English. The language of the original is indicated. Each entry consists of an identification number, author(s), including affiliations, the bibliographic description of the source, the abstract itself and the number of references.

The entries are arranged in accordance with the IEE's own subject-classification scheme (see Table 2.3). Each issue has an author index and there are separate indexes for bibliographies, books and conference proceedings. Indexes are cumulated twice yearly. Cumulated indexes for the four-year periods 1955–1959, 1960–1964, 1965–1968, 1969–1972, 1973–1976, 1977–1980, 1981–1984, 1985–1988 and 1989–1992 are also available. The subject index, arranged alphabetically, contains many more terms and more specific terms than those mentioned in the subject classification. Each abstract is assigned several indexing terms besides the keywords in the titles. Names of elements, their compounds and a few special compounds are included as indexing terms. The journal is useful for both current-awareness and retrospective searching.

Physics Briefs/Physikalische Berichte started as *Fortschritte der Physik 1845–1918* then continued as *Physikalische Berichte* during the period 1920–1978. Since 1979 it has been published jointly by AIP and Physik Verlag and has 24 issues each year. The abstracts are entered in a database PHYS which is available online through STN International. It now contains about 1.5 million records and the annual update is about 125 000 entries. The journals and series publications regularly abstracted are listed in *PHYS Database Reference Series* (published by FIZ, 1992).

Bulletin Signalétique is now part of the INIST database *PASCAL* which is available online from INIST Diffusion (CNRS). It was originally published as *Bulletin Analytique* during the period 1940–1955. Issues contain subject and author indexes; the following sections are of particular interest:

- 120 (astronomie, physique spatiale, géophysique);
- 130 (physique mathématique, optique, acoustique, mécanique – chaleur);
- 145 (électronique);
- 160 (physique de l'état condensé);
- 161 (structure de l'état condensé – cristallographie);
- 165 (atomes et molécules – plasmas).

The database covers science, technology and medicine and now includes over eight million references with French publications especially well represented. Physics represents about one million entries (13.5%).

Referativnyi Zhurnal – Fizika (VINITI) started in 1954 and now produces about 100 000 items annually. It is the hard copy version of the VINITI database which is held at the Russian National Public Library for Science and Technology in Moscow. The database is not yet generally available online beyond Eastern Europe. Over 20 000 serials are abstracted regularly as well as tens of thousands of books, standards and patents. A useful comparison of the growth of the *Fizika* section of *Referativnyi Zhurnal* with the growth of *Physics Abstracts*, was made by Vlachy (1983).

The growth in the contents of these abstracts of physics is a good indication of the growth of physics research in the world. A comparison between these four is shown in Table 2.1, covering the years 1952 to 1992.

In Table 2.2 we compare the distribution of journals abstracted, by country of publication, for these four services in 1992. There is inevitably a bias in favour of the native language of the service but this bias is remarkably insignificant, as these data show. For this reason it is believed that the following more detailed analysis of *Physics Abstracts*, from data supplied by the staff of INSPEC, is likely to provide a valid world picture of physics publications.

Physics Abstracts

Since 1973 *Physics Abstracts* has been included as *Section A* of the INSPEC database which includes also: *Section B: Electrical Engineering and Electronics*, *Section C: Computer and Control* and *Section D: Information Technology*. Many topics included in

TABLE 2.1 Growth of the four major physics abstracting services 1952 to 1992.

Year	PA	PB	BS	RZ
1992	152 000	130 000	65 000	100 000
1982	118 000	110 000	60 000	80 000
1972	85 000	70 000	70 000	70 000
1962	24 000	20 000	35 000	35 000
1952	9 000	7 000	—	—

Key PA *Physics Abstracts*
PB *Physics Briefs*
BS *Bulletin Signalétique*
RZ *Referativnyi Zhurnal*

TABLE 2.2 Percentage of physics journals abstracted in 1992, distributed by country of publication, for the four major physics abstracting services.

Country	PA	PB	BS	RZ
France	1	3	7	2
Germany	6	20	11	6
Great Britain	23	14	23	12
Japan	5	3	2	4
Netherlands	12	10	13	5
Russia	7	5	2	21
USA	32	22	30	25
Others	12	23	12	25

Key PA *Physics Abstracts*
PB *Physics Briefs*
BS *Bulletin Signalétique*
RZ *Referativnyi Zhurnal*

Sections B and *C* are of interest to physicists, although the central subject areas of classical and quantum physics and condensed matter are included in *Section A*. The contents of this database are published in hard copy as *Science Abstracts*.

Abstracting and indexing services in physics are discussed fully by R. J. Wyatt in Chapter 3. For the present discussion it will be useful to refer to the broad subject areas within the INSPEC classification, and these are listed in Table 2.3. The four abstracting publications

TABLE 2.3. Outline of the INSPEC classification

SECTION A: PHYSICS
00   General
10   The Physics of Elementary Particles and Fields
20   Nuclear Physics
30   Atomic and Molecular Physics
40   Classical Areas of Phenomenology
50   Fluids, Plasmas and Electric Discharges
60   Condensed Matter: Structure, Thermal and Mechanical Properties
70   Condensed Matter: Electronic Structure, Electrical, Magnetic, and
     Optical Properties
80   Cross-disciplinary Physics and Related Areas of Science and
     Technology
90   Geophysics, Astronomy and Astrophysics

SECTION B: ELECTRICAL ENGINEERING AND ELECTRONICS
00   General Topics, Engineering Mathematics and Materials Science
10   Circuit Theory and Circuits
20   Components, Electron Devices and Materials
30   Magnetic and Superconducting Materials and Devices
40   Optical Materials and Applications, Electro-Optics and
     Optoelectronics
50   Electromagnetic Fields
60   Communications
70   Instrumentation and Special Applications
80   Power Systems and Applications

SECTION C: COMPUTER AND CONTROL
00   General and Management Topics
10   Systems and Control Theory
30   Control Technology
40   Numerical Analysis and Theoretical Computer Topics
50   Computer Hardware
60   Computer Software
70   Computer Applications

SECTION D: INFORMATION TECHNOLOGY
10   General and Management Aspects
20   Applications
30   General Systems and Equipment
40   Office Automation Communications
50   Office Automation Computing

shown in this table comprise the INSPEC database which is available for online searching on several information retrieval services such as DIALOG and STN. It is also available on CD-ROM from 1989 onwards (UMI-ProQuest) in three versions:

INSPEC ONDISC which holds the complete database, with each year published on one disc;
INSPEC – PHYSICS ONDISC which corresponds to the printed *Physics Abstracts*;
INSPEC – ELECTRONIC & COMPUTING ONDISC which corresponds to the printed versions of *Electrical & Electronics Abstracts*, *Computer & Control Abstracts* and *Information Technology*.

The analysis of physics periodicals and documents, presented in Table 2.4, shows those items abstracted during a twelve-month period, October 1991 to September 1992, distributed between the five main subject groups in physics (as defined by the Institute for Information Science). The most significant feature, shown by these data, is the very large proportion of journal issues in the total number of items; also worth noting is the rather small number of books. There is a wide variation between subject groups in the number of

TABLE 2.4. The five information scientist groups for physics and the coverage of *Physics Abstracts* for the period 1 October 1991 to 30 September 1992

IS group[a]	Journals	Journal issues	Conferences[b]	Books	Reports
A	334	2 051	78	35	63
I	453	2 586	91	22	63
M	259	1 505	26	9	286
N	324	2 051	133	16	375
S	225	1 829	108	24	29
Total	1595	10 022	436	106	816

[a]   Key to IS group:
   A — *Astrophysics and Geophysics*
   I — *Interdisciplinary Physics: Electromagnetism, Heat, Mechanics, Materials Science, Biophysics*
   M — *Atomic and Molecular Physics: Fluid Dynamics, Fluid and Plasma Physics, Physical Chemistry*
   N — *General Physics: Acoustics, Elementary Particle Physics, Nuclear Physics, Energy Research*
   S — *Condensed Matter Physics*
[b]   *Includes conferences reported in journals*

reports extracted, Group N (which includes nuclear and particle physics and energy research) contributed nearly 50 per cent of the total number of reports.

The number of discrete physics items entered in the INSPEC database annually since 1969 is shown in Table 2.5. The total number of physics items added annually shows a steady growth, increasing by a factor of three during the 24-year period. In only five of the 24 years (1973, 1977, 1981, 1984 and 1988) was there even a slight reduction in the previous year's output of abstracts. However, the whole database has grown even more rapidly from 973 000 items in 1969 to 5 981 000 in 1992 thus, the proportion of physics items is diminishing. This is a result of the growing research activity in applied physics, computing and technology generally. This growth has had the inevitable result of rapidly increasing the costs of the

TABLE 2.5 Growth of *Physics Abstracts* database 1969–1992

Year	Annual input	Years	Total input	Cumulative total
		to 1968		547 384
1969	49 619			
1970	79 830			
1971	84 332			
1972	85 185	1969–72	298 966	846 350
1973	81 352			
1974	83 364			
1975	87 636			
1976	94 940	1973–76	347 292	1 193 642
1977	91 677			
1978	96 582			
1979	101 233			
1980	109 577	1977–80	399 069	1 592 711
1981	108 359			
1982	117 845			
1983	121 808			
1984	115 847	1981–84	463 859	2 056 570
1985	127 325			
1986	128 824			
1987	146 131			
1988	143 030	1985–88	545 310	2 601 880
1989	144 991			
1990	157 051			
1991	151 994			
1992	151 677	1989–92	606 036	3 207 916

hard copy abstracts journals and it may become necessary to sub-
sidize them if the service to the user is not to decline. Also, the
increasing availability of abstracts online has considerably reduced
the demand for hard copy. This has inevitably affected the market,
since income is increasingly derived from database charges instead
of sales of the hard copy printed abstracts journals.

Analysis of the language of publication of the items abstracted
shows an overwhelming and growing proportion in English. Data
supplied by INSPEC entered in Table 2.6 show the remarkable
increase since 1972 in the proportion of primary physics items
which are published in the English language. During this period
several foreign language journals changed to publishing in English in
order to satisfy their readers. To demonstrate that the preponderance of English-language items is not unduly influenced by a bias
due to the country of origin we list in Table 2.7 the distribution by
country of origin of the items abstracted in 1992.

TABLE 2.6 Analysis by language of publications abstracted in *Physics
Abstracts* in 1972, 1982 and 1992

Language	1972 (%)	1982 (%)	1992 (%)
English	68	90	95.7
Russian	17	3.8	0.7
German	7	1.4	0.5
French	6	1.4	0.6
Others	<3	2.8	2.4

TABLE 2.7 Analysis by country of origin of author (or affiliation of first
named author) of abstracts published in 1992

Country	Number of items	Percentage of total
USA	68 783	45.3
UK	25 357	16.7
Russia	7 288	4.8
Germany	7 136	4.7
Japan	5 466	3.6
France	2 581	1.7
Remainder	35 378	23.3

Nevertheless, there is a substantial output of primary publications in the Russian language, as has been mentioned above. The interest in, and importance of this, is demonstrated by the number of cover-to-cover translations into English of Russian-language primary journals now produced in the USA. In the list of the most important primary journals in physics abstracted by INSPEC (listed in Appendix A) it will be noted that eight of the first one hundred are US translations of Russian journals. These are shown in Table 2.8. In 1982/3 these titles were listed in the first 50 and there was also one Russian original *Metallofizika* then ranking 49, but now ranked at 156.

The journals listed in Appendix A are those which contributed the largest number of items to the INSPEC database for those subjects in the five Information Science subject groups detailed in the key to Table 2.4. The list comprises 415 titles which together contributed 90 per cent of all items selected from those issues received during the period 1 October 1991 to 30 September 1992. (Note: this selection is clearly not the same as that published in *Physics Abstracts* during the same period.) Eight of the journals listed changed title during the year and the computer output from which the list has been produced included separate entries for the old and new titles. A note correcting this appears at the end of the Appendix A.

TABLE 2.8 US translations of Russian journals listed in first one hundred of the most important primary physics journals according to INSPEC

Rank order in 1991/2	Title	Rank order in 1982/3
26	*Soviet Physics – Solid State*	18
43	*Soviet Technical Physics Letters*	50
57	*Soviet Physics – Technical Physics*	29
59	*Soviet Journal of Nuclear Physics*	261
62	*Soviet Physics – Semiconductors*	36
68	*Soviet Journal of Quantum Electronics*	46
71	*Soviet Physics – JETP*	129
95	*Soviet Physics – Doklady*	33

International organizations

International Union of Pure and Applied Physics

The flow of physics information is significantly influenced by the activities of international organizations which have become established during the twentieth century. From the Renaissance until the end of the nineteenth century such co-ordination of scientific activity as was deemed necessary was found possible through informal contacts between members of the various national academies. However, after the trauma and devastation of the First World War, which led to the breakdown, in Europe at least, of all the established academic links between scientists, a world movement was started towards a more formal international co-operation. In 1919 an International Research Council was created for this purpose and at its meeting in Brussels in 1922 an Executive Committe with Sir William Bragg as President took the steps necessary to establish an International Union of Pure and Applied Physics (IUPAP). The first Secretary-General was Henri Abraham, and the other members of the founding committee, besides the President, were M. Brillouin, O. M. Corbino, M. Knudsen, M. Leblanc, H. A. Lorentz, R. A. Millikan, H. Nagaoka and E. van Aubel. A Charter was promulgated at a General Assembly held in Paris in 1923 at which 16 countries were represented. Membership of the Union is by country through its national academy, research council or government, and it now constitutes a world federation of official national representatives which can act and proclaim authoritatively for the profession. In 1992, 42 countries were members of IUPAP. Thus the international character of physics is given a formal structure and recognition. IUPAP aims to obtain agreement on symbols, notations, constants, terminology and standards. Also, it co-ordinates the work of preparation and publication of abstracts and tables of constants, and generally promotes international co-operation among physicists. The most significant feature of IUPAP has been the convening of General Assemblies (listed in Table 2.9) whose reports cover a range of major topics.

IUPAP Commissions

The work of IUPAP in controlling physics nomenclature has been largely carried out through International Commissions. The first two were established at the Third General Assembly in 1932 to deal with:

- Publications (formerly Bibliography and Publications)
- Symbols, Units and Nomenclature.

TABLE 2.9 Year and venue for IUPAP General Assemblies

Year	Location	Year	Location	Year	Location
1923	Paris	1954	London	1975	Munich
1925	Brussels	1957	Rome	1978	Stockholm
1932	Brussels	1960	Ottawa	1981	Paris
1934	London	1963	Warsaw	1984	Trieste
1947	Paris	1966	Basel	1987	Washington
1951	Copenhagen	1969	Dubrovnik	1990	Dresden
1954	London	1972	Washington	1993	Nara

The latter commission, known by the acronym SUN, has been active ever since and has been responsible for recommendations on the definition of the absolute scale of temperature based on the triple point of water, the adoption of the metre-kilogram-second (MKS) system of units, on the designation of the newton as the MKS unit of force and on the rationalization of the equations for the electromagnetic field. Recently it was combined with the Commission on Atomic Masses and Fundamental Constants (1960) and now has the acronym SUNANMCO. Reference to these and other activities of SUN and SUNANMCO will be found in later chapters.

Other IUPAP Commissions with their dates of establishment include the following:

- Optics (1948)
- Statistical Physics (formerly Thermodynamic Notations) (1945)
- Cosmic Rays (1947)
- Low Temperature Physics (1949)
- Acoustics (1951)
- Structure and Dynamics of Condensed Matter (formerly Solid State Physics) (1960)
- Magnetism (1957)
- Particles and Fields (formerly High Energy Physics) (1957)
- Semiconductors (1957)
- Atomic Masses and Fundamental Constants (1960)
- Physics Education (1960)
- Nuclear Physics (1960)
- Atomic and Molecular Physics and Spectroscopy (1966)
- Plasma Physics (1969)
- General Relativity and Gravitation (1972)
- Quantum Electronics (1975)

- Mathematical Physics (1981)
- Physics for Development (1981)
- Astrophysics (1984)
- Biological Physics (1990).

An interesting account of the first fifty years of IUPAP activity was published in 1973 (Brown, 1973). The 50th Anniversary brochure produced for the Assembly in Washington in 1972 was updated by Cécile Balaux and re-issued in 1992 in preparation for the 23rd General Assembly held in Nara (Japan) in 1993.

International Council of Scientific Unions

The political influence of scientists has been greatly strengthened by the formation of the International Council of Scientific Unions (ICSU). This Council was created in 1931 to encourage international scientific activity for the benefit of mankind and now has 70 member countries. One of its best-known activities was the International Geophysical Year (IGY) in 1957, which led to the establishment of a huge international network of World Data Centres for Geophysical Observations. Other ICSU activities of importance to physics are those of the following bodies:

COSPAR   Committee on Space Research
COSTED   Committee on Science and Technology in Developing Countries
CODATA   Committee on Data for Science and Technology
SCOPE   Scientific Committee on Problems of the Environment
ICSTI   International Council for Scientific and Technical Information
IUBS   International Union of Biological Sciences (Biophysics)

The publications of these bodies frequently contain definitive statements of considerable importance to information exchange in physics, for example:

UNISIST: *Study Report on the Feasibility of a World Science Information System* (Unesco and ICSU, Paris, 1971).

ICSTI: *International Classification Scheme for Physics* (3rd edn, AIP 1991)

UNISIST aimed originally to create a world scientific information system, but this proved impossible to achieve. A modified objective was to co-ordinate an international network of information services in science (and technology) by fostering voluntary co-operation. A major task has been the definition of standards for machine-readable records.

Further information will be found in the *ICSU Yearbook*, which is published annually by ICSU in Paris and also in the *ICSU Annual Report.*

References

Annals of the International Geophysical Year (1959) Volume 1. Oxford: Pergamon.

Brown, S. C. (1973) *Physics 50 years later*. NAS.

Coblans, H. (1975) *Use of physics literature*. London: Butterworths.

IGY (1957) *IGY Bulletin 1957–*. NAS.

NAS (1972) *Physics in perspective*, Volume 1, Chapter 13. NAS.

Royal Society (1992) The scientific information system in the 1990s. *Royal Society News*, **6**(7), 6.

Singleton, A. K. J. (1978) *The communication system of physics. Final report of the Physics Information Review Committee*. BLRD Report, no. 5386. London: British Library.

Vlachy, J. (1983) World publication output in nuclear physics. *Czechoslovak Journal of Physics*, **33**(6), 709–712.

Weidner, R. T. (1974) *The New Encyclopaedia Britannica in 30 Volumes, Macro-paedia*, **14**, pp. 424–429. Chicago: Encyclopaedia Britannica Inc.

Tables 2.1 to 2.8 inclusive reproduced with kind permission of INSPEC.

CHAPTER THREE

Science libraries, reference material and general treatises*

R. J. WYATT

Research libraries

The libraries available to the physicist are of four types: national, university, public and special. The largest collections are held by libraries which enjoy the privilege of receiving their country's publishing output under legal deposit legislation or otherwise fulfil national functions. In the UK, there are six libraries of this type which are of interest to the physicist. The Science Reference and Information Service of the British Library (SRIS), at present on two sites in central London, will soon be unified at the new British Library building in St. Pancras. It has almost a quarter of a million monographs and nearly 30 000 current serials. The Radcliffe Science Library, dependent library of the Bodleian Library, Oxford, has approximately the same number of monographs and 30 000 British and foreign serials, of which at least 6 500 are currently received. Like the Radcliffe Science Library, Cambridge University Library purchases much non-British scientific literature to complement its British legal deposit receipts. The Scottish Science Library (a branch of the National Library of Scotland in Edinburgh) and the National Library of Wales (at Aberystwyth) also enjoy the copyright privilege. The Science Museum Library, South Kensington, which built up a magnificent collection between the world wars and acted as the British clearinghouse for interlibrary loans of scientific material in the 1930s and 1940s, now concentrates on the history and public understanding of science and technology, and has among

* based on the second edition chapter written by P. J. R. Warren

its 600 000 volumes 21 000 periodical titles, of which 2500 are currently received.

The national library of the United States, the Library of Congress in Washington, D.C., is the world's largest library and contains over 27 000 000 items. Its Science and Technology Division has more than 3 000 000 books and 60 000 serial titles, and also houses over 3 500 000 documents in its Technical Reports Section.

CISTI, the Canada Institute for Scientific and Technical Information, in Ottawa, is the national scientific library for Canada, and has about 500 000 books and 50 000 serials.

Universities where physics is taught will have good physics collections in their main libraries. Most notably, the John Crerar Library, the science library of the University of Chicago, has over 350 000 books and 10 000 current serials. Some of them will also have a library within the physics department which supplements or duplicates stock in the main library. The largest British public libraries, such as those in Birmingham, Liverpool and Sheffield, though they are small in comparison with the New York Public Library, which has in all 10 000 000 volumes and 40 000 periodical titles, have good collections in their scientific and technical departments. Government establishments (e.g. the United Kingdom Atomic Energy Authority), private industrial concerns, and learned societies (such as the Institute of Electrical Engineering (IEE), with 50 000 books, 20 000 reports and 1000 current periodicals) have their own special libraries to support the researches of their staffs or members. An unusual special library is the independent, non-profit Linda Hall Library in Kansas City, Missouri, a scientific and technological library containing 600 000 volumes and 38 000 serial titles.

The same fourfold division into national, university, public and special libraries is found in other countries. Further information is provided in *The world of learning* (annual, Europa), which gives worldwide coverage; the *Aslib directory of information sources in the United Kingdom* (7th edn., 1992) and *Guide to libraries and information units in government departments and other organisations* (annual, British Library), for the UK; *Guide to libraries in Western Europe: national, international and government libraries*, ed. P. Dale. (BL, 1991) for European coverage; and, for North America, the *American library directory* (annual, Bowker) and *Subject directory of special libraries* (16th edn., 1992, Gale).

Libraries differ in the terms on which they allow access, and also, obviously, in the sizes of their collections. Physicists with only a small library at their disposal, however, need not despair.

First, as noted in previous editions (Friend, 1975 and Warren, 1985), a small library with a well chosen collection of periodicals can be very effective. In the first edition of this book (Friend, 1975) it was pointed out that large libraries are not the only effective ones. Within any current subject literature there is a core of key journals which contain a very high number of significant papers. The Law of Scattering (Bradford, 1948) states that 'if scientific journals are arranged in order of decreasing productivity of articles on a given subject, they may be divided into a nucleus of periodicals more particularly devoted to the subject and several groups or zones containing the same number of articles as the nucleus, when the numbers of periodicals in the nucleus and succeeding zones will be as $1:n:n^2$ . . .'. It has been written about in many papers in documentation journals and need not detain us here, but its practical application means that the small specialist library can, provided its journals have been selected with care, supply a considerable proportion of the appropriate current literature to its users.

Secondly, the British Library Document Supply Centre (BLDSC) at Boston Spa, West Yorkshire, gives access to the world's physics literature by loans and photocopies provided from its own stock of seven million items, including 56 000 current serials, from the stocks of its back-up libraries, and from other sources at home and abroad.

Thirdly, fierce competition is breaking out among information providers – libraries, publishers, subscription agents, abstracting services and others – in the field of document delivery.

Since Peter Warren (1985) wrote the equivalent chapter for the previous edition of this book, libraries have seen a revolution in the services and the information sources available to physicists. Although indexing and abstracting journals were available as online services, provided by commercial or non-profit agencies ('hosts') from their mainframe computers, in general the medium in which reference works were produced was print.

Today, some reference works can be consulted online, and both abstracting services and reference works are available on compact disc (CD). It is not merely a theoretical possibility but a practical reality for a reader to travel the world electronically from the comfort of a seat at a workstation, consulting library catalogues and information services at will, and switching quickly from one to another. These developments seem quite remarkable, even by the standards of 1985, but it is probable that by the time as many years have again elapsed some of the services described here will seem outdated. CD is sometimes considered a transitional medium, and the cost and capacity of the means of computer storage are likely to improve, through such developments as 'flash-chip' technology (McSeán, 1991).

Whichever library he or she is using, and whether it is large or small, the physicist reader is likely to have one or more of seven intentions:

- to find a single definition, value or property;
- to begin work on a new topic;
- to find more literature on a topic;
- to find the most recent literature on a topic;
- to follow up earlier published research;
- to find a particular work from a full reference;
- to find a particular work from a partial or inaccurate reference.

Quick reference queries: dictionaries and tables

The *McGraw-Hill dictionary of scientific and technical terms*, 4th edn. (McGraw-Hill, 1989) contains definitions of about 100 000 terms from all branches of science and technology, each perhaps 100 words in length. Together with the more discursive *McGraw-Hill concise encyclopedia of science and technology* (7th edn., 1992) it is available on the MCGRAW-HILL CD-ROM SCIENCE AND TECHNICAL REFERENCE SET. *Dictionary of physics*, 3rd edn., edited by H. J. Gray and A. Isaacs (Longman, 1991) contains in a single volume about 7000 terms with definitions and explanations varying in length from a few words to more than a page. Account has been taken of advances since the previous edition, particularly in particle physics, nuclear physics, solid-state physics and astrophysics. The rather older *Concise dictionary of physics and related subjects*, edited by J. Thewlis (2nd edn., Pergamon, 1979), contains short definitions (about 30 words) of single concepts, again about 7000 in all.

The encyclopedic collection of tables, 'a compilation of all the certain results of physical, chemical, and technological research, characterized by both the greatest possible completeness and a critical attitude', is Landolt-Börnstein, *Zahlenwerte und funktionen aus naturwissenschaften und technik = Numerical data and functional relationships in science and technology* (published by Springer). This is an immense and continuing work of some complexity, so it is important to appreciate its history and structure. The 6th edition, published between 1950 and 1980, is in 28 parts, arranged in four volumes (vol. 1, Atomic and nuclear physics; vol. 2, Properties of matter in its aggregated states; vol. 3, Astronomy and geophysics; vol. 4, Technology). The predetermined plan of publication proved too restrictive in the event, so after only 11 years, in 1961, the New Series was begun. This has now reached 148 parts, organized, with one exception, into seven groups (group 1, Nuclear and particle physics; 2, Atomic and molecular physics; 3, Crystal and solid state physics; 4, Macroscopic and technical properties of matter; 5, Geophysics and space research; 6, Astronomy, astrophysics and space research; 7, Biophysics). The exception is *Einheiten und fundamentalkonstanten in physik und chemie = Units and fundamental constants in physics and chemistry*, edited by J. Bortfeldt and B. Kramer, of which the first subvolume was published in 1991. The first volume of the *Gesamtregister = Comprehensive index* to the 6th edition and the New Series appeared in 1987, listing all volumes and the indexes contained therein, and indexing the tables by keywords. Further index volumes were issued in 1988 and 1990.

Six decades since its publication, the seven-volume *International critical tables of numerical data, physics, chemistry and technology* (McGraw-Hill, 1926–33) is still of great value. This work was edited by E. W. Washburn for the National Research Council of the USA at the instigation of the International Union of Pure and Applied Chemistry (IUPAC). There are two indexes. The earlier, in English, French, German and Italian, ran to only 10 pages for each language. It was replaced in 1933 by a 300-page index in English.

Of more manageable size, and subject to considerable annual revision, is the Chemical Rubber Company's *CRC handbook of chemistry and physics*, which reached its 73rd edition in 1992 (CRC Press, Boca Raton, 1992). A somewhat shortened student edition of the work, edited by R.C. Weast, was issued by the same publisher in 1988. The *Composite index for CRC handbooks*, 3rd edition (1991) is available in print and also (updated annually) in CD format.

Tables of physical and chemical constants and some other mathematical functions, originally compiled by G. W. C. Kaye and T. H. Laby, is now in its 15th edition (Longman, 1986). It consists in the main of three substantial sections of approximately equal length (general physics, chemistry, and atomic and nuclear physics) and is completed by a very short section on mathematical functions and statistical methods for the treatment of experimental data.

The *American Institute of Physics handbook* (McGraw-Hill), aimed at scientists and engineers, has not been revised since its 3rd edition in 1972, but contains a remarkable range of tables from 'Density of woods' (oven-dry) to 'Nonlinear optical coefficients'.

The serials *Physik Daten = Physics Data*, published by Fachinformationszentrum Energie, Physik, Mathematik, and *Property Data Update* (Hemisphere) publish data compilations.

Beginning a new topic: encyclopedias, general works, review articles

The physicist seeking to begin work in a new and unfamiliar field will want to start with authoritative and scholarly overviews of topics. The first resource for these is the group of reference material which includes works indiscriminately entitled 'encyclopedia', 'handbook' or (occasionally and misleadingly) 'dictionary'.

The most general work of this type is the *New Encyclopaedia Britannica*, now in a revised version of the 15th edition (Encyclopaedia Britannica Inc., 1985). This runs to 32 volumes, the first 12 of which (the *Micropaedia*) provide over 100 000 short (up to 750 words) articles. Almost 700 longer, signed, articles on broader topics, which provide valuable select bibliographies, are found in the 17 volumes of the *Macropaedia*. Further means of access to the information are given by the two-volume *Index* and the single-volume *Propaedia*, a conspectus of human knowledge with references to the articles in the main text. Users of this encyclopedia (and indeed of any general encyclopedia) should check carefully which edition they are using, and should also be aware that the date of printing of the version in hand may well not be the date of the latest revision of each of the individual articles it contains.

The 18-volume *Encyclopedia of physical science and technology* (Academic Press, 2nd edn., 1992) is aimed at students, research scientists and engineers and contains over 700 articles, typically about 20 pages in length, which include definitions and glossaries of terms.

The *Handbuch der physik* (= *Encyclopedia of physics*), edited by S. Flügge (Springer, 1956–88) is a general overview of physics. There were 54 volumes in the original conception, published up to 1962, some written entirely or partly in German or French, but many in English. Supplementary parts have augmented some earlier volumes. The whole work was completed and enhanced in 1988 by the publication of an index to authors and subjects as volume 55.

Nothing has been added since 1975 to the original nine volumes and the 10 annual supplementary volumes of the *Encyclopaedic dictionary of physics*, edited by J. Thewlis (Pergamon, 1961–75). The work contains articles of up to 3000 words, and bibliographies. It is not to be confused with Thewlis's one-volume *Concise dictionary of physics and related subjects* (2nd edn., Pergamon, 1979), which is based on the larger work, but contains only short definitions.

A more recent work (in a single volume) is *The encyclopedia of physics*, 3rd edn., edited by R.M. Besançon (Van Nostrand

Reinhold, 1985). The more than 300 articles vary in their technical level from straightforward introductions with definitions of concepts to more detailed treatments.

The long-running *Methods of experimental physics*, edited by L. Marton (Academic Press, 1959–) allows physicists to find information on experimental methods outside their own fields. Marton conceived the work as a way of avoiding literature searches, which at the time were manual and thus time-consuming. By 1992 it had reached 25 volumes, and two of the early volumes are now in a second edition.

For an up-to-date view of a topic, review articles summarizing recent developments will prove very useful, not least because of the extensive bibliographies they contain. Review articles are published in ordinary physics journals, but there are also serials which specialize in such work. Titles currently published include:

Advances in Atomic, Molecular and Optical Physics
Advances in Chemical Physics
Advances in Electronics and Electron Physics
Advances in Geophysics
Advances in Heat Transfer
Advances in Magnetic and Optical Resonance
Advances in Multi-photon Processes and Spectroscopy
Advances in Nuclear Physics
Advances in Nuclear Science and Technology
Advances in Optical and Electron Microscopy
Advances in Physics
Annual Review of Astronomy and Astrophysics
Annual Review of Biophysics and Biophysical Chemistry
Annual Review of Earth and Planetary Sciences
Annual Review of Heat Transfer
Annual Review of Materials Science
Annual Review of Nuclear and Particle Science
Annual Review of Physical Chemistry
Contemporary Physics
Critical Reviews in Solid State and Materials Sciences
Fortschritte der Physik = Progress of Physics
Nuclear Science Applications. Section B. In-depth Reviews
Physics Reports (Amsterdam)
Progress in Biophysics and Molecular Biology
Progress in Low Temperature Physics
Progress in Nuclear Energy. New Series
Progress in Nuclear Magnetic Resonance Spectroscopy
Progress in Particle and Nuclear Physics
Progress in Quantum Electronics

Progress in Surface Science
Progress of Theoretical Physics. Supplement
Reports on Progress in Physics
Reviews in Modern Astronomy
Reviews of Geophysics
Reviews of Modern Physics
Reviews of Plasma Physics
Solid State Physics: Advances in Research and Applications

Springer Tracts in Modern Physics is a major monographic series in which whole volumes are devoted to reviews of single subjects.

Finding more literature on a topic: classification and subject catalogues

A question familiar to the staffs of scientific libraries is 'Where are the books on physics?'. In fact, the questioner will often have a particular work in mind – of which more later – but what lies behind the phrasing of this question is the idea that if the physics section can be found, then the work, or one like it, can be traced by browsing the shelves. Good classification schemes make effective browsing a realistic proposition by bringing close on the shelves works which are related in subject matter. The most practical schemes are those which are based on 'literary warrant', that is, at their detailed level they classify literature as it exists, rather than knowledge in general.

Classification schemes employ numerical or alphabetical notation to signify the place a work occupies in the whole sequence of works. The same notation can be used on bibliographic descriptions – catalogue entries – and so it is possible to browse at a remove from the shelves.

Large libraries in particular find it very difficult to reclassify their entire stocks, so some of them continue to use classification schemes of their own devising. However, three schemes dominate the scene: the Dewey Decimal Classification (DDC), used by most public libraries, the Universal Decimal Classification (UDC), largely the preserve of special libraries, and the Library of Congress classification scheme (LC), predominantly found in academic libraries.

DDC was first published in 1876, the work of Melvil Dewey, and is now in its twentieth edition (*Dewey Decimal Classification and Relative Index*, 20th edn., Forest Press, 1989). In 1988 both DDC and its publisher, Forest Press, were taken over by OCLC, the Online Computer Library Center. Knowledge is divided into 10 classes, which are traditional academic disciplines (000 Generalities; 100 Philosophy & psychology; 200 Religion; 300 Social sciences; 400 Language; 500 Natural sciences and mathematics; 600 Technology (Applied sciences); 700 The arts; 800 Literature & rhetoric; 900 Geography & history). Each class is made up of 10 divisions. Natural sciences & mathematics is divided as follows:

500	Natural sciences & mathematics [i.e. generalities]
510	Mathematics
520	Astronomy & allied sciences
530	Physics
540	Chemistry & allied sciences
550	Earth sciences
560	Paleontology. Paleozoology
570	Life sciences
580	Botanical sciences
590	Zoological sciences

Each division is further split into 10 sections:

530	Physics [i.e. generalities]
531	Chemical mechanics. Solid mechanics
532	Fluid mechanics. Liquid mechanics
533	Gas mechanics
534	Sound & related vibrations
535	Light & paraphotic phenomena
536	Heat
537	Electricity & electronics
538	Magnetism
539	Modern physics

Analysis of sections into constituent topics is achieved by the addition of further digits after a decimal point, each digit indicating a further downward step in a hierarchy of topics:

537	Electricity and electronics
537.6	Electrodynamics and thermoelectricity
537.62	Conduction and resistance
537.622	Semiconductivity and semiconductors
537.6226	Interactions in and specific properties of semiconductors

Standard subdivisions are used to denote, for example, special forms of works (dictionaries, periodicals, etc.). Thus, dictionaries and encyclopedias of electronics are classed at 537.03; periodicals on semiconductors are at 537.62205.

Classes 531–538 deal with the topics of classical physics: mechanics, sound, light, heat, electricity and magnetism. Modern physics has been forced rather awkwardly onto the structure, partly in 530 (mathematical physics at 530.1; quantum mechanics of specific states of matter at 530.4) and partly in 539 (structure of matter at 539.1; radiations at 539.2; molecular physics at 539.6; and atomic and nuclear physics at 539.7) (Dewey, 1989).

Because classification schemes have to deal with the literature of the past as well as publications of the present, they are inevitably rather conservative. Reclassification of stock is still labour-intensive, even in the age of the computer, and that means that extensive revisions of schedules are not undertaken lightly.

UDC was originally an expansion of the 5th edition of DDC undertaken to classify a projected universal subject bibliography and published in 1905 as *Manuel du répertoire bibliographique universel*. Class 53, Physics, has appeared in six English editions published by the British Standards Institution: two full editions (BS 1000) in 1943 and 1974; three abridged editions (BS 1000A) in 1948, 1957 and 1961; and the international medium edition (BS 1000M) in 1985. Since January 1992, UDC has been the responsibility of the.UDC Consortium. A UDC database has been created, in depth similar to that of the international medium edition, and this will be the basis of all future revision. A new medium edition is imminent at the time of writing (McIlwaine, 1993, personal communication).

UDC, like DDC, divides knowledge into 10 main branches:

.0	Generalities
.1	Philosophy. Psychology
.2	Religion. Theology
.3	Social sciences
.4	(Vacant: formerly Philology. Languages)
.5	Mathematics and natural sciences
.6	Applied sciences. Medicine. Technology
.7	The Arts. Recreation. Entertainment. Sport
.8	Language. Linguistics. Literature
.9	Geography. Biography. History

These decimal fractions are further subdivided decimally, such that .5 is analysed into:

.51	Mathematics
.52	Astronomy. Astrophysics. Space research. Geodesy
.53	Physics
.54	Chemistry. Mineralogical sciences
.55	Earth sciences. Geology. Meteorology, etc.
.56	Paleontology, etc.

Further decimal subdivision follows. The initial decimal point is omitted in practice, but additional points separate groups of (usually three) digits. For example, 'ferroelectricity' is found at 537.226.4.

The main outline of the scheme as it concerns physics (and as shown in the 1985 medium edition) is as follows:

53 Physics

530.1	Basic principles of physics
531	General mechanics. Mechanics of solid and rigid bodies
532	Fluid mechanics in general. Mechanics of liquids
533	Mechanics of gases. Aeromechanics. Plasma physics
534	Vibrations. Acoustics
535	Optics
536	Heat. Thermodynamics
537	Electricity. Magnetism. Electromagnetism
538.9	Physics of condensed matter (in liquid state and solid state)
539	Physical nature of matter
539.1	Nuclear, atomic and molecular physics
539.2	Properties and structure of molecular systems
539.3	Elasticity. Deformation. Mechanics of elastic solids
539.4	Strength. Resistance to stress
539.5	Properties of materials affecting deformability
539.6	Intermolecular forces
539.8	Other physico-mechanical effects

The UDC medium edition contains three times as many class numbers for physics as are found in DDC, and allows much greater depth than even this would suggest, since it permits the construction of compound numbers. It does this in two ways: by the use of signs, including the plus ('and'), the stroke (denoting a range of numbers) and the colon ('in relation to'); and by the use of auxiliary subdivisions to represent recurrent concepts either of general or of more limited application.

536+537	Heat and electricity (i.e. both subjects covered)
531/534	Mechanics
530.1:140.8	Philosophical outlook of physics
53(031)	Encyclopedias of physics
539.12	Elementary and simple particles
539.121.2	Mass and charge of elementary particles
539.126	Mesons
539.126.12	Mass and charge of mesons

While DDC and UDC show the places of works in hierarchies of subjects as well as indicating the linear order of items on shelves, LC is not hierarchical, merely enumerative, and this is the result of its origins as the shelf order of the Library of Congress. The science schedules (class Q) of the LC scheme originated in 1905, and were last published in full in the 7th edition, 1989. This is the main outline of class QC, Physics:

QC	1–75	General
	81–114	Weights & measures
	120–168.85	Descriptive & experimental mechanics
	170–197	Atomic physics
	220–246	Acoustics. Sound
	251–338.5	Heat
	350–467	Optics. Light
	474–496.9	Radiation physics (general)
	501–766	Electricity and magnetism
	770–798	Nuclear and particle physics. Atomic energy. Radioactivity
	801–809	Geophysics. Cosmic physics
	811–849	Geomagnetism
	851–999	Meteorology. Climatology

As in the case of DDC, this is at root a scheme for classical physics, beginning with generalities, and proceeding via mechanics, sound, heat, light, electricity and magnetism to meteorology and climatology. Modern physics has been shoe-horned into this basic arrangement, so that atomic physics follows mechanics, radiation physics follows light, and nuclear and particle physics follow electricity and magnetism.

Whole numbers and their subdivisions have no recurrent meaning. They are used indiscriminately within the range of numbers for a more general topic to indicate types of literature (histories, popular works, etc.) and to bring together special topics. The sub-arrangement of special topics within this framework is alphabetical and is indicated in the notation by so-called Cutter numbers:

QC 793	Elementary particle physics. Periodicals, congresses, yearbooks
QC 793.2	Elementary particle physics. General works, treatises, and advanced textbooks
QC 793.3	Elementary particle physics. Special topics A-Z
QC 793.3.H5	High energy physics
QC 793.3.S6	Spin
QC 793.9	Nuclear interactions. Periodicals, congresses, yearbooks
QC 794	Nuclear interactions. General works, treatises and textbooks
QC 794.6	Nuclear interactions. Special topics, A-Z
QC 794.6.F4	Feynman diagrams

Users of classification schemes have two ways to find what they want. They can perform a top-down analysis of a broader subject to reach the topic they want, using the classification schedules; or they can refer to an alphabetical index to the schedules. The former approach is more suitable to a strictly hierarchical scheme like UDC, while the latter is preferable in the case of an enumerative scheme like LC. It is important to remember that the same term may appear at different places in the schedules, according to context. In DDC, for example, fluid mechanics as an aspect of physics is at 532, but as an aspect of engineering it is at 620.106. Particular care must be taken with the LC scheme in this regard because major classes are published and indexed separately, so topics related to physics but in Class T, Technology, may easily be overlooked.

The traditional British subject catalogue is arranged in classification order and has a separate alphabetical subject index. There is, however, another way of organizing a subject catalogue: alphabetically, using subject headings. Typically, North American subject catalogues have been arranged in this way. There are three associated difficulties: the first is that the arrangement depends simply on the vagaries of the language concerned, and is not internationally meaningful; the second follows, that conceptually related topics are separated and so browsing is impossible; the third is that many terms have synonyms or near-synonyms. It is advisable, therefore, before approaching the subject catalogue, for the user to refer to the full list of subject headings, where typically both preferred and rejected terms will be listed, with references from the latter to the former. The place of each term in a hierarchy of concepts may be indicated, by giving, for each preferred term, the broader term which subsumes it, and the narrower terms which are its component parts. Terms related otherwise than hierarchically may be shown to aid clarification of the search.

Although libraries may have their own lists of subject headings, and there are other general schemes, *Library of Congress Subject Headings (LCSH)* maintained, revised and expanded since 1898 by the Library of Congress in Washington, D.C., comprises the predominant scheme in the English-speaking world.

Having searched for the term 'Science' in *LCSH,* one finds that it is at the top of a hierarchy of terms. 'Physics' is unsurprisingly on this hierarchy's next level. Among the almost 70 terms on the next level down is 'Thermodynamics'. The 30 terms given as narrower terms to 'Thermodynamics' include 'Spin temperature'. A term narrower than 'Spin temperature' is 'Nuclear magnetic resonance', and at the lowest level on this path traced down the hierarchy comes 'Meson resonance'.

This example creates the impression of totally discrete pyramids or trees of terms; in fact, when examined in more detail, the picture is much more complicated, for some terms have more than one broader term above them and the pyramids are in consequence interwoven.

In all, *LCSH* contains well over 100 000 headings for topics. It is continually updated to reflect the subject matter of current publications acquired by the Library of Congress, and published annually in print and quarterly in microfiche. The related CD-ROM, CDMARC SUBJECTS, is issued quarterly.

These, then, are the two traditional modes of subject access in library catalogues, which developed in the world of the card catalogue and of its successor, the microfiche catalogue. Computer technology has for a number of years allowed the sharing of catalogue records by libraries, either through co-operative schemes or on a commercial footing. However, the computer has developed from a means of catalogue production to a medium for catalogue display and search. Online public access catalogues (OPACs) as integral parts of library computer systems have a number of important advantages over even computer-produced catalogues of a more traditional kind. The catalogue may be made available simultaneously at different locations via local, national and even international networks. OPACs increasingly offer such facilities as access to records for items on order; sight of receipt records for periodical parts; details of loan status and the possibility of reservation. Often, a greater variety of access points is available than in traditional catalogues, including for example, International Standard Book Number (ISBN) and International Standard Serial Number (ISSN). No longer does the searcher have to move from one type of catalogue to another. How does all this affect the subject approach?

The first task for the physicist using an OPAC will be to establish whether there is a subject approach, and of what type it is, since the terminology used on the screen may be confusing, and there is no typical system. It may be possible to search classification numbers or subject headings; a third possibility may be offered, that of a more general search for words of the user's choice anywhere in a bibliographic record. It is well worth bearing in mind that as many libraries share catalogue records a classification scheme other than that used for arranging items on the shelves may be offered for searching. It will be necessary to check, however, that where class numbers or subject headings are searchable they are applied consistently to all catalogue records, or else items will inevitably be missed.

It is also essential to discover exactly what to expect from searches in a particular OPAC. The user may be permitted to browse indexes before selecting a term. Truncation may be allowed, so that, for example, a search on a DDC number will also retrieve all the items classified by subdivisions of that number. Boolean searching may be offered, so that sets of results can be combined using the logical operators AND, OR and NOT. Any or all of these options may be available in the same catalogue.

The characteristics of the classification scheme available are of particular importance. The highly structured UDC is particulaly suited to online searching, but only if the catalogue's software has sophistication to match. On the other hand, the lack of hierarchy in the LC scheme may render laborious the inclusion of narrower aspects in a search for a broader topic.

OPACs may free users of subject headings schemes from the tyranny of alphabetical order and worries about whether a term such as 'fluid mechanics' should be given in direct order or inverted as 'mechanics, fluid', as words within headings are often searchable.

Neither complete classification schemes together with their indexes, nor full lists of subject headings with references to preferred terms and showing the structure of broader, narrower and related terms, yet appear online as integral parts of catalogues. Progress is being made towards this goal, but in the interim period recourse must be had to the printed schemes, which may or may not be available at the terminal where the user is sitting.

The best advice which can be given at present is to utilize the subject searching possibilities to the full, having due regard to the limitations of a particular catalogue. If both classification numbers and subject headings are searchable, search both (Chan, 1990). If 'keyword' searching is on offer, use that as well, with the proviso that careful attention must be paid to the possible existence of synonyms, on the one hand, and words of the same spelling but different meaning, on the other, and also to variants such as plurals and adjectival forms.

The classified or alphabetical subject catalogue and its OPAC equivalents will direct readers to books and periodicals within the library. However, analytical catalogues of articles within periodicals are unusual, certainly outside special libraries, and so for a fuller subject search the physicist will need to turn to one or more of the major indexing or abstracting services in the field (or to one of the more specific services described in other chapters). These, too, will employ classification schemes or subject headings.

Finding more literature on a topic: indexing and abstracting services

Indexing services collect bibliographic information on papers and other forms of literature and provide lists in a suitable arrangement.

Abstracting services additionally provide resumés of the items listed. Such services typically now have at their heart an electronic database, which may be made available in one or more of several formats.

All the abstracting journals described below are published in print. Most are also available online. Regular snapshots of some databases are issued in CD-ROM format for use either on a single microcomputer or on local area networks of personal computers within a campus or organization. Each of the three formats has advantages and disadvantages.

Print is relatively cheap and portable, though unwieldy in multiple volumes. Simple searches can readily be conducted whether the primary arrangement of a printed bibliography is a classification scheme or an alphabetical subject vocabulary. However, more sophisticated searching has to be done with continual reference to indexes, and information collected has to be recorded by hand.

Online searching allows much more complexity and refinement in strategy (by the use of Boolean logic to broaden or narrow searches) and is not inhibited by the arrangement of the records in the database. Free-text searching of whole records, including abstracts, if present, is possible. The ability to limit results by language and by year of publication is normal. Online databases often do not stretch as far into the past as their printed counterparts; however, as J. L. Hall indicated in the previous edition of this book, this becomes less and less important as time goes on (Hall, 1985). On the other hand, costs of online searching include telecommunications charges, which are time-related, and charges for record use. Although time spent online has been reduced by the ability to download records, which avoids printing while online, libraries normally pass costs on to users, in whole or in part. To put some limitation on these costs, trained library staff usually perform searches on readers' behalf, after an interview in which the search is clarified and its possible development explored.

There is no such financial consideration in the case of CDs, where the recurrent costs (subscriptions and/or network licences) are known in advance. Readers use discs themselves. They can print or download records for personal use. Helpful menus of commands can be viewed as searches progress. The initial drawback of the medium, that only one person could use a disc at a time, has been overcome by networking. However, the user only sees the database frozen at a moment in the past; how old the snapshot is depends on how frequently the CD is reissued.

The main English-language abstracting service for physics is the INSPEC service of the IEE and the Institute of Electrical and Electronic Engineering (IEEE). For *Physics Abstracts* (and its sister journals in the fields of electronics and electrical engineering and computing, information technology and control technology) INSPEC scans the contents of over 4000 journals and over 1000 monographs, theses, conference proceedings and reports, comprising a quarter of a million items, each year. 150 000 of these are summarized in *Physics Abstracts*.

The INSPEC database from 1969 onwards is available online from many of the major hosts and is also available in CD-ROM format from 1989. INSPEC ONDISC covers the whole breadth of the database, while INSPEC-PHYSICS corresponds to *Physics Abstracts*. Subject searching approaches include classification codes, index terms from the *INSPEC Thesaurus* (also available in print – biennially, latest issue 1993), additional free-language (uncontrolled) terms, and treatment terms which enable the user to limit a search to, for example, review articles. Several university departments and research institutes subscribe to the INSPEC tape service enabling the database to be searched locally and this can significantly reduce the cost of access. A further facility, which is not unique to INSPEC, is the online order service for the user to request a photocopy of an article to be delivered by post or fax.

The printed version began in 1898. Since 1977, its primary arrangement has been the ICSTI International Classification for Physics, the broad outline of which is as follows:

0000 General
1000 The physics of elementary particles and fields
2000 Nuclear physics
3000 Atomic and molecular physics
4000 Classical areas of phenomenology
5000 Fluids, plasmas and electric discharges
6000 Condensed matter: structure, thermal and mechanical properties
7000 Condensed matter: electronic structure, electrical, magnetic and optical properties
8000 Cross-disciplinary physics and related areas of science and technology
9000 Geophysics, astronomy and astrophysics

Each twice-monthly issue contains an alphabetical subject guide to the classification scheme. Every six months a subject index to the individual entries is published, using terms from the *INSPEC Thesaurus*. The names of personal authors, corporate authors and conferences are also indexed.

The international competitors to *Physics Abstracts* are three: *Physics Briefs* = *Physikalische Berichte* (1979–) (formerly *Fortschritte der Physik* 1845–1918; then *Physikalische Berichte*

1920–78) is published twice a month in co-operation with the American Institute of Physics by the Deutsche Physikalische Gesellschaft and Fachinformationszentrum Energie, Physik, Mathematik. About 120 000 items drawn from all types of literature are abstracted each year, and the service claims strong Eastern European coverage. The corresponding database, PHYS, is available online from the host STN International. The classification scheme, PACS (the Physics and Astronomy Classification Scheme), expands the ICSTI scheme to include classes 84 (Electromagnetic technology), 85 (Electrical and magnetic devices) and 89 (Other areas of general interest to physicists), and gives more depth in classes 02 (Mathematical methods in physics) and 43 (Acoustics).

PASCAL (1984–, formerly *Bulletin Analytique* 1940–55; then *Bulletin Signalétique* 1956–83) is published by CNRS. It contains half a million entries a year on all aspects of science. It is available online from ESA and DIALOG, and on CD from 1987 onward. Sections of particular interest to the physicist include:

- *PASCAL Explore*, parts 11 (physique atomique et moléculaire, plasmas), 12 (état condensé), 13 (structure des liquides et des solides, cristallographie), 32 (métrologie et appareillage en physique et physiochimie), 99 (congrès, rapports, thèses);
- *PASCAL Folio*, part 10 (mécanique et acoustique);
- *PASCAL Théma*, part 230 (énergie).

The physics section of the monumental Russian abstracting journal *Referativnyi Zhurnal* (VINITI, 1953–) contains c. 80 000 abstracts per year from the physics literature of the world.

A rather different service, *INIS Atomindex* (1970–), is produced under the auspices of the International Atomic Energy Agency by co-operation among member states of the Agency and international organizations. It is a cross-disciplinary work, in that it covers publications on all aspects of nuclear science and its peaceful applications. The basic subject approach in the twice-monthly printed and microfiche issues is the classification scheme, the main outlines of which are:

A 00.00	Physical sciences	
	A 10.00	General physics
	A 20.00	High-energy physics
	A 30.00	Neutron and nuclear physics
B 00.00	Chemistry, materials and earth sciences	
C 00.00	Life sciences	
D 00.00	Isotopes, isotope and radiation applications	
E 00.00	Engineering and technology	
F 00.00	Other aspects of nuclear energy [including e.g.	

F 30.00 Nuclear documentation and F 40.00 Safeguards and inspection]

Provisional semi-annual and complete annual indexes are published, which allow searching by personal and corporate authors and conference names, by report, standard and patent numbers, and by subjects, using terms drawn from the *INIS Thesaurus*. *INIS Atomindex* is available online from a number of hosts including ESA and STN, and also on CD-ROM (SilverPlatter).

For an exhaustive search the physicist may wish to use the indexing and abstracting services appropriate to other closely related subjects.

Mathematical Reviews (1940–), published by the American Mathematical Society (AMS), appears monthly and is arranged in AMS classification order. The classes of particular interest to the physicist (that is, beyond mathematical methods) are:

70 Mechanics of particles and systems
73 Mechanics of solids
76 Fluid mechanics
78 Optics, electromagnetic theory
80 Classical thermodynamics, heat transfer
81 Quantum mechanics
82 Statistical physics
83 Relativity
85 Astronomy and astrophysics
86 Geophysics

Perhaps 15% of the (critical) reviews published in a typical issue are in these classes, making in total about 9000 a year. Each issue has an author index, and there are annual author and classified indexes. *Mathematical Reviews* is available online from BRS, DIALOG and ESA, and on CD-ROM (SilverPlatter)

Another user of the AMS classification is *Zentralblatt für Mathematik und ihre Grenzgebiete* = *Mathematics Abstracts* (1931–), issued by the Heidelberger Akademie für Wissenschaften and the Fachinformationszentrum Energie, Physik, Mathematik. Of the 50 000 items abstracted in a year (from 2200 journals) some 10% fall into AMS classes 70–86. The online equivalent is MATH, available through STN.

Every second weekly issue of *Chemical Abstracts* (1907–, issued by the American Chemical Society) includes six sections of particular interest to the physicist (though much else, especially in the field of physical chemistry, will be of value):

69 Thermodynamics, thermochemistry and thermal properties
70 Nuclear phenomena

71   Nuclear technology
73   Optical, electron, and mass spectroscopy and other related
      properties
76   Electric phenomena
77   Magnetic phenomena

The papers, reports, patents, etc. abstracted in these sections amount to over 65 000 a year out of the total coverage of half a million items. *Chemical Abstracts* is widely available online.

Engineering Index Monthly and its annual cumulation *Engineering Index Annual* (1962–, formerly *Engineering Index* 1884–1962) are arranged in alphabetical order of subject, using terms drawn from *SHE, Subject Headings for Engineering* (1970–) and indexed by author. They abstract over 100 000 items from about 2500 serials and many conferences. The online equivalent *COMPENDEX* (1970–) can be accessed from BRS, Data-Star, DIALOG, ESA, Orbit and STN. DIALOG publishes a CD-ROM version, also called COMPENDEX.

Finding the most recent work on a topic: current awareness services

Current awareness services developed to bridge the gap between the publication of a paper and its recording in an abstracting journal.

About 85 000 papers per year are listed in *Current Papers in Physics* (1966–), a product of the INSPEC service. Bibliographic details are arranged in classified order. There are no abstracts, and no indexes.

Current Physics Index (1975–), published by the American Institute of Physics (AIP), includes 32 000 abstracts per year of papers published in the 53 journals, including translations, of AIP and its member organizations, and is thus able to claim coverage of 90% of American research and 50% of that published in the former Soviet republics. *Current Physics Index* is able to achieve its timeliness because the abstracts are submitted by the authors of papers as part of AIP's publication process. Like *Physics Briefs, Current Physics Index* uses the PACS scheme. The corresponding online service is SPIN, available on DIALOG.

Early notification of mathematical papers, including those not destined for abstracting in *Mathematical Reviews*, is provided by *Current Mathematical Publications* (1969–), which is arranged in AMS classification order and has an author index.

Chemical Titles (1961–), published by the American Chemical Society, is a current awareness tool from the Chemical Abstracts

Service. It is published every two weeks and is basically an index of significant words (shown in their context) from the titles of papers in over 750 chemical journals. Each issue also contains an author index and a contents list for each journal issue scanned.

Current Contents:Physical, Chemical & Earth Sciences (weekly, 1961–) published by the Institute of Scientific Information (ISI), publishes tables of contents for 800 journals and indexes them by author (with addresses), and title words. It is also available on weekly diskettes.

Following up published research

It is possible to discover where a particular piece of published research led by means of *Science Citation Index* (*SCI*), the brain-child of Eugene Garfield, which is published by ISI.

In the printed version, *SCI* has three main parts. More than 3000 journals in all scientific subjects, those most frequently cited in scientific literature, are scanned and all articles, review articles, editorials, book reviews, letters, corrections, etc., are alphabetically indexed by first author (with references from other authors). This single alphabetical list forms the *Source index*. The titles of all the articles, review articles, etc., are broken down into their significant constituent words. A single alphabetical list of these words is produced. Under every word appears each other word found in combination with it, and against each entry appear the author and date of the article with that combination. It is thus a sophisticated title-word index to the *Source index*, and is called the *Permuterm subject index*. Thirdly, in the *Citation index*, every citation of another work made in each of the articles, review articles, etc., is recorded and put into a single alphabetical sequence by first author. Under each citation is listed the author and abbreviated biblio-graphic reference of the article making the citation. This abbreviated reference can be filled out from the *Source index*. Thus by looking up a known article in the *Citation index* one can find bibliographic details of articles which subsequently referred to it, and some at least of those articles will have taken the subject further.

SCI is widely available online (including on the Bath Information and Data Services – BIDS – operated on the British Joint Academic Network – JANET) and also (1980 onwards) on CD-ROM. Several universities subscribe to the ISI tape service and provide informa-tion from the database on the campus network. For example, Caltech takes records selectively from *SCI*, to provide a document delivery service from the periodicals within the library system (Card, 1989).

This is a work of great imagination and enormous utility. However, users need to heed some warnings. First, the citations are only as accurate as they were in the source article, and may appear in variant forms. Secondly, a paper establishing an experimental method is likely to be cited in many subsequent papers which take the method itself no further. Thirdly, the *Permuterm subject index* indexes combinations of title words, not subjects, and remains the only approximation to subject searching in the printed version. This defect has been mitigated in the electronic versions by a variety of innovations. Searchable 'research fronts' are clusters of pairs of co-cited articles named from frequently occurring title phrases. 'Related records' are articles which have a group of references in common. The addition to the database, since 1991, of searchable abstracts, 'author keywords', selected therefrom, and 'KeyWords Plus' generated from words which recur in the titles of cited references, has provided further improvement (Snow, 1991). (It should be noted, however, that in the CD-ROM version research fronts are not searchable at all, and abstracts and keywords are only available on the enhanced SCI CDE WITH ABSTRACTS). Finally, it must not be forgotten that *SCI* is not exhaustive, since it covers only 400 serials likely to be of immediate interest to physicists. From Appendix A one may infer that the coverage is therefore about 90%.

Current Research in Britain. Physical Sciences (British Library, annual) is a useful guide to where research on a topic is being carried out in the UK. The first volume lists under the names of universities and other institutions, subdivided first by department and then by the name of the principal investigator, the research being conducted. The second volume indexes the information in the first by general and specific subject words, and enables the user to trace the personal names which appear there.

Finding a work from a full reference: name/title catalogues

The overwhelming majority of readers who ask the question 'Where are the physics books?' are actually looking for one or more specific works, to which they have more or less accurate bibliographic references. What they should be asking, and what, with luck, they will be told, is 'Where is the catalogue and how does it work?'. The catalogue is the second key to the library's stock (the classification scheme being the first). It also appears to be one of the most alarming aspects of any library. Cataloguers have been less than successful in demystifying their trade and catalogues are utilized less than a librarian would hope, even by library staff. It is worth, therefore, outlining the typical features of modern catalogues, and the best mode of attack by the reader.

Regardless of format, name/title catalogues will perform two main functions: through one or more points of access ('headings') they will describe an item briefly and they will locate it. Access points may be names of personal or corporate authors, or titles of individual works or series. Catalogues aim to distinguish pairs of individuals or corporate bodies with identical names and to bring works together under a single form where a person or corporate body varies in name. Names, titles and subjects may be interfiled, but more typically the subject catalogue, if it exists, is separate.

The history of cataloguing and cataloguing codes is long and complex, and can scarcely be done justice in a few words. However, a very brief attempt to summarize it may assist the physicist faced with a library catalogue.

In the nineteenth century, libraries' stocks consisted almost entirely of monographs, largely written by a single person, and periodicals, often the printed version of oral proceedings of learned societies. The main access points ('main entries') in such nineteenth-century catalogues as were not subject catalogues, therefore, were typically authors' names (for monographs) and the names of learned societies (for periodicals). 'Added entries' with abbreviated descriptions might provide access by editors' names or (for periodicals) titles. However, each extra entry cost time and money.

Today, many academic works (including conference proceedings and other works produced under editorial direction) are of multiple or corporate authorship and few periodicals now consist of transcripts of papers actually read to learned societies. Reprographic technology (carbon paper, the typewriter, punched tape, then the computer) has reduced the cost of proliferating added entries.

The effect of these developments may be seen in the most modern cataloguing code, the *Anglo-American cataloguing rules* (Library Association, 2nd edn., 1978; revised edn., 1988). *AACR2* is in general more generous with added entries than earlier codes and all works are accessible by title. The relative decline of the personal author has to some extent diminished the importance of the cataloguer's stock-in-trade, the establishment of the correct (sometimes merely socially correct) form of the author's name. In general, persons are now called what they choose to call themselves in their works and the same commonsense approach has been applied to corporate bodies. All this helps the reader and should speed the cataloguer.

AACR2 is now the code of choice for libraries in the English-speaking world, though many library catalogues drawn up under more restrictive codes still survive.

So much for the theory of name/title catalogues. Here are three typical examples of complete references:

E. J. Burge, *Atomic nuclei and their particles*, 1988

C. Leubner *et al.*, Elementary relativity with 'everyday' clock synchronization, *European Journal of Physics*, vol. 13, no. 4 (1992), 170–177

R. Merlin, Raman studies of Fibonacci, Thue-Morse, and random superlattices, in *Light scattering in solids v*, ed. by M. Cardona and G. Güntherodt (Topics in applied physics, vol. 66), 214–232

How should the reader look for these? The first appears to be a monograph written by a single author. The most likely access point in the catalogue will be the author's name, i.e. Burge, though the form of name given as the heading in the catalogue may not correspond to the form in the reference. The heading could be Burge, E. J., or Burge, Edward James, or even Burge, E. J. (Edward James).

The second evidently refers to an article in a periodical. It is unlikely that the catalogue will provide access to single articles in periodicals, so the correct point of access is the periodical title: *European Journal of Physics*. The library may have a separate catalogue for periodicals.

The third appears to be a chapter in a work produced under editorial direction. A typical library catalogue would have an entry under the title of the book, *Light scattering in solids v*. However, note that the work is a volume in a series. It is possible that the library keeps the series together, as if it were a periodical, so if the work is not found under the title or the first editor, Cardona, it is worth checking for the series title in the catalogue (in the periodicals catalogue if that is separate).

Pitfalls of several kinds may be encountered. The first is the form in which names of persons are given. Some cataloguing codes insist on the researching of personal names to provide full forenames, and the user with a reference of surname and initials may in consequence be confused or misled.

The second is confusion of similar names (Brown/Browne; Andersen/Anderson/Andersson; Hofman/Hofmann/Hoffman/Hoffmann) and of title words with variants (encyclopaedia/encyclopedia; high temperature/high-temperature).

The third is in the filing order, which may be crucial in a large catalogue. The main issues faced by the framer of filing rules, and thus by the user of the catalogue, are: whether to count every word or to ignore minor parts of speech; whether to count a space between

words as a character which files before A ('word by word' filing) or
as no character ('letter by letter' filing); whether to treat a hyphen as a
character; whether to ignore diacritical signs such as the German
umlaut or whether to give them a value; how to interfile names and
titles; and the treatment of Mc and Mac. These issues are of much
less significance in an online catalogue, since accurate entry of the
search term should be sufficient to locate the correct part of the catalogue (although diacritics may cause trouble). However, users of
earlier types of catalogue would be well advised to enquire about
filing rules before beginning a catalogue search, so as to avoid need-
less frustration.

In large card catalogues, it may be difficult to find works whose
authors have common surnames or which have titles of a general
nature that have been used often. For example, M. S. Smith, *Modern
physics* (Longman, 1960) might be hard to trace on both counts. If
the catalogue gives the choice of searching under either author or
title, whichever is the more unusual should be chosen. If there is no
choice, it may sometimes save time to consult a bibliography for
more information, to discover, for example, that M. S. Smith's forenames are Michael Seaton.

Periodicals cause particular problems to library users and to
librarians because of their changing relationships with issuing bodies
and because of their frequent changes of title, mergers and splits. For
example, *Bulletin of the Institute of Physics* (1950–60) became
Bulletin of the Institute of Physics and the Physical Society (1961–
67). The title was then changed to *Physics Bulletin* (1968–88).
Physics Bulletin merged with *Physics in Technology* to form *Physics
World* (1988–). In the past, periodicals were typically described in
terms of either the earliest or the most recent manifestation, with
notes on, and added entries for, related titles. The norm today is
'successive entry', i.e. a separate description is made for each
manifestation, with references forward and/or back in each description. However, the reader should be aware of the continued existence of the older types of catalogue.

Finding a work from a partial or inaccurate reference

When the physicist has searched the library catalogue unsuccessfully
and suspects that a reference is deficient or inaccurate in some
respect, the available options will depend on the size and resources
of the library.

It would be possible at this point to list a large number of general
reference works and bibliographies, but I shall instead recommend

in passing the two classic bibliographies of reference works, *Walford's guide to reference material,* (5th edn., 1989–1991, Library Assoc.) (Vol. 1 covers Science and technology) and *Guide to reference books*, edited by E. P. Sheehy (10th edn., 1986, ALA, with a supplement covering 1985–90, edited by R. Balay, 1992), and focus on a relatively small number of items which will help to deal with the inadequate reference.

One of the causes of difficulty most frequently encountered in tracing a work from a reference is confusion as to the meaning of an abbreviation of a periodical title. It is often a cause of surprise that there is no universally accepted abbreviation for every title. There is a British and international standard (BS4148, ISO4, *Specification for abbreviation of title words and titles of publications*) but few journal publishers insist on its use. A most useful work, therefore, is *Periodical title abbreviations: by abbreviation*, edited by L.G. Alkire (7th edn., 1989, Gale). This does not attempt to prescribe a correct form of abbreviation, but it does list variant forms where they occur, and is based on commonly used abstracting and indexing services. Its sources include *INIS Atomindex,* INSPEC, *Mathematical Reviews, Science Citation Index, Physics Briefs,* and *Surface and Vacuum Physics Index.* Excluded from its coverage, however, is *Chemical Abstracts Service Source Index (CASSI).* This lists around 18 000 current serials, as many again that have ceased or have changed title, as well as many conference proceedings and other collections of papers. Titles are listed alphabetically in full, with the ISO4 abbreviations highlighted within them, so *CASSI* is not always easy to use if one is starting with an abbreviation.

A title keyword approach to imperfectly known serial titles is available with the British Library's microfiche *Keyword Index to Serial Titles (KIST)*, which covers 440 000 titles, including the 72 000 currently received by BLDSC and SRIS and a quarter of a million non-current serials held in the British Library. The CD version is BOSTON SPA SERIALS ON CD-ROM.

The Serials Directory (annual, Ebsco) and *Ulrich's International Periodicals Directory* (annual, Bowker) each cover about 150 000 current periodicals. In the printed versions, the arrangement is by subject, with title and ISSN indexes (*Ulrich's* additionally has useful indexes to serials available online, arranged by title and by vendor, and to serials available on CD-ROM). However, in the CD versions a much wider range of search options is possible, including keyword searching.

For title (but not keyword) access to details of older physics periodicals, the best work, both for coverage and for ease of use, is *World list of scientific periodicals* (4th edn., Butterworths, 1965), though it should be noted that it does not cover periodicals which ceased publication before 1900. There are supplements which cover 1961–68 (together) and 1969–1980 (separately). Most needs will be met by *World list, KIST* and one of the two current periodicals directories mentioned above.

The 600 000 machine-readable bibliographic records for serials collected under the CONSER (Co-operative Online Serials Program) scheme, originally a resource for cataloguers, are now becoming available for reference use. The Library of Congress issues the whole database on CD-ROM as CDMARC SERIALS (quarterly, cumulative, 1992–), offering a wide range of search options; the same records are to be found in the online union catalogue of OCLC, the Online Computer Library Center; 220 000 CONSER records for current serials are searchable on CARL, the Colorado Alliance of Research Libraries network; and the CD version of THE SERIALS DIRECTORY contains CONSER records in addition to its primary, more conventional, descriptions of over 150 000 current titles.

A third group of sources will enable references to monographs to be confirmed or corrected. Access to details of recent books is about to be transformed by the issue of a joint English-language books-in-print CD-ROM by Whitaker (publisher of *British Books in Print*) and Bowker (publisher of the American *Books in Print*) (Bowker, 1992).

Bowker's *Scientific and Technical Books and Serials in Print* (annual) combines the scientific and technical entries from *Books in Print* and *Ulrich's International Periodicals Directory*. It is available on CD, together with *American Men and Women of Science, Corporate Technology Directory* and the *Directory of American Research and Technology*, on a disc called SCITECH REFERENCE PLUS.

Only the largest libraries will have the gigantic *National Union Catalog* (*NUC*), which records the holdings of the Library of Congress and of many other major North American libraries. However, for the physicist, a much more compact and easily used substitute is *Pure & applied science books 1876–1982* (Bowker, 1982), arranged under Library of Congress subject headings but with author and title indexes. There are 220 000 entries in this latter work, including some 5000 published before 1876.

Many British libraries will have the *British National Bibliography* (*BNB*) (weekly, with various cumulations, 1950–), which is the record of the works received under legal deposit legislation by the Legal Deposit Office of the British Library and its predecessor, the Copyright Office of the British Museum. *BNB* is in classified order (DDC) with author and title and subject indexes. It now contains details, supplied by Whitaker, of forthcoming British books.

For earlier British books not found in *NUC* or *Pure & applied science books 1876–1982* the best source is the *British Library general catalogue of printed books to 1975* in CD-ROM format (Saztec Europe; distributed by Chadwyck-Healey) or in print (360 volumes and six supplements; Clive Bingley, K. G. Saur, 1979–88) (also found in different cumulations in earlier printed editions as the *British Museum general catalogue of printed books)*.

Volumes in monographic series can prove remarkably elusive, since references may omit either the volume title or the series information, and catalogues may treat each volume separately, with no series title access, or only list the series as a whole. Bowker has published three works which will assist in these circumstances: *Books in series, 1876–1949* (1982), *Books in series* (4th edn., 1985), which covers the period 1950–1984, and *Books in series, 1985–88* (1989). Each indexes volumes in monographic series numerically under the series title, then by author and by title. A CD-ROM is available.

Conference proceedings can be identified by keywords from titles and conference names in BLDSC's *Index of Conference Proceedings* (monthly, 1964–; annual cumulation; microfiche 1964–1988; available online on BLAISE; CD equivalent BOSTON SPA CONFERENCES ON CD-ROM). A quarter of a million conferences are now covered, some published as monographs, others as parts of periodicals. BLDSC shelfmarks are given. Those shelfmarks which have a decimal point after the first four digits belong to serials, whose titles can be discovered by looking up their shelfmarks in *KIST*.

ISTP, ISI's *Index to Scientific and Technical Proceedings* (1978–), allows conference proceedings and the articles in them to be traced by authors, editors, title words, sponsors, meeting locations and subject categories. *ISTP* is available online from Orbit and (to British academic users) on BIDS. Other papers will best be sought in one of the abstracting services mentioned above, or in *SCI*.

The reader with a faulty reference is best advised to decide whether, in all the circumstances, the details refer to a monograph, a journal paper or a conference paper, and to act accordingly. *In extremis* help can often be obtained from the *Citation index* of *SCI*, where it may be possible to find an alternative version of the reference in hand which will clarify matters.

Last, but not least, there is one often underestimated resource: the library staff. Few librarians will claim to be expert in physics. However, they will all be familiar with the tools of their own trade and, perhaps equally importantly, they will be aware of the idiosyncrasies of their own library. Most librarians are only too glad to wrestle with seemingly intractable references on behalf of their readers.

References

Bowker (1992) World's largest books-in-print disc. (1992) *Information World Review*, **75**, (November), 3.

Bradford, S. C. (1948) *Documentation*, pp. 86, 106–121. St. Albans: Crosby.

Card, S. (1989) TOC/DOC at Caltech: evolution of citation access online. *Information Technology and Libraries*, **8** (2), 146–160.

Chan, L. M. (1990) The Library of Congress classification system in an online environment. *Cataloging and Classification Quarterly*, **11** (1), 7–25.

Dewey Decimal Classification and Relative Index (1989) vol. **4**: Relative index manual, p. 885. Albany, NY: Forest Press.

Friend, P. D. (1975) Science libraries. In: *Use of physics literature*, ed. (from second edition text by PJRW) H. Coblans, p. 24. London: Butterworths.

Hall, J. L. (1985) Abstracting, indexing and on-line services. In *Information sources in physics*, 2nd edn., ed. D. F. Shaw, p. 68. London: Butterworths.

McSeán, T. (1991) CD-ROM and beyond, *Serials*, **4** (2), 47–52.

Snow, B. (1991) SCISEARCH changes: abstracts and added indexing. *Online*, **15** (5), 102–106.

Warren, P. J. R. (1985) Science libraries, reference material and general treatises. In: *Information sources in physics*, 2nd edn., ed. D. F.Shaw, p. 23. London: Butterworths.

CHAPTER FOUR

Atomic and molecular physics

JULIE M. HURD

Atomic and molecular physics makes up part of the field commonly referred to as 'modern physics'. The adjective *modern* is intended to contrast this specialization with the older field of classical physics, and emphasizes its relatively recent emergence during this present century. Atomic and molecular physics, in particular, can be considered to have developed from the Bohr theory of the hydrogen atom (1913), the discovery of electron spin by Goudsmit and Uhlenbeck (1925) and the statement of the Pauli exclusion principle (1926). Associated mathematical formulations attributed to such theoreticians as Dirac and Heisenberg form the basis for quantum mechanical calculations that enhance our understanding of experimentally observed atomic and molecular spectra and allow the prediction of various properties of atoms and molecules. The information sources described in this chapter will include those treating the following topics:

- the electronic structure of atoms and molecules;
- atomic spectra and the interaction of atoms with photons;
- molecular spectra and the interaction of molecules with photons;
- atomic and molecular collision processes;
- experimentally derived information on atoms and molecules.

Most of this chapter will deal with sources of information that are theoretical or computational rather than experimental in nature. Nonetheless, it must be recognized, as it indeed has been by Per-Olov Löwdin, that quantum mechanical calculations rest on some 150 years of experimental research in spectroscopy and related fields. The reader is also referred to other chapters in this same

volume covering chemical physics, statistical physics and thermodynamics (Chapter 5), computational physics (Chapter 7) and
optoelectronics, quantum optics and spectroscopy (Chapter 16).

The literature of physics has been subjected to extensive study by
historians of science, sociologists of science, librarians and information scientists. Using various bibliometric techniques they have
examined the scatter of the literature, citation practices, the nature
of information sources cited by physicists, obsolescence and other
information-related aspects of physics. Common to many of these
studies is the finding that physicists desire very current information
and that journal articles reporting primary research in physics are
'used' most, to the extent that use can be measured by citations in the
writings of other scientists, in a two-to three-year period following
their publication. Some of the sources treated in this chapter will
address the physicist's need for the most up-to-date information
available on the topic of interest.

Contrasted with this need for very current materials is another
requirement for information that surveys a specialized area of
research over a period of time, for example, nuclear magnetic
resonance or microwave spectroscopy. In some circumstances a
scientist (who may be a graduate student beginning research or a
researcher from one specialization hoping to apply concepts
developed in another field) will find it useful to consult a book that
provides basic theory, computational or methodological techniques,
analyses of experimentally derived data and, perhaps, bibliographies of significant earlier writings and tables of numerical data. In
atomic and molecular physics a reader will note the number of
monographic treatises serving this type of information need that are
regarded as 'classics'. These works appear to see sustained use as
texts by graduate students and as reference works by research
scientists for long periods of time following their publication. They
tend to be revised and reprinted as well. In the pages that follow,
treatises of this sort will be listed, some of which were first
published over twenty years ago. The ongoing utility of such
sources is frequently verified by the many citation analyses
reported by information scientists working with the large bibliographic databases of journal articles created by the Institute for
Scientific Information (ISI). In reports published by Eugene
Garfield and his co-workers one finds ranked lists of very heavily
cited books, and included on such lists are quite a number of the
'classic' works identified later in this chapter. Additional evidence
of the significance of many of these titles can be acquired by visiting
science libraries, where the worn and tattered state of these books
on the shelves provides silent evidence of their use by numerous
information seekers.

To conclude these introductory remarks it seems appropriate to acknowledge the interdisciplinary nature of atomic and molecular physics. The earliest publications in this specialization were written by scientists who were considered primarily to be physicists and who, if academicians, were based in departments or institutes of physics. They published in journals of physics or, perhaps, applied mathematics. These disciplinary boundaries soon shifted, however, as chemists began to share the same concerns, to utilize the findings of this research and to work on the same types of research problems themselves. Such scientists who might have been called 'chemical physicists' or 'physical chemists' were active in founding the *Journal of Chemical Physics* in 1933, and this publication continues today as a widely read and highly regarded bridge between the disciplines of physics and chemistry. More recently, biological scientists have come to apply quantum mechanical methods to study complex, naturally occurring molecules. This further broadening of the discipline base of atomic and molecular physics will be noted in several conferences and symposia to be cited below.

As to the level of information sources discussed here, most are intended for use by the specialist rather than the lay person. Some are considered suitable for advanced undergraduates in physics or chemistry, but more are comprehensible only by the graduate student or research scientist. An effort will be made to note any exceptions to this generalization.

Primary journals

As in almost all other scientific fields, much informal communication of research findings in atomic and molecular physics takes place prior to formal publication in a primary journal. That earlier exchange of scientific information at seminars and conference presentations and through preprint exchanges is treated in considerable detail in the discussion of grey literature by J. Chillag in Chapter 19. Herein we shall consider the more formal channels of communication, whether these be printed, as in journals or books, or electronic (or magnetic), as in bibliographic databases.

Research in atomic and molecular physics sees publication in a variety of journals ranging from more general physics journals with large circulations to smaller more specialized periodicals. In the former category are the highly regarded general physics journals which have long been published by leading professional societies.

Il Nuovo Cimento (Società Italiana di Fisica, 1855–), biweekly. Articles are accepted in English, French, German, Italian and

Spanish, although all have English summaries. Part D covers condensed matter, atomic, molecular and chemical physics and biophysics and most articles now appear to be in English.

Journal of Chemical Physics (AIP, 1931–), semi-monthly. The stated purpose of this journal is to bridge the gap between journals of physics and journals of chemistry. Of particular interest here are the articles published in the sections on spectroscopy and light scattering; molecular interactions and reactions, scattering, photochemistry; and quantum chemistry, theoretical, electronic and molecular structure.

Journal of Physics B: Atomic, Molecular and Optical Physics (Institute of Physics, Bristol, UK, 1968–). This section appears twice-monthly and publishes research articles on the study of atoms, ions and molecules and their interaction with radiation and other particles. Articles are accepted in English, French or German; all have English summaries. This journal also includes 'Letters to the Editor' (shorter communications, comments and notes).

Physical Review A: General Physics (AIP, 1970–). This section, published monthly, contains articles on atomic and molecular spectroscopy, interaction of atoms and molecules with radiation, atomic and molecular structure and collision processes. In addition to research articles are included comments, errata, rapid communications and brief reports. The parent journal began publication in 1893.

Other journals in which articles of interest to scientists working in this field are published are of a more specialized nature. These journals tend to be of a more recent origin and to be products of commercial publishing houses rather than of professional societies. Their authorship is international, although English is the predominant language of publication. John Wiley, based in New York, publishes the *Journal of Computational Chemistry* (quarterly since 1990) and the *International Journal of Quantum Chemistry* (monthly since 1966), both of which contain purely theoretical papers. Academic Press, also in New York, has been publishing in the field of atomic and molecular physics for some time. Their *Journal of Computational Physics* (monthly since 1966) includes both review and research articles on mathematical techniques developed in the solution of data-handling problems related to the description of physical phenomena. The *Journal of Magnetic Resonance* (15 issues per year since 1969) and the *Journal of Molecular Spectroscopy* (12 issues per year since 1957), both also from Academic Press, combine the theoretical and the experimental in their articles which provide theoretical analyses of experimentally measured data.

Several primary journals in the field are published in Amsterdam for an international audience. North-Holland Press, a subsidiary of Elsevier Scientific Publishing, offers the *Journal of Luminescence* which appears in 18 issues per year and has text in English, French and German with summaries in English. Elsevier's products also include the *Journal of Molecular Structure* and its subsection *THEOCHEM*, devoted to discussions of applications of quantum mechanics to organic, inorganic and biological problems; the *Journal of Electron Spectroscopy and Related Phenomena*, treating theoretical and experimental studies in photoelectron spectroscopy and ultraviolet and X-ray-induced electron-impact energy-loss spectroscopy; and the *International Journal of Mass Spectrometry and Ion Physics*, with a primarily experimental focus. All these Elsevier publications accept research papers in English, French or German.

A group of journals with quite a narrow subject focus are published by John Wiley & Sons, Ltd. in Chichester, UK:

JRS: Journal of Raman Spectroscopy (1973–), monthly
OMS: Organic Mass Spectrometry (1968–), monthly
XRS: X-ray Spectrometry (1972–), quarterly

Still others are the products of a diverse array of publishers in various countries:

Atomic Spectroscopy (Perkin-Elmer, 1962–), bimonthly
Mass Spectrometry (Royal Society of Chemistry, 1971–), biennial
Molecular Physics (Taylor and Francis, 1958–), 18/year
Spectrochimika Acta. Part A: Molecular Spectroscopy and *Part B: Atomic Spectroscopy* (Pergamon, 1939–), monthly
Spectroscopy: An International Journal (Pare, 1992–), bimonthly
Theoretika Chimica Acta: A journal for structure, dynamics and radiation (Springer, 1962–), monthly

Except for the last title, these journals tend to combine theory and experiment in their articles. English is the primary language of publication, although some also accept manuscripts in French or German. All of the journals mentioned in the foregoing discussion contain articles that report in considerable detail theoretical and/or experimental research in atomic and molecular physics. Many of them also publish shorter articles frequently categorized as 'Letters to the Editor' including communications of very significant new findings, research notes, comments on previously published work and errata correcting earlier articles. The type of short article commonly called a 'communication' is of particular interest to those active at a scientific research frontier. This brief report serves not only to communicate, with a minimum time lag, research in a rapidly developing area but also to establish a priority claim for the author.

Refereeing of such articles tends to be both rigorous and rapid, as the journal editor is desirous of sharing the communication with readers as speedily as possible. In an effort to minimize further the time lag from submission to publication, a number of publishers have, in recent years, begun to produce separate communications journals devoted entirely to these brief articles. Frequently these journals may photo-reproduce author-submitted typescripts rather than typeset manuscripts. *Physical Review Letters*, published by the American Institute of Physics, was a pioneer publication of this type; now there are many serving various specialized scientific fields. Examples of journals publishing communications in the area of atomic and molecular physics include the following:

Chemical Physics Letters (North-Holland, 1967–), weekly
Europhysics Letters (Editions de Physique, 1986–), 24 issues/year
Letters in Mathematical Physics (Kluwer, 1975–), 12 issues/year
Physics Letters. Section A: General, Atomic and Solid State Physics
   (North-Holland, 1962–), 90 issues/year
Spectroscopy Letters (Dekker, 1968–), 10 issues/year

During the decade of the 1960s and thereafter, the applications of computers to the publishing of scientific journals contributed to the development of online bibliographic databases. Now computer-assisted literature searching is commonplace, whether it is carried out by a scientific information specialist or the end user. More recently technology has produced both multi-media and electronic journals, and the emerging electronic journals promise even more rapid dissemination of research findings than the letters and communications journals described above. The first of these to appear seem to have been the electronic equivalents of the paper versions and were created in the process of typesetting and producing the more familiar format. Database vendors such as STN International have made available files of online journals that include many of the titles listed in this chapter; these journals can be searched full-text to improve retrieval of information not necessarily accessible through indexes or abstracts. At the present time a number of institutions and organizations are engaged in projects designed to explore means for improving access and delivery of scientific information through use of electronically-stored journals.

More innovative types of electronic journals have a mix of electronic and paper components and these represent a newer phase of development. One of these which includes articles in molecular physics is a multi-media publication: *Tetrahedron Computer Methodology* and is an international journal published bimonthly since 1988 by Pergamon. It consists of a collection of articles in the standard paper format, but is accompanied by a diskette which includes the full-text of the articles in ASCII as well as executable programs, source code, datafiles, and parameter sets. Subscribers specify whether the disk is sent in Macintosh or IBM-compatible format. A reader of an article in this journal may follow a text discussion of computation of a molecular structure, while simultaneously viewing on an adjacent terminal a rotating picture representing the molecule. Subsequently, the programs may be used by the reader for additional computations, and the datafiles provided as input for other research. In 1991, *Spectrochimica Acta, Part B* began publishing an electronic supplement, *Spectrochimica Acta Electronica*, to allow simultaneous publication of executable computer programs and data in a medium convenient for both authors and readers. The disks are supplied, with a looseleaf binder, to subscribers; users are expected to consult the hardcopy papers for specific instructions on use of the programs and data.

Yet other electronic journals do not exist at all in paper unless their readers choose to print copies from their computers. These journals consist of articles, with or without tables or other graphics, which are distributed over computer networks such as the Internet or on CD-ROM. By eliminating any dependence on paper publication, the publishers of these journals hope to reduce publication delays further. Both scientists and information specialists are examining the implications of electronic journals and speculating on the impact they will have on scientific communication. It is certain that readers will be called upon to change some comfortable habits as electronic journals are constrained by the technology that supports them and, at this time, that equipment is not as fully portable as a paper issue of a journal. The peer review process, on which journal publication rests, may also see changes that go beyond the mere speeding-up of the editorial processing of submitted articles. The reward structure, particularly for academic scientists, in turn relies on peer review, and how new forms of scientific publication will be evaluated by promotion and tenure committees will likely influence acceptance of electronic journals. Electronic journals are just beginning to appear and many questions related to their production and use are only now being explored; resolution will come at some future date when the situation is more fully evolved.

Symposia and conferences

Conferences and meetings serve several purposes for the scientists who attend them. These gatherings provide opportunities for scientists to present preliminary reports of their research findings and to obtain helpful criticism and evaluation from peers. Although conference papers may later appear in print as published proceedings, studies have shown that much of this research subsequently appears in primary journals as well in more complete, less speculative and perhaps even revised form. Scientists also engage in less formal interchange of information at conferences as they gather to discuss papers in meeting rooms and hallways and at meal times. The speakers and other visible participants are identified as specialists in a particular field, and this serves to facilitate discussion with those who share their interests. Conferences provide significant opportunities for members of an 'invisible college' to meet; among attendees may be potential new members as well. Many conferences are open to all who choose to register; those large international meetings sponsored by professional societies are of this type. Others, however, function as invited meetings and are only open to a small, select group. Any published proceedings from this latter type of gathering will be of particular interest to those who were not among the invited attendees.

In atomic and molecular physics there are a number of active groups whose conference presentations appear in published form. Sometimes these can be troublesome to locate, however, as the official titles of the meetings may vary from year to year, as may locations. In addition, the scientists participating may use their own familiar designations in referring to meetings. The Löwdin Symposia meetings on Sanibel Island, Florida are an example:

International Symposium on Atomic, Molecular and Solid-State Theory, Collision Phenomena and Computational Methods. Proceedings, Per-Olov Löwdin (ed.) (Wiley, 1967–) These articles are also published in the *International Journal of Quantum Chemistry.* Individual meetings have been held to honour noted scientists: John C. Slater, Henry Eyring, E. Bright Wilson, John H. van Vleck, etc.

Other significant published meetings include:

Computational Techniques in Quantum Chemistry and Molecular Physics (NATO Advanced Study Institute) (Reidel).
International Conference on the Physics of Electronic and Atomic Collisions (North-Holland. 5th, 1967), irregular. *International Congress of Quantum Chemistry.* The first congress was held in 1973 and the proceedings were published in 1974. Subsequent congresses have followed at approximately three-year intervals. The first three proceedings were published as a series by Kluwer; later proceedings appeared either as separate monographs or

were included in the *International Journal of Quantum Chemistry*. The publication pattern for this particular meeting illustrates a number of the difficulties encountered with the conference literature.

Jerusalem Symposia on Quantum Chemistry and Biochemistry (Reidel, 1969–).

NATO Advanced Study Institutes Series: Series B, Physics and *Series C, Mathematical and Physical Sciences* (various publishers), irregular.

Secondary services

The literature of most scientific disciplines includes several types of secondary publications that provide access to and evaluation of the primary research literature. We shall examine here indexing and abstracting services, review journals and reviews of progress. Also noted are the increasingly numerous electronic versions of these publications in both online and CD-ROM formats.

Indexes and abstracts

The indexing and abstracting services providing the most comprehensive and timely coverage of the research literature in atomic and molecular physics are those serving the entire discipline of physics. Unlike many other fields, physics has a number of partially overlapping secondary services including the major abstracts described below.

Bulletin Signalétique (CNRS, 1940–), monthly (former title (1940–1955): *Bulletin Analytique*) was recently reorganized and renamed as the *PASCAL* journals by its producer CNRS, (Centre de Documentation Scientifique et Technique). This service indexes journal articles, theses, reports, conference proceedings, books, and some patents in many disciplines, including atomic and molecular physics, and provides bibliographic citations and indicative abstracts in French. Titles of articles not originally in French are translated into that language with a notation to the original language of publication. There are subject and author indexes in each issue and in annual cumulations. Users of this service should note that section titles and scope change fairly frequently; as this was written there appeared to be 79 separate sections. Questel, a database vendor based in France, offers online access to the *PASCAL* files which provide multi-disciplinary coverage of over four million articles back to 1973. The Questel system may be searched with either a French or English command language and the *PASCAL* records contain both French and English titles and descriptors.

Records from 1977 forward also have Spanish keywords. The online version of this information source thus provides easier access for those who lack language skills in French. Dialog Information Services also offer access to the *PASCAL* files.

Current Physics Index (AIP, 1975–), quarterly. This service covers some 45 journals including those published by the American Institute of Physics and its member societies; 17 journals are the AIP's cover-to-cover translations of Russian-language equivalents. The information contained in this publication is timely, as much of it is recorded during the computer-assisted photocomposition of the AIP journals. The abstracts are those supplied by the authors for publication in the primary journals. Abstracts are arranged according to the 99 classes of the Physics and Astronomy Classification Scheme and supplemented with author and subject indexes. Atomic and molecular physics falls into classes 30–36. *Current Physics Index* covers some 90 per cent of the physics literature of the United States and 50 per cent of the Soviet literature (through translation journals). It is also available as part of an online database, SPIN (Searchable Physics Information Notices), marketed by Dialog Information Services, Inc., and in that format is updated monthly.

Physics Abstracts (Science Abstracts. Section A) (IEE, 1898–), 26 issues/year. Approximately 2000 journals are scanned by this source which publishes over 120 000 indicative and informative abstracts annually arranged according to the Physics and Astronomy Classification Scheme (PACS). Of particular value are the numerous indexes in each issue which provide access by subject, author and corporate author and assist in locating bibliographies, books, conferences, patents and reports. *Physics Abstracts* is available online from Bibliographic Retrieval Service, STN International, and Dialog Information Services, Inc. as part of the INSPEC database.

Physics Briefs/Physikalische Berichte (AIP, in co-operation with the Deutsche Physikalische Gesellshaft, 1979–), semi-monthly. This publication supersedes *Physikalische Berichte*, which has provided comprehensive coverage of the physics literature since 1920. Abstracts are now published in English and are arranged according to the Physics and Astronomy Classification Scheme. In addition to journals, coverage is also provided for books, patents, reports, theses and conference papers, and an effort is made to emphasize non-conventional (i.e. non-journal) literature and materials in Eastern languages. This abstracting service is also available as an online database from STN International.

Referativnyi Zhurnal: Fizika (VINITI, 1954–), monthly. This Russian-language abstracting service provides indicative and informative abstracts and includes a section on atomic and molecular physics and magnetic resonance. Titles and bibliographic citations are in the original language but this source will be most useful to those who read Russian. It is a very important one, however, for scientists who desire access to literature published in the Eastern countries.

Other indexes and abstracts worthy of mention here include *Chemical Abstracts*, published by the American Chemical Society (ACS), which is very likely the most comprehensive scientific abstracting service in English, and the interdisciplinary *Science Citation Index*, published by the Institute for Scientific Information (ISI). Both of these services are also produced in machine-readable form and are available from several vendors of search services. Because of the scatter of atomic and molecular physics literature these publications are valuable in comprehensive literature searching. The most timely coverage of core journals in the field will no doubt be provided by ISI's *Current Contents (Physical, Chemical, and Earth Sciences)*, which reproduces the title pages of selected journals and is often in a subscriber's hands before the journals it indexes arrive. *Current Contents* is especially useful to those who wish to write to authors for reprints, as author addresses are provided. *Current Contents* may also be searched as an online database.

Very much narrower in their focus are a number of abstracts which endeavour to pull together applied and experimental literature dealing with specific spectroscopic techniques. PRM in London publishes the following:

Atomic Absorption and Emission Spectrometry Abstracts (1969–), bimonthly. (Former title: *Atomic Absorption and Flame Emission Spectroscopy Abstracts*).
Electron Spin Resonance Spectroscopy Abstracts (1973–), quarterly.
Laser Raman and Infrared Spectroscopy Abstracts (1971–), bimonthly. (Formed by the merger of: *Laser-Raman Spectroscopy Abstracts* and *Infrared Spectroscopy Abstracts*).
Mössbauer Spectroscopy Abstracts (1978–), quarterly.
Nuclear Magnetic Resonance Spectrometry Abstracts (1971–), bimonthly: Preston has published *Nuclear Magnetic Resonance Literature – Abstract and Index* (formerly *Nuclear Magnetic Resonance Abstract Service*) since 1964.

All these very specialized secondary services may be of more value to individuals and their research groups than to most libraries. The latter will more frequently invest in the more costly but comprehensive general services such as *Physics Abstracts* and *Physics Briefs*. As is evident from the number of abstracts and indexes identified above, only the largest and best-funded libraries are likely to subscribe to all of them.

As noted in the above descriptions of the various indexes and abstracts, the larger, more comprehensive services are now searchable as online databases through arrangement with vendors of online services such as Dialog, BRS, and STN International. Online services have become well-established in most full-service science research libraries and reflect both the time-saving features and the enhanced retrieval capability of the computer-stored files. Within the last five years several newer forms of electronic access have been developed and these have led to a decentralization of databases. Both the CD-ROM database and the locally-mounted database use hardware controlled by the institution which subscribes to, or leases, the data. CD-ROM databases are stored on equipment attached to microcomputers and searched with software designed for the database and usually marketed with it; depending on the hardware configuration utilized, these may be searched by one or several users simultaneously. Locally mounted databases are versions of the files stored on an institution's mainframe computer where they may be accessed by many simultaneous users; the search software may be developed by the institution or purchased commercially. Users of indexes and abstracts now have choices previously unavailable and may select the mode of access most appropriate for their environment, whether paper copies, use of remote online search services, or locally managed databases on a CD-ROM or mainframe computer. The library literature includes extensive discussions of the merits and drawbacks of each type of access as well as descriptions of information services developed using one or another format.

Review journals

Evaluative comments on the published literature of atomic and molecular physics are provided by review journals. These service scientists who wish to survey the literature in a specialized area. The scope of some titles is quite broad.

Reviews of Modern Physics (AIP, 1929–), quarterly. This widely read journal contains perspectives and tutorial articles in rapidly developing fields in physics. It also includes listings of review articles published elsewhere and so can usefully be scanned regularly to identify reviews of interest. The quality of articles is high and all submissions, even when solicited, are refereed.

Journal of Physical and Chemical Reference Data (AIP, co-sponsored with the ACS and the United States National Bureau of Standards, 1972–), quarterly. This unique journal reviews the research literature of physics and chemistry and provides critically evaluated data and techniques documented as to origin and criteria of evaluation. Contributions from the National Standard Reference

Data System of the US Department of Commerce result from the co-ordinated efforts of data-evaluation centres in universities and in industrial and government laboratories. This journal is valuable for its compilations of data on atomic and molecular properties and spectra.

Other review-type journals are more narrowly focused on specialized fields, as for example:

Applied Spectroscopy Reviews (Marcel Dekker, 1964–), quarterly. This journal provides information on principles, methods, and applications of spectroscopy and discusses the relation of physical concepts to chemical applications.

Comments on Atomic and Molecular Physics (Gordon and Breach, 1969–), bimonthly. This journal provides critical discussion of current literature and suggests directions that future research might take.

Magnetic Resonance Review: a Quarterly Literature Review Journal (Gordon and Breach, 1972–), quarterly.

Reviews in Mathematical Physics (World Scientific, 1989–), quarterly. This journal offers survey and expository articles in mathematical physics.

Still other review articles are to be found in primary journals, conference proceedings and other serial publications. ISI's *Index to Scientific Reviews* provides access to those appearing in the journals scanned in producing its *Science Citation Index*. Subject searching is possible using pairs of title words in a Permuterm index.

Reviews of progress

Another type of information source, the review of progress, is similar to the review journal in that it attempts to provide a critical summary, usually by a recognized expert, of developments in a specialization over a defined period of time. These works are generally published in a series for which a library may place a 'standing order', to be assured of the prompt arrival of each new volume. A typical period of time for a review of progress to span is one year, and one refers to these as 'annual reviews', although longer or more irregular publication schedules also exist. Volumes in these series may contain articles covering several aspects of a field and the articles are usually solicited by the series editor. Reviews of the literature of atomic and molecular physics will be found in the following reviews of progress:

Advances in Atomic, Molecular and Optical Physics (Academic, 1965–), irregular.

Advances in Chemical Physics (Interscience, 1958–), irregular.

Advances in Magnetic Resonance (Academic, 1965–1990), irregular. Continued by: *Advances in Magnetic and Optical Resonance* (1991–)

Advances in Quantum Chemistry (Academic, 1964–), irregular.

Advances in Theoretical Physics (Academic, 1965–), irregular. See the *Proceedings of the Landau Birthday Symposium* (Pergamon, 1990) published as part of this series.

Annual Review of Physical Chemistry (Annual Reviews, 1950–), annual.

Electron Spin Resonance (Royal Society of Chemistry, 1971–), every 18 months.

Mass Spectrometry (Royal Society of Chemistry, 1971–), biennial.

Methods in Computational Physics (Academic, 1968–), irregular. Each volume also has a distinctive title.

Methods of Experimental Physics (Academic, 1959–). Each volume treats a specific topic such as *Molecular physics* (volume 3), *Atomic and electronic physics* (volumes 4 and 7), and *Spectroscopy* (volume 13).

Nuclear Magnetic Resonance (Royal Society of Chemistry, 1971–), annual.

Perspectives in Quantum Chemistry and Biochemistry (Wiley, 1976–), irregular.

Dictionaries, encyclopedias and handbooks

Tertiary sources such as dictionaries, encyclopedias and handbooks further distill the information first published in the primary sources. Unlike the indexes and abstracts discussed above, the tertiary sources can stand alone and serve certain types of information needs without requiring consultation of the primary source. For such sources treating the entire field of physics the reader is referred to Chapter 3 by R. J. Wyatt on reference materials. Herein we mention only those specific to atomic and molecular physics.

R. C. Denney's *Dictionary of spectroscopy* was published in a second edition in 1982 by Wiley. It has an A-Z arrangement of terms used in the literature of spectroscopy with brief (approximately 100–word) definitions. This information source, unlike most others in this chapter, is suitable for use by non-specialists or undergraduate students. References supplying additional information are provided. Other specialized lists with definitions of terms have been produced as pamphlets by the American Institute of Physics (AIP). For example, George L. Trigg prepared a *Glossary of terms frequently used in quantum mechanics*, which was published in 1964 by the AIP. Another useful publication of this type is *Acronyms and abbreviations in molecular spectroscopy: an encyclopedic dictionary* by Detlef A. W. Wendisch, published in 1990 by Springer.

Encyclopedias provide more information than dictionaries, although they may display a similar alphabetic arrangement of terms. Dated but still useful is the *Encyclopedia of spectroscopy*, edited by G. L. Clark and published in 1960 by Reinhold and Chapman and Hall. Contributions from more than 100 specialists are classified into 23 major subjects which follow an alphabetic sequence. The main topics are then further subdivided, with the narrower topics also arranged A-Z within the broader section. Some 185 of the contributions are signed essays. This source overlaps with the *Encyclopedia of chemistry*, edited by C. A. Hampel and G. E. Hawley, published by Van Nostrand (3rd edn., 1973) and with the *Encyclopedia of physics*, edited by Rita Lerner and George L. Trigg, published by VCH (2nd edn., 1991). These encyclopedias would serve beginners in the discipline by providing a simplified explanation or overview. Although called an encyclopedia, and discussed here for that reason, the *Handbuch der physik/Encyclopedia of physics,* edited by S. Flügge and published by Springer, is what might more accurately be called a systematic treatise. In fact, it is the only one which currently covers all of physics. A second edition of 54 volumes began appearing in 1955 and was completed in 1988. It has several volumes devoted to topics in atomic and molecular physics including:

Volume 5: *Principles of quantum theory* (1958)
Volume 27–28: *Spectroscopy* (1964, 1957)
Volume 32: *Structural research* (1957)
Volume 35–36: *Atoms I and II* (1957, 1956)
Volume 37(1): *Atoms III – Molecules I* (1959)
Volume 37(2): *Molecules II* (1961).

Contributions are in English, French and German by recognized experts and the subject indexes in each volume can serve as French and German glossaries; the lengthy articles in this work are intended for use by specialists and supply a state-of-the-art survey of each topic at the time of writing.

The handbooks published by the CRC Press are ubiquitous in the sciences, and their *CRC practical handbook of spectroscopy* is typical of other members of the family. Edited by J. W. Robinson and published in 1991, it contains sections prepared by specialist contributors with coverage of all major subfields such as nuclear magnetic resonance, infra-red, Raman, ultra-violet, ESCA photoelectron, appearance potential, and X-ray spectroscopy, mass spectrometry and atomic absorption. Diagrams, charts, graphs and tables provide detailed data, and references are supplied to the primary literature.

The data compilation is similar to the handbook in that it draws together in a single convenient reference work numerical data that may initially have been scattered throughout the primary literature. Such data, although derived from experimental measurements, may be required even in purely theoretical research in atomic and molecular physics. Spectral and structural data for atoms and molecules are to be found in the following sources:

Stanley Baskin and John O. Stoner, Jr. *Atomic energy levels and grotrian diagrams* (North-Holland, 1975–), irregular.

S. Fraga, J. Karwowski and K. M. S. Saxena (1976) *Handbook of atomic data* (Elsevier).

S. Fraga, K. M. S. Saxena and J. Karwoski (1979) *Atomic energy levels: data for parametric calculations* (Elsevier).

K. P. Huber and G. Herzberg. *Molecular spectra and molecular structure: vol IV: constants of diatomic molecules* (1979) (Van Nostrand).

A. L. McClellan. *Tables of experimental dipole moments* (1963) (Freeman).

C. E. Moore. *Atomic energy levels as derived from analyses of optical spectra. Circular 467, volumes I, II, III* (NBS, 1949, 1952, 1958). This has been reprinted in 1971 by the US Government Printing Office.

As workstations are developed for use by physicists and chemists, there will be an increased need for data compilations in electronic format that permit efficient access to numerical and spectral data. An example is the MASS SPECTRAL DATABASE from the Mass Spectrometry Data Center at the US National Institute of Standards and Technology in Gaithersburg, Maryland. The Center provides data for over 62 000 compounds on magnetic tapes and diskettes suitable for use with personal computers. Other efforts to convert existing data compilations to electronic format are underway; in addition numerous new databanks are growing as the use of complex instrumentation results in data collection in electronic form. Libraries may expect to see a shift toward this type of compilation and will need to plan for appropriate hardware to access the data. The potential benefits for both librarian and scientist will include the opportunity to save space in crowded facilities, more timely distribution of data, and much enhanced retrieval capabilities.

Treatises, monographs and texts

Books in the field of atomic and molecular physics include those categorized variously as treatises, monographs and texts. Treatises endeavour to cover a topic comprehensively in a systematic fashion and may describe the historical development of the topic under consideration, drawing together information previously scattered in years of primary literature. A subject is discussed exhaustively in a treatise and full bibliographic citations refer the reader back to the original literature. The preparation of such works is by nature painstaking and lengthy, and hence they cannot be expected to deal with the very most recent research in an area; for that information reviews of progress, mentioned above, can supplement the background provided by the treatise.

Monographs resemble treatises in several ways, in fact, some works may be difficult to classify in one or the other category. The significant difference lies in the narrower subject focus of the monograph and the more contemporary emphasis which normally avoids coverage of the historical or background material. Like the treatise, however, the monograph also contains full bibliographic citations in its many references to the primary literature. Monographs in many scientific fields are frequently offered as publisher's series for which libraries may place standing orders.

Both treatises and monographs are publications intended for use by specialists, i.e. scientists engaged in research in a field. As graduate students advance in their courses they too make use of such works. Earlier, beginning a course of graduate study they are more likely to read textbooks. The primary function of the text is to teach the reader by elucidating principles and providing descriptions, explanations and examples. Material selected for inclusion will be representative of the principles illustrated rather than comprehensive. Textbooks that serve teachers will endure for a number of years; the best will see several revisions to reflect a major change in the field.

Following is a selection of treatises, monographs and texts in atomic and molecular physics classified according to subject.

Electronic structure of atoms and molecules

Included here are works that are primarily theoretical in focus that discuss the quantum mechanical calculations that provide a framework for the understanding of electronic transitions in atoms and molecules, chemical binding and molecular interactions.

Bartlett, Rodney J. (ed.) (1985) *Comparison of ab initio quantum chemistry with experiment for small molecules: the state of the art* (D. Reidel.)

Bethe, Hans (1986) *Intermediate quantum mechanics* 3rd edn. (Benjamin/Cummings).

Buckingham, A. D. (ed.) (1975) *MTP international review of science, Physical Chemistry, Series Two. Volume 2. Molecular structure and properties* (Butterworths)

Davidson, E. R. (1976) *Reduced density matrices in quantum chemistry* (Academic).

Fano, U. and Fano, L. (1972) *Physics of atoms and molecules: an introduction to the structure of matter* (U. of Chicago Pr.).

Hartree, D. R. (1957) *The calculation of atomic structures* (Wiley)

Löwdin, Per-Olov (ed.) (1966) *Quantum theory of atoms, molecules and the solid state: a tribute to John C. Slater* (Academic). A Festchrift of collected papers by former students and colleagues of John C. Slater recognizes the influences of his contributions to the quantum theory of matter.

Pauling, Linus (1960) *The nature of the chemical bond and the structure of molecules and crystals*. 3rd edn. (Cornell U. Pr.). This work is intended for use by students in courses covering the structure of molecules and crystals and the theory of the chemical bond. Although dated, it continues to provide a valuable introduction to these topics.

Paunz, R. (1967) *Alternant molecular orbital method* (Studies in Physics and Chemistry, 4) (Saunders). This monograph assumes an understanding of basic quantum mechanics in providing details, with examples, of the alternant molecular orbital method and the use of projection operators.

Slater, John C. (1979) *The calculation of molecular orbitals* (Wiley). The manuscript of this book was found in Slater's papers after his death in 1976 and is the last volume in the author's seven-book series on the quantum theory of matter. It outlines, although incompletely, his last research on the exact solution to the self-consistent field problem.

Slater, John C. (1960) *Quantum theory of atomic structure* , 2 vols (McGraw-Hill). The first volume of this set is intended as an introductory text for use by first-year graduate students with problems illustrating principles and references to sources of additional information. The second volume serves as a reference work on the more advanced aspects of atomic structure for use by chemists, theoretical physicists, metallurgists and electrical engineers. This set surveys the historical development of modern physics.

Slater, John C. (1963–1974) *Quantum theory of molecules and solids*, 4 vols (McGraw-Hill). This multi-part treatise on molecular quantum mechanics describes both the techniques of pure wave mechanics and various semi-empirical methods with an emphasis on the former.

Atomic spectra

The works included here are primarily concerned with the theoretical interpretation of atomic spectra. Since any theory represents an effort to interpret experimentally observed spectra, these works will necessarily treat, to varying extents, observational aspects of atomic spectroscopy. They do not, however, provide details on the design of experiments or apparatus.

Bethe, H. A. and Salpeter, E. E. (1977) *Quantum mechanics of one-and two-electron atoms* (Plenum). This reprinting of the 1957 publication by Springer testifies to the ongoing importance of a classic monograph.

Condon, E. U. and Odabasi, H. (1980) *Atomic structure* (CUP). The stated purpose of this book is to interpret the line spectra of atoms semi-quantitatively using quantum mechanics and the nuclear atom model.

Fano, Ugo and Rau, A. R.(1986) *Atomic collisions and spectra* (Academic).

Herzberg, G. (1944) *Atomic spectra and atomic structure*, 2nd edn. (Dover). Despite the age of this translation of a work originally published in German, Herzberg's text continues to provide an elementary introduction to principles and the calculations based upon them.

Kuhn, H. G. (1969) *Atomic spectra*, 2nd edn. (Longman and Academic). This introductory treatment goes beyond texts on modern physics in its reconciliation of observed facts with classical concepts. Readers desirous of more mathematical detail are referred to E. U. Condon and G. H. Shortley, *Theory of atomic spectra* (1935) (CUP; repr. with corrections in 1951) or the works of Slater.

Shore, Bruce W. (1990) *The theory of coherent atomic excitation* (Wiley).

Shore, B. W. and Menzel, D. H. (1968) *Principles of atomic spectra* (Wiley).

Sobel'man, I. I. (1979) *Atomic spectra and radiative transitions* (Springer Series in Chemical Physics, 1) (Springer).

Sobel'man, I. I. (1972) *Introduction to the theory of atomic spectra* (Pergamon).

Woodgate, G. K. (1970) *Elementary atomic structure* (2nd edn., 1990 Clarendon).

Molecular spectra

Under this heading we list works that detail the theory and analysis of radio-frequency, microwave, infrared, Raman, Rayleigh, visible, ultra-violet, X-ray, nuclear magnetic resonance, electron paramagnetic resonance and nuclear quadrupole resonance spectroscopy. Frequently a work will be devoted to only one or two of these topics.

Abragam, A. (1983) *The principles of nuclear magnetism* (Clarendon) (International Series of Monographs of Physics).

Allen, H. C. and Cross, P. C. (1963) *Molecular vib-rotors: the theory and interpretation of high resolution infrared spectra* (Wiley).

Bauman, R. P. (1962) *Absorption spectroscopy* (Wiley). This introductory text is intended to provide undergraduates with an understanding of basic techniques for analysis of spectra.

Cotton, F. A. (1971) *Chemical applications of group theory* 2nd edn. (Wiley-Interscience).

Dykstra, Clifford E. (1991) *Quantum chemistry and molecular spectroscopy* (Prentice Hall).

Gordy, W. and Cook, R. L. (1984) *Microwave molecular spectra* 3rd edn. (Wiley-Interscience).

Graybeal, Jack D. (1988) *Molecular spectroscopy* (McGraw-Hill).

Gribov, L. A. and Orville-Thomas W. J. (1988) *Theory and methods of calculation of molecular spectra* (Wiley).

Harmony, M. D. (1972) *Introduction to molecular energies and spectra* (Holt-Reinhart).

Herzberg, G. (1971) *The spectra and structures of simple free radicals: an introduction to molecular spectroscopy* (Cornell U.Pr.). For a more advanced treatment of the subject, see this author's three-volume set:

Herzberg, G. *Molecular spectra and molecular structure*, 3 vols. Volume I: *Spectra of diatomic molecules* (2nd edn., 1950). Volume II: *Infrared and raman spectra of polyatomic molecules* (1945). Volume III: *Electronic spectra of polyatomic molecules* (1966) (Van Nostrand). This set supports multiple-level use by students and advanced research scientists through the use of varying typefaces. Each volume can stand alone because needed background information is repeated and all volumes provide numerous references to original papers.

King, G. W.(1964) *Spectroscopy and molecular structure* (Holt-Rinehart). This undergraduate-level text provides examples and references on the analysis and interpretation of spectroscopic measurements.

Papousek, D. and Aliev, M. R. (1982) *Molecular vibrational-*

rotational spectra: theory and applications of high resolution infrared, microwave, and raman spectroscopy of polyatomic molecules (Elsevier).

Steinfeld, J. I. (1974) *Molecules and radiation – an introduction to modern molecular spectroscopy* (Harper & Row). This text was developed to support a graduate course at Massachusetts Institute of Technology, and to introduce students to the research literature and the more detailed treatments found in the works of Herzberg, Townes and Schawlow, and Condon and Shortley.

Struve, Walter S. (1989) *Fundamentals of molecular spectroscopy* (Wiley).

Townes, C. H. and Schawlow, A. L. (1955) *Microwave spectroscopy* (International Series in Pure and Applied Physics) (McGraw-Hill): this still-useful reference work was the first to cover definitively the field that had just developed prior to its publication.

Wang, Zu-Geng (1991) *Molecular and laser spectroscopy* (Springer).

Wilson, E. B., Decius, J. C. and Cross, P. C. (1955) *Molecular vibrations: the theory of infrared and raman vibrational spectra* (McGraw-Hill).

Wollrab, J. E. (1967) *Rotational spectra and molecular structure* (Academic). This state-of-the-art review is intended for use by physicists, chemists and engineers.

Atomic and molecular collisions

These works treat the theory of atomic, molecular, and ionic interactions in elastic or inelastic scattering or in the diffusion of particles in gases.

Electronic and Ionic Impact Phenomena, 2nd edn. (Clarendon). Volume 1: Massey, H. S. W. and Burhop, E. H. S. (1969) *Electronic collisions with atoms*. Volume 2: Massey, H. S. W. (1969) *Electron collisions with molecules and photo-ionization*. Volume 3: Massey, H. S. W. (1971) *Slow collisions of heavy particles*. Volume 4: Massey, H. S. W. and Gilbody, H. B. (1974) *Recombination and fast collisions of heavy particles*. Volume 5: Massey, H. S. W., Burhop, E. H. S. and Gilbody, H. B. (1974) *Slow positron and muon collisions – and notes on recent advances*.

Huxley, L. G. H. and Compton, R. W. (1974) *The diffusion and drift of electrons in gases* (Wiley).

Mason, E. A. (1988) *Transport properties of ions in gases* (Wiley).

McDaniel, E. W. (1989) *Atomic collisions: electron and photon projectiles* (Wiley).

McDaniel, E. W. (1964) *Collision phenomena in ionized gases* (Wiley).

McDaniel, E. W. and Mason, E. A. (1973) *The mobility of diffusion of ions in gases* (Wiley).

Mott, N. F. and Massey, H. S. W. (1965) *The theory of atomic collisions*, 3rd edn. (Clarendon).

It should be emphasized that the treatises, monographs and texts chosen for inclusion here are selected from a much larger body of works covering these topics. An effort has been made here to identify those books that have proved over the years to be reference sources of enduring value and, in addition, to note some more recent publications that may eventually supplant them.

CHAPTER FIVE

Chemical physics, statistical physics and thermodynamics

ROBERT C. MICHAELSON

Although the term 'chemical physics' has been used for well over a century at least (Miller, 1855; this book is chiefly known for its third edition of 1863 containing the first publication of Thomas Andrews' discoveries on the critical region of carbon dioxide), there does not seem to be a coherent, generally accepted definition of the field or a sharp differentiation from the related area of 'physical chemistry'. Thus the *Journal of Chemical Physics*, published by the American Institute of Physics, describes the field broadly as research which is as much chemistry as physics, while the UK Chemical Society in 1972 split its *Faraday Transactions* into Part I (physical chemistry) and Part II (chemical physics), implicitly limiting chemical physics to those topics published in Part II: 'Theoretical papers, especially those on valence and quantum theory, statistical mechanics, inter-molecular forces, relaxation phenomena, spectroscopic studies . . . leading to assignments of quantum states, and fundamental theory, and also studies of impurities in solid systems, etc.'. (The sections were reunited in 1990).

The blurred usage of the terms 'physical chemistry' and 'chemical physics' may be understood in historical perspective. The first English-language periodical in this boundary area between physics and chemistry was the *Journal of Physical Chemistry*, founded in 1896 by Wilder Bancroft in Ostwald's tradition of what was variously called physikalische, allgemeine, or theoretische chemie (Servos, 1982). Bancroft's conception of the field was rather limited, however, and his journal published papers on colloid science and on applications of Gibbs' phase rule but avoided contributions having much physical or mathematical content. Since A. A. Noyes, G. N. Lewis, L. Pauling, H. Urey and other proponents of modern physicochemical research were unable to bring Bancroft's journal towards their position, the way was open for the creation of the *Journal of Chemical Physics*, which began publication in January 1933. When John Slater wrote of the range of study common to both physics and chemistry, 'for want of a better name, since Physical Chemistry is already pre-empted, we may call this common field Chemical Physics' (Slater, 1939) he undoubtedly had in mind the *Journal of Physical Chemistry*.

In this chapter 'chemical physics' will be understood in the broad sense used by Slater and by the *Journal of Chemical Physics*, i.e. the range of study common to both physics and chemistry. The reader should note, however, that many subjects included under this interpretation and generally considered as chemical physics are treated in other chapters of this book. In particular, attention is directed to the previous chapter on atomic and molecular physics, which treats sources of information on electronic structure of atoms and molecules, atomic and molecular spectra and interaction with photons, atomic and molecular collision processes, and experimentally derived information on atoms and molecules. For the experimental techniques of spectroscopy, see Chapter 16 (optoelectronics, quantum optics and spectroscopy); and there are separate chapters on experimental heat and low-temperature physics (Chapter 11) and on crystallography (Chapter 8).

General chemical physics

No single monograph provides an adequate introduction to the entire field of chemical physics, but a basic introduction to the earlier literature is given by J. C. Slater (1939) *Introduction to chemical physics* (McGraw-Hill), and an impressively broad, but in many places now much-outdated treatment, is found in J. R. Partington (1949–1954) *An advanced treatise on physical chemistry*, 5 vols (Longman). There have been two projects to provide comprehensiveness by means of collections of individual monographs: the first, by E. A. Guggenheim *et al.* (eds.) (1960–1975) *International encyclopedia of physical chemistry and chemical physics*, 30 vols (Pergamon), began with an elaborately planned group of 22 topics to be treated in about 100 separate volumes. This collection produced a number of important brief monographs but seems to have been abandoned far short of its original goal. Another series, by H. Eyring *et al.* (eds.) (1967–1975) *Physical chemistry, an advanced treatise*, 11 vols, in 15 parts (Academic), covers 11 fairly broad topics, each in one or two volumes of articles by several authors. The theoretical areas of chemical physics are reviewed in the series of volumes by W. H. Miller (ed.) (1976–1977) *Modern theoretical chemistry*, 8 vols (Plenum), and many experimental methods are treated by B. W. Rossiter and J. F. Hamilton (eds.) (1986–) in *Physical methods of chemistry*, 6 vols in 7 (2nd edn., Wiley) and in several volumes of A. Weissberger (ed.) (1971–) *Techniques of chemistry* (Wiley).

Several publishers have issued monographic series on chemical physics, including the Wiley Monographs in Chemical Physics, the Springer Series in Chemical Physics and Academic's Physical Chemistry, a Series of Monographs; only the Springer series is still published. Academic still issues Theoretical Chemistry, a Series of Monographs. The principal journals treating the entire field of chemical physics are the *Journal of Chemical Physics* and the *Faraday Transactions* mentioned above, and *Chemical Physics, Chemical Physics Letters* and *Molecular Physics*. Serial publications containing important review articles include *Annual Review of Physical Chemistry* (1950–), *Advances in Chemical Physics* (1958–), which is published in several volumes per year and frequently as single topic volumes, *Annual Reports on the Progress of Chemistry, Section C: Physical chemistry* (1979–), *Theoretical chemistry* (1974–1981), one of the Chemical Society Specialist Periodical Reports, and *Theoretical chemistry: advances and perspectives* (1975–1981). Also of importance are the annual conferences on single topics published as the *Faraday Discussions* and the *Faraday Symposia* of the Chemical Society; since 1986 the latter have been incorporated into the *Faraday Transactions*.

The *Journal of Physical and Chemical Reference Data*, published co-operatively by the American Chemical Society, the American Physical Society and the US National Institute of Standards and Technology, publishes critically reviewed compilations of data on a wide variety of topics in chemical physics. An important monographic series publishing critically reviewed data is the US National Bureau of Standards *National Standard Reference Data Series* (1964–1987), also known as the NSRDS-NBS Series.

By far the most comprehensive indexing of the recent literature in the subject areas of chemical physics is that provided in *Chemical Abstracts (CA)*. *Chemical Abstracts* is available on computer tape as well as in printed form. Files of these tapes for the period from 1967 onwards are made available by several commercial vendors for online searching, including Chemical Abstracts Service itself which offers the database CAS ONLINE through their database vendor STN International. The 12th Collective Index to *CA*, covering 1987–1991, and the abstracts for that period, are being offered on CD-ROM from 1993. Since there is a delay of several months from the time of publication of an article until it is abstracted in *Chemical Abstracts*, use may be made of *Chemical Titles*, also produced by Chemical Abstracts Service but indexing only about 800 journals and using indexing by title keyword and by author, thus indexing much more current literature than *CA*. The French abstracts journal *PASCAL* (formerly *Bulletin Signalétique*) currently covers some chemical physics in its *PASCAL* folio F17, *Chimie Générale, Minérale et Organique*, but other sections may be important for specific subject areas, and the section numbering and organization are changed at irregular but fairly frequent intervals. *PASCAL* is also searchable in an online version. For a comprehensive search of a topic in chemical physics, it is necessary also to consult *Physics Abstracts*, either in printed form or as part of the INSPEC online database; the INSPEC database is also available on CD-ROM covering 1989–.

When physical properties are needed for specific inorganic compounds, it is useful to check the *Gmelins Handbuch der Anorganischen Chemie* (1924–), particularly if the data needed is likely to have been obtained some time ago. Volumes since 1984 are in English and bear the title *Gmelin Handbook of Inorganic and Organo-metallic Chemistry*. Some volumes are more up to date than others, because of the irregular publication of supplements. An online GMELIN file includes critically evaluated data and bibliographic citations from the handbook, and also selected data from more recent literature.

Since CRC Press publishes over 300 handbooks in addition to the well-known *CRC handbook of chemistry and physics*, it may be useful to consult their (1991) *Composite Index for CRC Handbooks*, 3 vols (3rd edn., CRC) which includes a CD-ROM version with the set of printed volumes.

Molecular interactions and reactions

One of the most active topics in chemical physics is the study of reactions of atoms and molecules. This area of research has expanded so much recently that it may no longer be possible for a single monograph to have the comprehensive coverage of the still very useful book by S. W. Benson (1960) *The foundations of chemical kinetics* (McGraw-Hill), or even thorough coverage of narrower topics as given in the classic books on gas-phase reactions by V. N. Kondrat'ev (1964) *Chemical kinetics of gas reactions* (Pergamon) and on photochemistry by J. G. Calvert and J. N. Pitts, Jr. (1966) *Photochemistry* (Wiley). Something approaching comprehensive coverage can only be expected from a series of monographs, the method used in the treatise edited by C. H. Bamford and C. H. M. Tipper (eds.) (1969–) *Comprehensive Chemical Kinetics*, Vols 1–10, 12–14, 14A, 15–33 (Elsevier).

This rapid development has come about because of progress in both experimental and theoretical techniques. The use of molecular beams, lasers, sensitive photon and particle detectors and digital electronic techniques have allowed the experimentalist to approach the goal of preparing reactants in specific quantum states, bringing them together with known relative orientation and translational energy, and observing the products in equal detail. Theoreticians have similarly developed methods, facilitated by computers, allowing them to make detailed calculations on such processes. Fortunately, R. B. Bernstein (1982) *Chemical dynamics via molecular beam and laser techniques* (OUP) is a monograph that is a superb guide to many of these developments, as is J. I. Steinfeld *et al.* (1989) *Chemical kinetics and dynamics* (Prentice-Hall).

Detailed information on molecular beam techniques may be found in the classic work by N. F. Ramsey (1956) *Molecular beams* (OUP), the compilations edited by J. Ross (1966) *Molecular beams*, vol. 10 of *Advances in Chemical Physics* (Wiley), Ch. Schlier (1970) *Molecular beams and reaction kinetics* (*Proceedings of the International School of Physics 'Enrico Fermi', Varenna, Italy. Course XLIV*) (Academic) and G. Scoles (ed.) (1988–1992) *Atomic and molecular beam methods*, 2 vols (OUP), and in the monographs by H. S. W. Massey (1971) *Slow collisions of heavy particles*, vol. III of H. S. W. Massey *et al.*, (1969–1974) *Electronic and ionic impact phenomena*, 2nd edn., 5 vols (OUP) and by M. A. D. Fluendy and K. P. Lawley (1973) *Chemical applications of molecular beam scattering* (Chapman and Hall). Laser techniques and applications are treated by A. H. Zewail (ed.) (1978) *Advances in laser chemistry* (Springer), J. Steinfeld (ed.) (1981) *Laser-induced chemical processes* (Plenum), D. L. Andrews (1986) *Lasers in chemistry* (Springer) and in the series edited by C. B. Moore (1974–1977) *Chemical and Biochemical Applications of Lasers*, 5 vols (Academic).

For many years the standard theoretical approach to chemical rate problems was through the 'activated complex' theory, which has its definitive treatment in S. Glasstone *et al.* (1941) *The theory of rate processes: the kinetics of chemical reactions, viscosity, diffusion and electrochemical phenomena* (McGraw-Hill) and which is still important, particularly for dealing with complex systems (R. D. Gandour and R. L. Schowen (eds.) (1978) *Transition states of biochemical processes* (Plenum)). However, the detailed experiments now being done require detailed calculations of molecular collision processes. H. S. Johnston (1966) *Gas phase reaction rate theory* (Ronald) reviews the relation between activated-complex and collision theories. R. D. Levine and R. B. Bernstein (1987) *Molecular reaction dynamics and chemical reactivity* (2nd edn., OUP) provide an excellent introduction to theories of molecular reactions. The monographs by R. D. Levine (1969) *Quantum mechanics of molecular rate processes* (OUP) and E. A. Nikitin (1974) *Theory of elementary atomic and molecular processes in gases* (OUP) give more advanced treatments; and M. S. Child (1974) *Molecular collision theory* (Academic), W. H. Miller (ed.) (1976) *Dynamics of molecular collisions*, 2 vols, vols 1 and 2 of W. H. Miller (ed.) (1976–1977) *Modern theoretical chemistry* (Plenum), R. B. Bernstein (ed.) (1979) *Atom-molecule collision theory: a guide for the experimentalist* (Plenum) and D. Henderson (1981) *Theory of scattering: papers dedicated to Henry Eyring*, vol. 6 of *Theoretical Chemistry: Advances and Perspectives* (Academic), contain surveys of molecular collision theory and its applications to energy transfer and chemical reactions.

Photochemistry has progressed in a similar manner, since new experimental techniques allow preparation of specific excited states and detailed monitoring of subsequent products on a picosecond time scale, while new theoretical approaches are being developed to treat the time evolution of excited states. A good collection of articles on this new photoselective chemistry is J. Jortner *et al.* (eds.) (1981) *Photoselective chemistry*, vol. 47 of *Advances in Chemical Physics* (Wiley). Picosecond and faster methods are found in G. R. Fleming (1985) *Chemical applications of ultrafast spectroscopy* (OUP) and in R. J. H. Clark and R. E. Hester (eds.) (1989) *Time resolved spectroscopy*, vol. 18 of *Advances in Spectroscopy* (Wiley). An introduction to modern photochemistry of organic molecules is given in N. J. Turro (1991) *Modern molecular photochemistry* (University Science Books), and specialized articles on organic photophysics are in J. B. Birks (ed.) (1973–1975) *Organic molecular photophysics*, 2 vols (Wiley). K. P. Lawley (ed.) (1982) *Dynamics of the excited state*, vol. 50 of *Advances in Chemical Physics* (Wiley) and E. C. Lim (ed.) (1973–1982) *Excited states*, 6 vols (Academic) treat the whole question of the evolution of excited molecular states, and experimental techniques for excited state processes are reviewed in A. A. Lamola and W. R. Ware (eds.) (1971–1976) *Creation and detection of the excited state*, 4 vols (Dekker). Optoacoustic techniques, revitalized by the use of lasers, are very useful in the study of radiationless transitions and are treated in V. P. Zharov (1986) *Laser optoacoustic spectroscopy* (Springer). The photochemistry of small molecules, of intrinsic interest and also important to the study of planetary atmospheres, has been reviewed in

H. Okabe (1978) *Photochemistry of small molecules* (Wiley). Compilations of experimental information and techniques for photochemistry are given in S. L. Murov (1973) *Handbook of photochemistry* (Dekker) and J. F. Rabek (ed.) (1982) *Experimental methods in photochemistry and photophysics*, 2 vols (Wiley). Important review articles are to be found in the series *Advances in Photochemistry* (1963–), and in *Photochemistry* (1969–), a Royal Society of Chemistry Specialist Periodical Report.

Chemical reaction processes are fundamentally connected with molecular energy transfer; J. T. Yardley (1980) *Introduction to molecular energy transfer* (Academic) gives a good introduction to this subject and a thorough treatise has been edited by G. M. Burnett *et al.* (1969–1974) *Transfer and storage of energy by molecules*, 4 vols (Wiley). Other important topics include unimolecular reactions, reviewed in P. J. Robinson and K. A. Holbrook (1972) *Unimolecular reactions* (Wiley) and W. Forst (1973) *Theory of unimolecular reactions* (Academic); and ion-molecule reactions are reviewed in H. S. W. Massey (1971) *Slow collisions of heavy particles*, vol. III of H. S. W. Massey *et al.* (1969–1974) *Electronic and ionic impact phenomena*, 2nd edn., 5 vols (OUP) and in P. Ausloos (ed.) (1979) *Kinetics of ion-molecule reactions* (Plenum) and M. T. Bowers (ed.) (1979–1984) *Gas phase ion chemistry*, 3 vols (Academic). The phenomena of oscillatory chemical reactions present interesting problems for the theoretician (P. Gray (1990) *Chemical oscillations and instabilities: non-linear chemical kinetics* (OUP)) and have significant implications for biology (D. Henderson (ed.) (1978) *Periodicities in chemistry and biology*, vol. 4 of *Theoretical Chemistry: Advances and Perspectives* (Academic)). Sources dealing with surface reactions are discussed below with the section on surfaces.

There are critical compilations of data in Arrhenius form for gas-phase unimolecular reactions in S. W. Benson and H. E. O'Neal (1970) *Kinetic data on gas phase unimolecular reactions*, (vol. 21 of US National Bureau of Standards, *National Standard Reference Data Series* (US Government Printing Office)) and bimolecular and termolecular in J. A. Kerr (ed.) (1981–1987) *CRC handbook of bimolecular and termolecular gas reactions*, 3 vols in 5 (CRC). Evaluated data for high-temperature reactions have been compiled in D. L. Baulch *et al.* (1972–1977) *Evaluated kinetic data for high temperature reactions*, vols 1–3 (Butterworths); D. L. Baulch *et al.* (1981) *Evaluated kinetic data for high temperature reactions*, vol. 4, supplement 1 to vol. 10 of *Journal of Physical and Chemical Reference Data* (ACS). Liquid-phase reaction rate constants are to be found in E. T. Denisov (1974) *Liquid-phase reaction rate constant* (Plenum). Methods for estimating unknown Arrhenius reaction rate parameters for gas-phase reactions are given in S. W.

Benson (1976) *Thermochemical kinetics: methods for the estimation of thermochemical data and rate parameters* (2nd edn., Wiley).

Statistical physics and thermodynamics

Thermodynamics is a subject of central importance in both physics and chemistry; as such it has received many treatments from the points of view of the physicist and of the chemist. A classic introductory text for physicists is M. W. Zemansky (1968) *Heat and thermodynamics: an intermediate textbook* (5th edn., McGraw-Hill), but topics of importance for the physical chemist are neglected. Excellent standard advanced texts concentrating on applications of chemical interest include K. Denbigh (1981) *The principles of chemical equilibrium* (4th edn., CUP) and G. N. Lewis and M. Randall (1961) *Thermodynamics*, 2nd edn., revised by K. S. Pitzer and L. Brewer (McGraw-Hill); a rigorously formulated approach is J. G. Kirkwood and I. Oppenheim (1961) *Chemical thermodynamics* (McGraw-Hill), and M. L. McGlashan (1979) *Chemical thermodynamics* (Academic) emphasizes the experimental basis of chemical thermodynamics. The standard authoritative treatise for physicists and chemists is E. A. Guggenheim (1967) *Thermodynamics: an advanced treatment for chemist and physicist* (5th edn., North-Holland). A. B. Pippard (1957) *The elements of classical thermodynamics for advanced students of physics* (CUP) is an outstanding discussion of the principles of thermodynamics for advanced physics students, while a splendid recent textbook is J. R. Waldram (1985) *The theory of thermodynamics* (CUP). A different postulational approach, H. B. Callen (1985) *Thermodynamics and an introduction to thermostatistics* (2nd edn., Wiley) based on work by L. Tisza, has been found to be useful. The searching examination of the nature of thermodynamics in P. W. Bridgman (1941) *The nature of thermodynamics* (Harvard U. Pr.) is still worth reading.

Comprehensive presentations of the state of the art of experimental thermochemistry – the determination of heats of chemical reactions – have been produced under the auspices of the International Union of Pure and Applied Chemistry (IUPAC): the original volumes, F. D. Rossini (ed.) (1956) *Experimental thermochemistry*, vol. 1 (Interscience) and H. A. Skinner (ed.) (1962) *Experimental thermochemistry*, vol. 2 (Interscience), are being superseded by a new series, the first volume of which is S. Summer and M. Mansson (eds.) (1979) *Combustion calorimetry*, vol. 1 of *Experimental Chemical Thermodynamics* (Pergamon). Reviews of experimental methods in thermodynamics have also been issued by IUPAC (J. P. McCullough and D. W. Scott (eds.) (1968) *Calori-*

metry of non-reacting systems, vol. 1 of *Experimental Thermo-dynamics* (Butterworths); B. Vodar and B. LeNeindre (eds.) (1974) *Experimental thermodynamics of non-reacting fluids*, vol. 2 of *Experimental Thermodynamics* (Pergamon)). (Also see Chapter 11, 'Experimental heat and low-temperature physics', by R. Berman.)

Current experimental work in thermodynamics is indexed in the *Bulletin of Chemical Thermodynamics*, an IUPAC (International Union of Pure and Applied Chemistry) publication. This is an 'annual index, bibliography and review for published and unpub-lished research in the intersecting areas of thermodynamics and chemistry'. Sources of critically evaluated chemical thermodynamic data include M. W. Chase *et al.* (1985) *JANAF thermochemical tables*, 2 vols, 3rd edn., supplement 1 to vol. 14 of *Journal of Physical and Chemical Reference Data* (ACS), which gives enthalpies, entropies and Gibbs energies of formation for over 1100 (chiefly inorganic) species, a compilation designed to provide information on rocket propellants and thus including high-temperature data for pure substances; and D. D. Wagman *et al.* (1982) *The NBS tables of chemical thermodynamic properties: selected values for inorganic and C1 and C2 organic substances in SI Units*, supplement 2 to vol. 11 of *Journal of Physical and Chemical Reference Data* (ACS), with enthalpies of formation, Gibbs energies of formation, entropies and heat capacities for inorganic and small (one or two carbon atom) organic compounds at 25° C, both pure and in aqueous and organic solutions – superseding Series I of tables in F. D. Rossini *et al.* (1952) *Selected values of chemical thermodynamic properties*, US National Bureau of Standards *Circular 500* (US Government Print-ing Office). An extensive collection of critically evaluated data is in I. Barin (1989) *Thermochemical data of pure substances*, 2 vols (VCH). Enthalpies of formation of gas-phase ions are in the critical compilation, S. G. Lias *et al.* (1988) *Gas-phase ion and neutral thermochemistry*, supplement 1 to vol. 17 of *Journal of Physical and Chemical Reference Data* (ACS). Evaluated data for about 1000 commercially important chemicals is given by T. E. Daubert and R. P. Danner (1985) *Data compilation tables of properties of pure compounds*, looseleaf (AIChE). Critically evaluated thermo-dynamic data for a very large number of organic compounds may be found in TRC Hydrocarbon Project (1985–) *TRC thermodynamic tables. Hydrocarbons*, 12 vols, looseleaf, updated (Therm. Res. Cen.), and Thermodynamics Research Center (1985–) *TRC thermodynamic tables. Non-hydrocarbons*, 9 vols, looseleaf, updated (Therm. Res. Cen.). A critical compilation of standard enthalpies of formation and of vaporization for both organic and organometallic compounds at 25° C was made in J. D. Cox and G. Pilcher (1970) *Thermochemistry of organic and organometallic compounds* (Academic); J. B. Pedley, R. D. Naylor, and S. P. Kirby (1986) *Thermochemical data of organic compounds* (2nd edn., Chapman & Hall) brought that work up to date, including also enthalpies of combustion and of other reactions. Another useful compilation of organic thermochemical data is D. R. Stull *et al.* (1969) *The chemical thermodynamics of organic compounds* (Wiley). G. J. Janz (1967) *Thermodynamic properties of organic compounds. Estimation methods, principles and practice* (rev. edn., Academic) gives methods of estimating thermodynamic properties of organic compounds in the absence of experimental data.

The statistical mechanical foundations of thermodynamics and the molecular theory of matter have been treated by a large number of advanced monographs and it is possible to mention only some of them here. R. C. Tolman (1938) *The principles of statistical mechanics* (OUP) gives a classic treatment of the theoretical foundations of the subject, and a classic treatise on the applications of statistical mechanics is R. H. Fowler and E. A. Guggenheim (1939) *Statistical thermodynamics* (CUP). Standard advanced textbooks include K. Huang (1987) *Statistical mechanics* (2nd edn., Wiley) and J. E. Mayer and M. G. Mayer (1977) *Statistical mechanics* (2nd edn., Wiley). Any topic of theoretical physics is likely to be covered in penetrating fashion by a volume of the Landau and Lifshitz, *Course of theoretical physics*; statistical physics receives such treatment in this series from a two-part volume which in its most recent edition is E. M. Lifshitz and L. P. Pitaevskii (1978–1981) *Statistical physics*, 2 vols (3rd edn., Pergamon). An excellent exposition of modern methods and developments is given in L. E. Reichl (1980) *A modern course in statistical physics* (U. of Texas Pr.), and a rigorous treatment of the mathematical aspects of the subject is D. Ruelle (1969) *Statistical mechanics: rigorous results* (Benjamin). D. A. McQuarrie (1976) *Statistical mechanics* (Harper and Row) presents a fine modern account of areas of statistical mechanics relating to chemical physics, including transport processes. Also of importance are A. Münster (1969–1974) *Statistical thermodynamics*, 2 vols (Springer), R. Balescu (1975) *Equilibrium and nonequilibrium statistical mechanics* (Wiley), W. T. Grandy, Jr. (1987–1988) *Foundations of statistical mechanics*, 2 vols (Reidel) and the unique approach in R. P. Feynman (1972) *Statistical mechanics: a set of lectures* (Benjamin). The important topic of entropy is discussed in detail in J. D. Fast (1968) *Entropy. The significance of the concept of entropy and its applications in science and technology* (2nd edn., Philips), and more theoretically in J. Yvon (1969) *Correlations and entropy in classical statistical mechanics* (Pergamon). A recent history of the field by S. G. Brush (1983) *Statistical physics and the atomic theory of matter, from Boyle and Newton to Landau and*

Onsager (Princeton U. Pr.) includes as a final chapter a stimulating sample of expert opinion on current outstanding problems in statistical physics.

The only major journal which concentrates solely on statistical physics is the *Journal of Statistical Physics*, although *Physica, Section A: Statistical and Theoretical Physics* publishes chiefly in this area. Critical reviews of basic problems in statistical mechanics are presented at the International Summer School on Statistical Mechanics, held every several years; the proceedings are published as the series *Fundamental Problems in Statistical Mechanics* (1962–). Other significant reviews are published in the series *Studies in Statistical Mechanics* (1963–). *Statistical Mechanics* (1973–), a Chemical Society Specialist Periodical Report, topically covering the literature year by year, has unfortunately not been published since 1975.

There has been a great deal of interest in the theory of non-equilibrium processes during the past few decades; some standard works are S. R. de Groot and P. Mazur (1962) *Non-equilibrium thermodynamics* (North-Holland), I. Prigogine (1969) *Introduction to the thermodynamics of irreversible processes* (3rd edn., Wiley) and W. Yourgrau *et al.* (1966) *Treatise on irreversible and statistical thermophysics* (Macmillan). Some of the wide-ranging work of the Brussels school, emphasizing the role of fluctuations, is found in G. Nicolis and I. Prigogine (1977) *Self-organization in non-equilibrium systems: from dissipative structures to order through fluctuations* (Wiley). An unusually detailed and careful critical examination of the field is provided in H. J. Kreuzer (1981) *Non-equilibrium thermodynamics and its statistical foundations* (OUP). The rapidly developing theory of chaos may come to be of great significance in the study of processes far from equilibrium; the collection of reprints in P. Cvitanović (ed.) (1989) *Universality in chaos* (2nd edn., Hilger) gives an overview of some recent applications of this research.

The entire range of the application of molecular theory to the properties of gases and liquids, for both equilibrium and non-equilibrium cases, is treated in the massive and very influential monograph, J. O. Hirschfelder *et al.* (1964) *Molecular theory of gases and liquids*, corrected printing with notes added (Wiley) (unchanged, except for minor corrections and the addition of 29 pages of notes, from the 1954 edition). This is still a most useful work, but it should be noted that computations in it were done largely using Lennard-Jones 6–12 potentials for convenience; more recent work has shown that such potentials are often unrealistic, and use of this book should be supplemented with the modern treatment of potentials in G. C. Maitland *et al.* (1981) *Intermolecular forces: their origin and determination* (OUP). T. G. Cowling (1950)

Molecules in motion. An introduction to the kinetic theory of gases (Hutchinson) is a beautifully written elementary introduction to the kinetic theory of gases; a somewhat more mathematical introduction is J. Jeans (1940) *An introduction to the kinetic theory of gases* (CUP). The classic reference work on the kinetic theory of transport properties in gases is S. Chapman and T. G. Cowling (1970) *The mathematical theory of non-uniform gases* (3rd edn., CUP), and a comprehensive review of the theory and applications of the Boltzmann equation is C. Cercignani (1985) *The Boltzmann equation and its applications*, vol 67 of *Applied Mathematical Sciences* (Springer). The theory of transport for dense gases and for liquids is not as yet satisfactory; however, in many cases these properties can be estimated and correlated by semi-empirical methods such as those in R. C. Reid *et al.* (1987) *The properties of gases and liquids* (4th edn., McGraw-Hill). Measured transport properties of fluids are included in N. B. Vargaftik (1983) *Handbook of physical properties of liquids and gases: pure substances and mixtures* (2nd edn., Hemisphere) and in P. E. Liley *et al.* (1988) *Properties of inorganic and organic fluids*, vol. V-1 of *CINDAS Data Series on Material Properties* (Hemisphere). Also for viscosity data see Y. S. Touloukian *et al.* (1975) *Viscosity*, vol. 11 of *Thermophysical properties of matter* (Plenum); diffusion data is also in Y. S. Touloukian *et al.* (1973) *Thermal diffusivity*, vol. 10 of *Thermophysical properties of matter* (Plenum) and in T. R. Marrero and E. A. Mason (1972) 'Gaseous Diffusion Coefficients', *Journal of Physical and Chemical Reference Data*, **1**, 3–118. Also see K. Schäfer (ed.) (1968–1969) *Transportphänomene*, 2 vols, vol. 2, sections 5a and 5b of Landolt-Börnstein, *Zahlenwerte und funktionen* (6th edn., Springer).

The equation of state for gases is often conveniently represented by a virial expansion, as discussed in E. A. Mason and T. H. Spurling (1969) *The virial equation of state* (Pergamon). J. H. Dymond and E. B. Smith (1980) *The virial coefficient of pure gases and mixtures: a critical compilation* (OUP) includes critically evaluated virial coefficients. Detailed equation of state data for, so far, 11 important fluids has been compiled under the auspices of IUPAC in S. Angus *et al.* (eds.) (1972–) *International Thermodynamic Tables of the Fluid State*, vol. 1, Argon; vol. 2, Ethylene; vol. 3, Carbon dioxide; vol. 4, Helium; vol. 5, Methane; vol. 6, Nitrogen; vol. 7, Propylene; vol. 8, Chlorine; vol. 9, Oxygen; vol. 10, Ethylene; and vol. 11, Fluorine. (Pergamon; except Blackwell Scientific for the latest volume)

Liquids

The study of simple liquids has made significant progress in recent years (H. L. Frisch and Z. W. Salsburg (eds.) (1968) *Simple dense fluids* (Academic), J. P. Hansen and I. R. McDonald (1986) *Theory of simple liquids* (2nd edn., Academic)). One of the best introductions to the modern theory of liquids is J. S. Rowlinson and F. L. Swinton (1982) *Liquids and liquid mixtures* (3rd edn., Butterworths); a volume of reviews of modern liquid state physics is E. W. Montroll and J. L. Lebowitz (eds.) (1982) *The liquid state of matter: fluids, simple and complex*, vol. 8 of *Studies in Statistical Mechanics* (North-Holland). A special field in liquid state physics which has been the subject of much interest is the study of liquid metals (S. Z. Beer (ed.) (1972) *Liquid metals, chemistry and physics* (Dekker), T. E. Faber (1972) *Introduction to the theory of liquid metals* (CUP), N. H. March (1990) *Liquid metals: concepts and theory* (CUP)). Another special area, in which much remains to be understood but which is of great significance for our understanding of biological phenomena, is the study of water. An excellent introduction is D. Eisenberg and W. Kauzmann (1969) *The structure and properties of water* (OUP), and F. Franks (ed.) (1972–1982) *Water – a comprehensive treatise*, 7 vols (Plenum) is a multi-volume treatise giving a comprehensive review. Often the best calculations on properties of fluids are done using computer simulation rather than by analytical methods. Sources for the techniques of computer simulation are R. W. Hockney and J. W. Eastwood (1988) *Computer simulation using particles* (Hilger) and M. P. Allen and D. J. Tildesley (1987) *Computer simulation of liquids* (OUP). A range of recent examples applying these methods may be found in K. Binder (ed.) (1987) *Applications of the Monte Carlo method in statistical physics*, vol. 36 of *Topics in Current Physics* (Springer).

Another area of significant current interest is the study of phase transitions. The best introduction to the field is H. E. Stanley (1971) *Introduction to phase transitions and critical phenomena* (OUP) and H. E. Stanley (ed.) (1973) *Cooperative phenomena near phase transitions: a bibliography with selected readings* (MIT). A modern treatment is given by J. M. Yeomans (1992) *Statistical mechanics of phase transitions* (OUP). The topic is covered comprehensively in C. Domb and M. S. Green (eds.) (1972–) *Phase Transitions and Critical Phenomena*, vols 1–3, 5 (in 2 parts), 6–14 (Academic). Data on temperatures, enthalpies, entropies and heat capacities relating to phase transitions are given in Series II of F. D. Rossini *et al.* (eds.) (1952) *Selected values of chemical thermodynamic properties*, US National Bureau of Standards, *Circular 500* (US Government

Printing Office). The literature on changes of state in solids is reviewed in N. B. Hannay (ed.) (1975) *Changes of state*, vol. 5 of N. B. Hannay (ed.) (1973–1976) *Treatise on solid state chemistry* (Plenum). The most dramatic recent theoretical advances in the study of phase transitions have been made following K. G. Wilson's development of renormalization group theory, as described in D. J. Amit (1984) *Field theory, the renormalization group, and critical phenomena* (2nd edn., World Scientific), S.-K. Ma (1976) *Modern theory of critical phenomena* (Benjamin), and J. J. Binney *et al.* (1992) *The theory of critical phenomena: an introduction to the study of the renormalization group* (OUP).

Liquid crystals

Liquid crystals, i.e. liquids whose component molecules have a degree of order approaching that of solid crystals, exhibit fascinating properties and present interesting theoretical problems. Excellent treatments of the physics of these problems are presented by P. G. de Gennes (1974) *Physics of liquid crystals* (OUP) and S. Chandrasekhar *Liquid crystals* (CUP, 1992). Information on the basic science, properties and applications of liquid crystals may be found in H. Kelker and R. Hatz (eds.) (1980) *Handbook of liquid crystals* (Verlag Chemie).

Polymers

The classic introduction to polymer science is still P. Flory (1953) *Principles of polymer chemistry* (Cornell U. Pr.); other fine texts are C. Tanford (1961) *Physical chemistry of macromolecules* (Wiley) and F. A. Bovey and F. H. Winslow (eds.) (1979) *Macromolecules: an introduction to polymer science* (Academic). *Encyclopedia of polymer science and engineering* (1985–1990), 19 vols (2nd edn., Wiley) is an excellent source of information on all areas of polymer science, and J. Brandrup and E. H. Immergut (eds.) (1989) *Polymer handbook*, (3rd edn., Wiley) is a very useful source of data on polymers. Many properties can be estimated using the approach of D. W. van Krevelen (1990) *Properties of polymers: their correlation with chemical structure, their numerical estimation and prediction from additive group contributions* (3rd edn., Elsevier). Modern experimental techniques of polymer science are conveniently compiled in J. F. Rabek (1980) *Experimental methods in polymer chemistry. Physical principles and applications* (Wiley). The important monographic series *High Polymers* (1940–1977) treats many significant topics in polymer physics and chemistry. Useful sources of review articles include the series *Macromolecular Chemistry* (1980–), *Report on Progress in Polymer Physics in Japan* (1958–) and *Advances in Polymer Science (Fortschritte der Hochpolymeren-Forschung)* (1958–), and the journals *Progress in Polymer Science, Progress in Colloid and Polymer Science, Journal of Macromolecular Science C: Reviews in Macromolecular Chemistry and Physics* and *Journal of Polymer Science D: Macromolecular Reviews.*

The statistical mechanics of polymers is treated in a number of important monographs, including M. Volkenstein (1963) *Configurational statistics of polymeric chains* (Interscience), T. M. Birshtein and O. B. Ptitsyn (1966) *Conformation of macromolecules* (Interscience) and P. Flory (1969) *Statistical mechanics of chain molecules* (Wiley). Much of the recent work on polymer theory uses techniques of renormalization group theory, mentioned in the section above on statistical mechanics. A fine introduction to this approach is P. G. de Gennes (1979) *Scaling concept in polymer physics* (Cornell U. Pr.); a recent work of major importance is M. Doi and S. F. Edwards (1986) *The theory of polymer dynamics* (OUP). R. T. Bailey *et al.* (1981) *Molecular motion in high polymers* (OUP) show how the molecular dynamics of polymers affects macroscopic phenomena. A classic treatment of polymer solutions is H. Morawetz (1965) *Macromolecules in solution* (Interscience). One of the best accounts of the modern theoretical treatment of polymer solutions is H. Yamakawa (1971) *Modern theory of polymer solutions* (Harper and Row). Important recent works in this area include J. Des Cloizeaux (1990) *Polymers in solution: their modelling and structure* (OUP), and H. Fujita (1990) *Polymer solutions* (Elsevier). S. Rice and M. Nagasawa (1961) *Polyelectrolyte solutions* (Academic) is the standard monograph on the theory of polyelectrolyte solutions. R. B. Bird *et al.* (1987) *Dynamics of polymeric liquids*, 2 vols (2nd edn., Wiley) give an authoritative treatment of polymeric liquids, including fluid mechanics, rheology and kinetic theory.

Surfaces

Phenomena at the interface of two phases are of interest in physics, chemistry and biology, and are also of considerable importance in areas of technology such as the fabrication of integrated circuits. Excellent introductions to the study of interfaces of gases, liquids and solids are given in G. Somorjai (1972) *Principles of surface chemistry* (Prentice-Hall), F. MacRitchie (1990) *Chemistry at interfaces* (Academic), D. Myers (1991) *Surfaces, interfaces and colloids*

(VCH) and A. W. Adamson (1990) *Physical chemistry of surfaces* (5th edn., Wiley); a somewhat more advanced treatment is S. R. Morrison (1990) *The chemical physics of surfaces* (2nd edn., Plenum).

Important sources of review articles on surface science include the journals *Surface Science Reports* and *Progress in Surface Science*, and the annuals *Surface and Colloid Science* (1969–), *Progress in Surface and Membrane Science* (1964–1981), *Advances in Colloid and Interface Science* (1967–) and *Chemical Physics of Solids and Their Surfaces* (1970–). An extensive index to surface science literature, containing 14 500 references for the period 1956–1977, has been compiled in E. Umbach *et al.* (1982) *Surface science index 1956–1977*, no. 24–1 of *Physik Daten* (FIZ).

Monographs on the science of liquid-liquid, liquid-gas, liquid-solid and solid-solid interfaces are J. T. Davies and E. K. Rideal (1963) *Interfacial phenomena* (Academic) and R. D. Vold and M. J. Vold (1983) *Colloid and interface chemistry* (Addison-Wesley); liquid interfaces are treated in C. A. Croxton (ed.) (1986) *Fluid interfacial phenomena* (Wiley); the solid-liquid interface, and particularly crystal growth at that interface, is described in D. P. Woodruff (1973) *The solid-liquid interface* (CUP). The definitive account of the theory of capillarity and of the liquid-gas interface is given in J. S. Rowlinson and B. Widom (1983) *Molecular theory of capillarity* (OUP).

A brief introduction to the physics of solid surfaces is given in M. Prutton (1983) *Surface physics* (2nd edn., OUP), and in A. Zangwill (1988) *Physics at surfaces* (CUP). Solid surfaces are reviewed in N. B. Hannay (ed.) (1976) *Surfaces*, vol. 6 of N. B. Hannay (ed.) (1973–1976) *Treatise on solid state chemistry* (Plenum). F. Garcia-Moliner and F. Flores (1979) *Introduction to the theory of solid surfaces* (CUP) treat the electronic structure of solid surfaces. Techniques for studying solid surfaces are covered in D. Briggs and M. P. Seah (eds.) (1990–) *Practical surface analysis* (2nd edn., Wiley), J. C. Riviere (1990) *Surface analytical techniques* (OUP), and J. M. Walls (1989) *Methods of surface analysis* (CUP). Important specific techniques include electron diffraction (G. Ertl and J. Küppers (1985) *Low energy electrons and surface chemistry* (2nd edn., VCH)), field ion and field emission microscopies (T. T. Tsong (1990) *Atom-probe field ion microscopy* (CUP)) and electron microscopy (J. I. Goldstein *et al.* (1992) *Scanning electron microscopy and X-Ray microanalysis* (2nd edn., Plenum)). The exciting new scanning microscopies are introduced by D. Sarid (1991) in *Scanning force microscopy* (OUP). The interaction of gases with the solid surface is reviewed in M. N. R. Ashford and C. T. Rettner (eds.) (1991) *Dynamics of gas-surface interactions* (RSC) and in

W. A. Steele (1974) *Interaction of gases with solid surfaces* (Pergamon); the theory of gas-surface scattering is treated in F. O. Goodman and H. Y. Wachman (1976) *Dynamics of gas-surface scattering* (Academic). S. C. Saxena and R. K. Joshi (1989) *Thermal accommodation and adsorption coefficients of gases*, vol. II-1 of *CINDAS Data Series on Material Properties* (Hemisphere) critically review data on thermal accommodation and adsorption coefficients in gas-surface interactions.

For a molecule to participate in a surface reaction, it must first be 'chemisorbed', that is, it must form a chemical bond with the surface. Very useful discussions are in F. C. Tompkins (1978) *Chemisorption of gases on metals* (Academic), in T. N. Rhodin and G. Ertl (eds.) (1978) *The nature of the surface chemical bond* (North-Holland), J. R. Smith (ed.) (1980) *Theory of chemisorption* (Springer), and in N. H. March (1986) *Chemical bonds outside metal surfaces* (Plenum). G. Somorjai (1981) *Chemistry in two dimensions: surfaces* (Cornell U. Pr.) presents a splendid study of solid surfaces and the gas-solid interface as they relate to chemical catalysis; he includes lengthy tables of data compiled on the adsorption of molecules onto well-characterized surfaces and on kinetic parameters of surface reactions. Other treatments of the closely related fields of surface science and heterogeneous catalysis are in E. Drauglis and R. I. Jaffee (eds.) (1975) *The physical basis for heterogeneous catalysis* (Plenum) and D. A. King and D. P. Woodruff (eds.) (1981–1990) *The chemical physics of solid surfaces and heterogeneous catalysis*, 5 vols in 6 (Elsevier), and in M. W. Roberts and C. S. McKee (1979) *Chemistry of the metal-gas interface* (OUP). Review articles on this subject are found in the annual series *Catalysis* (1977–) and *Advances in Catalysis* (1948–) and in the journals *Catalysis Reviews* and *Catalysis Today*.

Electronic information sources

Electronic sources of bibliographic information are mentioned above in the section on general chemical physics, together with their print counterparts. However, a large and increasing number of electronic sources of numerical data are also available. In addition to the numerical databases listed below, it is possible to find sources of numerical data through the CODATA REFERRAL DATABASE (CRD) produced by CODATA (Committee on Data for Science and Technology). The database on this diskette derives from CODATA directories and the *UNESCO inventory of data referral sources in science and technology*. Searches can be done for keywords combined with Boolean logic, and the results are names and locations of organizations which can provide the needed data, including databases these organizations produce. Another useful source of information is the online CUADRA DIRECTORY OF DATABASES, corresponding to the print title *Gale directory of databases*.

DETHERM – SDC. Produced by DECHEMA (Deutsche Gesellshaft für Chemisches Apparatwesen). Available online, as diskette, and magnetic tape. Data for calculating thermophysical properties and phase equilibria of about 550 pure substances and their mixtures in the fluid state.

DETHERM – SDR. Produced by DECHEMA. Available online, as diskette, and magnetic tape. Data on thermophysical properties of about 5700 industrially important compounds and their mixtures.

DIPPR DATA COMPILATION OF PURE COMPOUND PROPERTIES. Produced by the American Institute of Chemical Engineers (AIChE); a print version (1985) is *Data compilation tables of properties of pure compounds* (see above in the section on statistical physics and thermodynamics). Available online, as diskette, and magnetic tape. Includes thermodynamic and transport data for over 1100 commercially important chemicals.

JANAF THERMOCHEMICAL TABLES. Produced by US National Institute of Standards and Technology; print version is (1985) *JANAF thermochemical tables* (see above). Available online; for diskette, see NIST STRUCTURES AND PROPERTIES DATABASE AND ESTIMATION PROGRAM, below.

NISTFLUIDS. Produced by US National Institute of Standards and Technology. Available online. Calculates thermophysical and transport properties over a wide range of temperatures and pressures for 12 cryogenic fluids.

NIST STRUCTURES AND PROPERTIES DATABASE AND ESTIMATION PROGRAM. Produced by US National Institute of Standards and Technology, as *NIST Standard Reference Database 25*. Diskette. Contains thermochemical data for about 4900 compounds or species from NIST POSITIVE ION ENERGETICS DATABASE, NIST CHEMICAL KINETICS DATABASE, and NIST JANAF THERMOCHEMICAL TABLES DATABASE. Also includes data estimation software.

NISTTHERMO. Produced by US National Institute of Standards and Technology; print version (1982) as *NBS tables of chemical thermodynamic properties* (see above). Available online. Critically evaluated chemical thermodynamic properties of over 8000 substances (chiefly inorganic).

PPDS (Physical Property Data Service). Produced by Institution of Chemical Engineers. Available online, and as magnetic tape. Physical properties of 880 compounds at temperatures up to 1000° C; allows calculation of properties of mixtures of up to 20 components. Allows calculation of vapor-liquid equilibrium conditions. Contains equations of state prepared under the auspices of IUPAC.

TRCTHERMO. Prepared by Thermodynamics Research Center; print version as *TRC thermodynamic tables. Hydrocarbons* and *TRC thermodynamic tables. Non-hydrocarbons*. (see above). Available online. Selected thermodynamic properties of pure compounds in about 350 000 records.

TRC VAPOR PRESSURE DATAFILE. Produced by Thermodynamics Research Center. Available online, as diskette. Vapor pressure and boiling point data for about 8700 organic compounds.

References

Miller, W. A. (1855) *Elements of chemistry. Part I. Chemical physics*. London: Parker.

Servos, J. W. (1982) A disciplinary program that failed: Wilder D. Bancroft and the *Journal of Physical Chemistry*, 1896–1933. *Isis*, **73**, 207–229.

Slater, J. C. (1939) *Introduction to chemical physics*. New York: McGraw-Hill.

CHAPTER SIX

Classical optics

A. L. RODGERS

Introduction

Optics is one of the oldest scientific subjects, dating back to the Ancient Greeks. Important milestones in the second millenium, from the invention of spectacles by Bacon in the 13th century, have been the creation of the laws of refraction by Snell and Descartes and the development and application of the astronomical telescope by Galileo in the 16th century; the observation of diffraction by Grimaldi and the investigation of wave fronts and propagation of light by Huygens and Fermat in the 17th century; and the publication of *Opticks* by Newton in 1704, covering refraction, dispersion, interference, diffraction and polarization of light. Discoveries in the 19th century were the establishment of the transverse nature of light by Young and Fresnel and the investigation of the relationship between optical and magnetic phenomena by Faraday. The culmination of these discoveries appears in the creation of electromagnetic theory by Maxwell and of physical optics by Lorentz, laying the foundation of classical optics. Twentieth century milestones are the discovery of electromagnetic waves by Hertz, the evolution of quantum optics by Planck and the derivation of the hydrogen atom spectra by Bohr. The subject has produced Nobel Laureates, the most recent awards to Lamb (1955) for work on the fine structure of the hydrogen spectrum; to Cherenkov (1958) for discovery of radiation named after him; to Toomes, Prokhorov and Basov (1964) for experimental verification of the laser; to Kastler (1966) for studies of resonances in atoms by optical methods; to Gabor (1971) for discovery of holography; and to Bloembergen and Schawlow (1981) for work in laser spectroscopy.

The subject of optics has been described as the study of phenomena associated with the generation and detection of electromagnetic radiation within the wavelength range extending from the
near ultraviolet to the far infrared, though other definitions have
adopted a broader spectral range.

This chapter is concerned mainly with classical optics which is
defined as the study of rays and electromagnetic waves, interference
and diffraction, scattering and polarization, optical properties of
materials, optical radiation measurements and optical systems
design. Sections on colour, vision and image processing are included
although these could equally be at home in the associated Chapter
16 entitled 'Optoelectronics, quantum optics and spectroscopy.'

Brief summaries to references are depicted where made available,
otherwise author, title and publisher details are given. References
are classified within subject headings but there is overlap and attention is drawn to associated material in Chapter 16.

In a retrieval exercise of this type it is inevitable that source
information is missed, and omissions are regretted. Full consultation
with the particular publication is recommended and is essential in
order to confirm interest if considering a purchase.

Historical

A contemporary trilogy for science historians up to the middle of the
nineteenth century is A. I. Sabra (1982) *Theories of light from
Descartes to Newton* (CUP), A. E. Shapiro (ed.) (1983) *The optical
papers of Isaac Newton. Vol. 1–The optical lectures 1670–1672*
(Pergamon), giving the complete text of his lectures, and G. N.
Cantor (1983) *Optics after Newton: theories of light in Britain 1704–
1840* (MUP), a new interpretation of British optical theories.

L. Mendel and E. Wolf (eds.) (1970) *Selected papers on the coherence and fluctuations in light* (Dover) is a historical review covering the years 1850–1960 with bibliography.

Bairerlein R. (1992) *Newton to Einstein. The trail of light* (CUP) is
an undergraduate text requiring no previous knowledge of physics
and only limited mathematical ability. It helps understanding by
presenting many exercises, with clues, before answers are revealed.

Journals, series and conference proceedings

The following lists are also part of Chapter 16 bibliography.

Journals

Journals within the general field of optics are:

Applied Optics, Optics Letters and Optics News (OSA)
Applied Spectroscopy (Society for Applied Spectroscopy)
Journal of Optical Society of America, Part A: Optics and Image Science, Part B: Optical Physics (OSA)
Journal of Physics B, Atomic, Molecular and Optical Physics (IOP)
Journal of Optical and Quantum Electronics (Chapman and Hall)
Journal of Quantitative Spectroscopy and Radiative Transfer (Pergamon)
Journal of Quantum Electronics (IEEE)
Journal of Modern Physics – formerly *Optica Acta* (Taylor and Francis)
Infra-Red Physics (Pergamon)
Journal of Lightwave Technology (IEEE/OSA)
Journal of Plasma Physics (CUP)
Inform D'Optique (FSIO)
Physica D: Nonlinear Phenomena (North Holland)
Physical Review A, Atomic, Molecular and Optical Physics (APS)
Physical Review Letters (APS)
Physics Letters A (North Holland)
Plasma Devices and Operations (Gordon and Breach)
Plasma Physics and Controlled Fusion (IOP)
Progress in Atomic Spectroscopy, Parts A, B and C (Plenum)
Quantum Optics (IOP)
Journal of Optics (Masson)
Journal of European Optical Society, Part A, Pure and Applied Optics, and Part B, Quantum Optics (IOP)
Review of Scientific Instruments (AIP)
Reviews of Modern Physics (APS/AIS)
Engineering Optics (IOP)
Optical Engineering (SPIE)
Optics Communications (North Holland)
Opto and Laser Europe (IOP)
Optics and Laser Technology (Butterworth and Heinemann)
Optik (Wissenschaftliche, Verlagsgesellschaft, MbH, Stuttgart)
Laser and Particle Beams (CUP)
Nonlinear Optics (Gordon and Breach)
Advances in Optical and Electron Microscopy (Academic)
Contemporary Physics (Taylor and Francis)
Physics of Quantum Electronics (Addison-Wesley)
Soviet Physics-JETP (AIP)
Journal of Raman Spectroscopy (Wiley)

Journal of Optical Computing (Wiley)
Microwave and Optical Technology Letters (Wiley)
Advanced Materials for Optics and Electronics (Wiley)
Laser Therapy (Wiley)
Lasers in Surgery and Medicine (Wiley)
Soviet Lightwave Communications (IOP with Russian Academy of Science)
Soviet Journal of Optical Technology (OSA/SPIE)
Applied Physics B, Photophysics and Laser Chemistry (Springer)
Journal of Nonlinear Science (Springer)
Meteorology and Atmospheric Physics (Springer)
Physics and Chemistry of Minerals (Springer)
Zeitschrift für Physik D, Atoms, Molecules and Clusters (Springer)
Zeitschrift für Physik B, Condensed Matter (Springer)
Journal of Fluorescense (Plenum)
Journal of Soviet Laser Research (Plenum)
Radiophysics and Quantum Electronics (Plenum)
Journal of The Illuminating Engineering Society (IES, New York)
Journal of Luminescence (North Holland)

Series

A selection of book series is given below:

Series on Optics and Optoelectronics (Hilger)
Series on Plasma Physics (Hilger)
Series in Pure and Applied Optics (Wiley)
Series on Sensors (Hilger)
Series in Modern Optics (CUP)
Topics in Applied Physics (Springer)
Topics in Current Physics (Springer)
Series in Optical Sciences (Springer)
Reviews of Infra-red and Millimetre Waves (Plenum)
Technology Transfer Series (Pergamon)
Series in Materials Science (Springer)
Modern Problems in Condensed Matter Series (North Holland)
Malvern Physics Series (IOP)
Manchester Physics Series (MUP)
Ettore Majorana International Science Series: *Physical Science* e.g. vol. 35 *Laser science and technology*, vol. 49 *Nonlinear optics* and vol. 54 *Optoelectronics and environmental science* (Plenum)
Optical Engineering (Marcel Dekker)
Lasers, Photonics and Electro-optics (Plenum)
NATO Advanced Science Institute Series B, Physics (Plenum)
Optical Physics and Engineering (Plenum)
Cambridge Lecture Notes in Physics (CUP)

American Institute of Physics Conference Proceedings (IOP)
Institute of Physics Conference Series (IOP)
Medical Science Series (IOP)
Scottish Universities Summer School in Physics (IOP)
Analytical Spectroscopy Library (Elsevier)
Electromagnetic Waves (Elsevier)
Laser Handbook (Elsevier)
Optical Wave Sciences and Technology (Elsevier)
Progress in Electromagnetic Research (Elsevier)
Progress in Optics (Elsevier)
SPIE Optical Engineering Press (SPIE)
Physics-Based Vision (Jones and Bartlett)
Montroll Memorial Lecture Series in Mathematical Physics (CUP)
Advanced Topics in Interdisciplinary Mathematical Series
    (Addison-Wesley)
Handbook Series on Laser Science and Technology (CRC)
IEEE Conference Publication Series (IEEE)
Transactions in Communications (IEEE)
NATO Advanced Study Institute Series (Plenum)
Electronic and Electrical Research Studies in Optoelectronics
    (Wiley/Research Studies)
McGraw-Hill Science Reference Series (McGraw-Hill)
Advanced Science and Technology Series (McGraw-Hill)
Optoelectronics Library Series (Artech)
Optical Radiation Measurements (Academic)
Applied Optics and Optical Engineering (Academic)
Monographs on Applied Optics (Hilger)
Advances in Optical and Electron Microscopy (Academic)
Methods in Experimental Physics (Academic)

Conference proceedings

In order to limit the number of sections both in this chapter and in
Chapter 16, conference proceedings are listed with books within the
subject classification.

Though different publishers are responsible for conference proceedings, a prolific source of information in all branches of contemporary optics, optical and electro-optical engineering and
technology are the publications by the International Society for
Optical Engineering and the Society of Photo-optical Instrumentation Engineers (SPIE). Every branch of these subjects is regularly
covered at an advanced research level in publications, annual meetings and conferences run by these societies under SPIE. The size of
the output can be appreciated by the statistics within *Publication
Index* vol. 1515 for 1990, listing 225 proceedings and books, 7300

technical papers involving 16 900 authors for that year alone. The volume of this output cannot be properly accommodated within this chapter and Chapter 16, which list selected publications for the years 1990 to 1992, and the reader is referred to earlier publication lists and programme announcements.

Fundamental optics

Books and proceedings

GENERAL TEXTS

Wolfe, E. (ed.) *Progress in Optics* (Elsevier/North-Holland) is a comprehensive source of information in all branches of optics covering theoretical and experimental aspects with applications. Articles are contributed by leading scientists and annual volumes reflect contemporary progress. The latest is vol XXX 1992.

Handbook of optics (OSA, McGraw-Hill, 1978) is another authoritative reference work, specializing in optical materials, components and systems.

Levi, L. *Applied optics*, vol. 1 (1968) and vol. 2 (1980) (Wiley) is another substantial review, specializing in optical systems design and optical data, which is aimed at the graduate engineer and experienced practitioner.

Born, M. and Wolfe, E. (1980) *Principles of optics: electromagnetic theory of propagation interference and diffraction of light*, 6th edn. (Pergamon) is a significant source and reference book giving a rigorous account of wave physics and physical and geometrical optics at an advanced level.

Jenkins, F. A. and White, H. E. (1976) *Fundamentals of optics*, 4th edn. (McGraw-Hill) has been updated to include lasers and holography.

Lipson, S. G. and Lipson, H. (1981) *Optical physics*, 2nd edn. (CUP) is another advanced book with a full discussion of statistical optics and the contributions of lasers to physical optics.

Garbung, M. (1970) *Optical physics* (Academic) covers the dynamic interactions between light and matter from which structural information is derived.

Other general texts include:

Ditchburn, R. W. (1976) *Light*, 3rd edn. (Academic) and *Light* by Deeson, E. (John Murray, 1975), Welford, W. T. (1988) *Optics*, 3rd edn. (OUP), Mathieu, J. P. (1975) *Optics*, within the International Series in Natural Philosophy (Pergamon) and associated *Problem in optics* by the same author with Rousseau, M., and Brill, T. (1980) *Light* (Plenum).

Longhurst, R. S. (1973) *Geometrical and physical optics*, 3rd edn. (Longman, reprinted 1990).

Smith, F. G. and Thompson, J. H. (1988) *Optics*, in the Manchester Physics Series, 2nd edn, (Wiley).

Hecht, E. (1974) *Optics* (McGraw-Hill). The second edition (Addison-Wesley, 1987) brings in new developments in optical technology.

Fowles, G. R. (1968) *Introduction to modern optics* (Holt-Rinehart).

The following have been published since the second edition of this book:

Welford, W. T. (1988) *Optics*, 3rd edn. (OUP)

Iizuka, K. (1987) *Engineering optics*, 2nd edn. (Springer).

Freeman, M. H. (1990) *Optics*, 10th edn. (Addison-Wesley) gives the fundamentals and recent technological advances.

Parker, S. P. (ed.) (1988) *Optics source book* (McGraw-Hill) is a general book within the McGraw Hill Science Reference Series.

Guenther, B. D. (1990) *Modern optics* (Wiley) gives a modern perspective to traditional methods and examines the basis of new optical devices.

Meyer-Arendt, J. R. (1989) *Introduction to classical and modern optics*, 3rd edn. (Prentice Hall).

Klein, M. V. and Furtak, T. E. (1986) *Optics*, 2nd edn. (Wiley).

Yu, F. T. S. and Khoo, I. (1990) *Fundamentals of optic engineering* (Wiley) expands from geometrical optics through to modern electro-optics covering lasers, holography, fibres and signal processing.

Bates, D. R. and Bederson B. (1991) *Advances in atomic, molecular and optical physics* (Academic) covers the control of atoms and molecules by light, along with non-optical topics.

Heavons, O. S. and Ditchburn, R. W. (1991) *Insight into optics* (Wiley) covers the general principles and applications of the whole field of optics in unique and condensed form. It is eminently readable and contains problems to secure knowledge.

Parker, S. P. (ed.) (1988) *Optics source book* (McGraw-Hill) is a comprehensive compendium of optical phenomena presented by experts, with precise descriptions of each subject matter and lists of references.

Bally, G. von (ed.) (1991) *Optics in life sciences* (Springer).

Goodman, J. W. (1991) *International trends in optics* (Academic) is a collection of essays providing an overview of the current state of knowledge in many branches of optics.

Strong, J. *Procedures in applied optics* (Marcel Dekker, Opt. Eng. Series vol. 17).

GEOMETRICAL OPTICS

Geometrical optics forms the basis of the design of many optical instruments:

Fry, G. A. (1969) *Geometrical optics* (Chilton) is written from the traditional ray-tracing aspects.

Stravroudis, O. N. (1972) *The optics of rays, wavefront and caustics*, vol. 38 within the Monographs on Pure and Applied Physics (Academic), is a detailed study of the mathematical basis of geometrical optics.

Zimmer, H. G. (1970) *Geometrical optics* (Springer) is an introductory text which develops the subject from the principle of conservation of radiated energy.

Jurek, B. (1976) *Optical surfaces* (North-Holland) examines image formation from reflecting surfaces.

Kline, M. and Kay, I. W. (1965) *Electromagnetic theory and geometrical optics* (Wiley-Interscience, repr. Krieger, 1979) interrelates the two subjects, as does Luneberg, R. K. (1984) *Mathematical theory of optics* (U. of Cal. Pr.).

Slyusorev, G. G. (1984) *Aberration and optical design theory* (IOP) is concerned with ray tracing techniques employing theory of aberration.

Welford, W. T. (1991) *Aberrations of optical systems* (Hilger) covers the elementary optics and the aberration theory of optical systems. Mainly based upon geometrical optics but some physical optics is invoked. Aimed for postgraduate students, optical designers in industry and researchers in relevant fields.

Klavtsov, Y. A. and Orlov, Y. I. (1990) *Geometrical optics of inhomogenous media* (Springer).

Of historical interest is *Geometrical Optics* by R. S. Heath (CUP, 1895) giving a 19th century review of the subject.

HAMILTONIAN OPTICS

There are two books which deal with Hamiltonian formalism:

Buchdahl, H. A. (1970) *An introduction to Hamiltonian optics* (CUP) is concerned with the treatment of aberration theory.

Blaker, J. W. (1984) *Hamiltonian optics* (Pergamon) employs the Hamilton function in deriving optical equations and in describing optical phenomena, specifically for final-year undergraduates taking optical options as well as for postgraduate.

PHYSICAL OPTICS

The wave aspects of light have significance in the performance of optical instruments and in most applications in optics.

Tolansky, S. (1973) *An introduction to interferometry* (Longmans) is a student text, as is Françon, M. (1966) *Diffraction: coherence in optics* (Pergamon). A more advanced work is Lipson, H. (1972) *Optical transforms* (AP), and Northover, F. H. (1971) *Applied diffracton theory* (Elsevier) gives a mathematical treatment with practical applications.

The theory and use of diffraction gratings are covered in Petit, R. (1980) *The electromagnetic theory of gratings* (Springer), which uses vector formalism.

Practical books are:

Guild, V. (1960) *Diffraction gratings as measuring scales: practical guide to the metrological use of Moiré fringes* (OUP).

Dyson, J. (1970) *Interferometry as a measuring tool* (Machinery).

Françon, M. and Mallick, S. (1971) *Polarisation interferometers: applications in microscopy and macroscopy* (Wiley).

Tolansky, S. (1960) *Surface microtopography* (Longmans) develops the technique of multiple-beam interferometry to study surface topography.

Cowley, J. M. (1981) *Diffraction physics* (North-Holland) gives a unified treatment for all branches, including optics.

Steel, W. S. (1984) *Interferometry* (CUP) is a monograph which has been updated to reflect the growth in laser techniques in remote sensing, speckle interferometry and holography. The latter two subjects are also covered in Jones, R. and Wyker, C. M. (1984) *Holography and speckle interferometry* (CUP), which contains a unified theoretical analysis of the basic principles and associated techniques.

Dainty, J. C. (ed.) (1975) *Laser speckle* (Springer) explains the principles of speckle formation by coherent and partially coherent light and describes applications in astronomy and metrology.

The following have been published since the second edition of this book:

Hariharan, P. (1992) *Basics of interferometry* (Academic) is an introductory book and the same author has written *Optical interferometry* (Academic, 1985) which gives a full review of the principles and techniques.

Vaughan, J. M. (1989) *The Fabry Perot interferometer* (Hilger) is a substantial review of the history, theory, practice and applications of this technique.

Hernandez, G. (1986) *Fabry Perot interferometers* (CUP) is a comprehensive review of the principles and design of these systems.

Belyakov, V. A. (1992) *Diffraction optics of complex structural periodic media* (Springer) is an advanced treatise on coherent interaction of electromagnetic and corpuscular radiation with regular, spatially periodic, media having complicated structures.

Taylor, C. A. (1987) *Diffraction* (IOP) covers Fraunhöfer and Fresnel diffraction, image limitation and applications.

Barakat, N. and Hanza, A. A. (1990) *Interferometry of fibrous material* (Hilger) describes the use of interferometric techniques to fibre analysis and research.

Wilcox, C. H. (1984) *Scattering theory for diffraction gratings* (Springer).

Hariharan, P. (ed.) (1991) *Interferometry* (SPIE, vol. MS28) is a selection of key discovery and development papers from the world literature.

Pryputniewicz, R. J. (1990) *Laser interferometry: quantitative analysis of interferograms* (SPIE, vol. 1162) is the proceedings of a conference held in San Diego in 1989 and covers laser diode interferometry, speckle interferometry, systems and applications.

Neito-Vesperinas, M. (1991) *Scattering and diffraction in physical optics* (Wiley) is a comprehensive treatment of wave fields, extinction theorems and applications; based upon a tutorial approach.

Mickelson, A. R. (1992) *Physical optics* (Van Nostrand Reinhold) derives from a one semester, first year, graduate level course and aims to present a unified quantitative treatment of physical optics at a deeper insight than for an undergraduate course.

STATISTICAL OPTICS

Frieden, B. R. (1991) *Probability statistical optics and data testing* (Springer) is a textbook describing the essential aspects of probability and statistics required for the study of optics.

Goodman, J. W. (1985) *Statistical optics* (Wiley) covers random variables and processes, propagation of light waves, coherence and partial coherence in imaging systems, imaging in the presence of random inhomogenous media, limits to photo-electric detection of light and Fourier transforms.

SCATTERING OF LIGHT

The scattering of light is of concern whenever light is propagated through a medium. Books on light scattering are:

Van de Hulst, H. C. *Light scattering by small particles* (Chapman and Hall, 1957; Dover, 1981).

Fabelinskii, I. L. (1968) *Molecular scattering of light* (Plenum).

Kerker, M. (1969) *The scattering of light and other electromagnetic radiation* (Academic).

Cardona, M. (1983) *Light scattering in solids*, 2nd edn. vol. 8 in Topics in Applied Physics (Springer), deals with inelastic light scattering in amorphous and crystalline solids with applications to their electronic and vibronic structure. The same author with G. Guntherodt is co-editor of *Light scattering in solids III recent results*, Topics in Applied Physics, 51 (Springer, 1982), which is concerned with applications of Raman and Brillouin scattering to various specific case studies and another book on the same theme is Birman, J. *et al.* (1979) *Light scattering in solids* (Plenum).

Hayes, W. (1978) *Scattering of light by crystals* (Wiley) is a study of inelastic light scattering and its use in obtaining information about solid-state excitation; experimental aspects are described in detail.

Degiorgio, V. (1980) 'Light scattering in liquids and macromolecular solutions' is reported in *The 1979 proceedings on quasielastic light scattering studies, Milan* (Plenum).

Baron, L. D. (1982) *Molecular light scattering and optical activity* (CUP) reviews classical and quantum methods to develop the theory of a variety of optical activity phenomena of a symmetrical and anti-symmetrical nature.

Bohren, C. F. and Hoffman, D. R. (1983) *Absorption and scattering of light by small particles* (Wiley) is a classical treatment of absorption and scattering of light using linear optics.

Other books are by Scherman, D. W. (1980) *Light scattering by irregularly shaped particles* (Plenum); Dahneke, B. E. (1983) *Measurement of suspended particles by quasi-elastic light scattering* (Wiley) and Wickramasinghe, N. C. (1973) *Light scattering functions for small particles* (Hilger).

Bates, H. P. (ed.) (1980) *Inverse scattering problems in optics* (Springer) discusses inverse scattering as a method to study different applications.

Additional books published since 1984 follow:

Brown, W. (ed.) (1992) *Dynamic light scattering* (OUP) addresses single photon correlation techniques and noise generation; applications to polymers in bulk and in solution, to polymer gels, rigid rods, simple fluids and biological systems.

Schmitz, K. S. (ed.) (1991) *Photon correlation spectroscopy* (SPIE) covers instrumentation, internal motions, polymers, gels and polyelectrolytes.

Chu, B. (1991) *Laser light scattering*, 2nd edn. (Academic) covers the basic principles and applications, incorporating fibre optics for instrumentation.

Zege, E. P., Ivanov, A. P. and Katsev, I. I. (1990), *Image transfer through a scattering media* (Springer) gives a detailed theoretical background of image generation and detection in scattering media and acts as a handbook for solutions in practical problems.

Richards, P. H. (1985) *Optical measurements in fluid mechanics* (Institute of Physics) is the proceedings of the Sixth International Conference on Photon Correlation and Other Techniques in Fluid Mechanics held at Cambridge, 1985.

Cardona, M. and Guntherodt, G. (eds.) *Light scattering in solids* (Springer, vol. 5, 1989 and vol. 6, 1991).

Lockwood, D. J. and Young, J. F. (1991) *Light scattering in semiconductor structures and superlattices* (Plenum).

Nussenzveig, H. M. (1992) *Diffraction effects in semiclassical scattering* (CUP) is the first book in the Montroll Memorial Lecture Series in Mathematical Physics.

ELECTROMAGNETIC WAVES

Of those books concerned with the applications of Maxwell's equations:

Read, E. H. (1980) *Electromagnetic radiation* (Wiley) provides a sound introduction covering classical and quantum treatments and introducing coherence and laser light.

Garbuny, M. (1965) *Optical physics* (Academic) covers the fundamental laws of propagation and interaction of radiation with matter.

Marcuse, D. (1982) *Light transmission optics*, 2nd edn. (Van Nostrand) is an advanced treatment; and Papas, C. H. (1965) *Theory of electromagnetic wave propagation* (McGraw-Hill) and Sander, K. F. and Read, G. A. L. (1986) *Transmission and propagation of electromagnetic waves*, 2nd edn. (CUP) includes transmission through optical fibre.

Stamners, J. J. (1986) *Waves in focal regions* (Hilger) examines image formation and focusing of electromagnetic waves and also examines acoustic and water waves.

Kritikos, H. N. and Jaggard, D. L. (eds.) (1990) *Recent advances in electro-magnetic theory* (Springer).

Petykiewicz, J. (1992) *Wave optics* (Kluwer) is based on a course of lectures held at the Department of Technical Physics and Applied Mathematics of the Warsaw University of Technology.

See further references in Chapter 16 under 'Optical communications'.

FOURIER OPTICS

The subject of Fourier optics is covered in:

Goodman, J. W. (1968) *Introduction to Fourier optics* (McGraw-Hill), in which linear system theory and Fourier analysis is used as a foundation for the theory of image formation, optical data processing and holography, the treatment being oriented towards electrical engineers.

Duffieux, P. M. (1983) *The Fourier transform and its application in optics* in the Wiley series of Pure and Applied Optics (Wiley) shows how concepts of Fourier optics may be developed using only Fourier series and integrals.

Steward, E. G. (1987) *Fourier optics: an introduction*, 2nd edn. (Ellis Horwood) is an introduction to Fourier image formation. This edition includes a chapter on image formation by computed tomography, several new appendices that broaden the scope of the book and contains problems with solutions and guidance.

Gaskill, J. D. (1978) *Linear systems Fourier transforms and optics* (Wiley) is another book in the Wiley series of Pure and Applied Optics.

Stark, H. (ed.) (1982) *Applications of optical Fourier transforms* (Academic) gives an in-depth study of the principal applications to the solution of some scientific and engineering problems.

Strennler, F. G. (1979) *Fourier transforms and applications* (Wisconsin).

The following additional books should be noted:

Reynolds, G. O., Detelis, J. B., Parrent Jnr., G. B. and Thompson, B. J. (1989) *The new physical optics notebook tutorials in Fourier optics* (SPIE and AIP/Hilger) is a comprehensive advanced text covering the underlying fundamentals and applications. It addresses Fourier transforms, theory of optical coherence, partial coherence, image formation and resolution, interferometry, holography and applications of Fourier analysis to filters, communications, optical computing, phase contrast imagery, synthetic aperture imaging systems and optical lithography.

Korner, T. W. (1993) *Exercises in Fourier analysis* (CUP) is a compilation which helps to bring out the usefulness of Fourier analysis as a tool for a wide variety of techniques and applications.

Two books which are concerned with the velocity of light are:

Froome, K. D. and Essen, L. (1969) *The velocity of light and radio waves* (Academic) and Sanders, J. H. (ed.) (1965) *The velocity of light* (Pergamon).

POLARIZATION OF LIGHT

Polarised light: production and use by W. A. Schurclift (Harvard U. Pr., 1962) is an introduction to the subject.

Polarised light and optical measurement by D. Clarke and J. F. Grainer (Pergamon, 1971) is an advanced text.

Beckmann, P. (1968) *The depolarisation of electromagnetic waves* (Golden Press) discusses the theory of change of state of polarization.

Ellipsometry and polarised light by R. A. Azzam, and N. M. Bashara (North-Holland, 1977) discusses this technique and lists over 300 references for research workers in the field.

Further titles are:

Robson, B. A. (1974) *The theory of polarization phenomena* in the Oxford Studies in Physics Series (Clarendon).

Konnen, G. P. (1985) *Polarised light in nature* (CUP) discusses the features of polarised light, methods of observing and occurrence in nature.

Kliger, D. S., Lewis, J. W. and Randall, C. E. (1990) *Polarised light in optics and spectroscopy* (Academic) emphasises the practical applications and computational methods to describe the interaction of polarised light with different types of optical element.

Robinson, P. C. and Bradbury, S. (1992) *Qualitative polarised light microscopy* (OUP) is a practical handbook describing the use of the polarised light microscope. It covers the basic concepts and observational phenomena.

Sowinski, J. and Vigdor, S. E. (1990) *Physics with polarised beams on polarised targets* (World Scientific) is the proceedings of a conference on this subject held in Spencer, Indiana, in October, 1989.

COLOUR AND VISION

Introductory texts are *Colour*, by R. W. Burnham *et al.* (Wiley, 1963), and *Light and colour*, by R. D. Overheim and E. L. Wagner (Wiley, 1982).

At a more advanced and comprehensive level is:

Wysecki, G. and Stiles, W. S. (1967) *Colour science—concepts and methods, quantitative data and formulae*, 2nd edn. (Wiley).

Kornerup, A. and Wonscher, J. H. (1967) *The Methuen book of colour*, 2nd edn. (Methuen) includes a dictionary of colour samples with the British Standard and Munsell equivalents.

Wright, W. D. (1969) *The measurement of colour* (Hilger) describes the trichromatic system of colour measurement.

Chamberlin, D. G. and Chamberlin, G. J. (1980) *Colour, its measurement, computation and application* (Heyden) describes the physical and physiological basis of colour perception and defects in colour vision.

Judd, D. B. and Wysecki, G. (1975) *Colour in business, science and industry*, 3rd edn. (Wiley) covers applications over a broad field.

The International Society Colour Council News, published by the Society of the same name, is a journal providing regular updates in this field.

Publications since the second edition of this book are:

Fletcher, R. and Voke, J. (1985) *Defective colour vision* (Institute of Physics) covers the fundamentals, diagnosis and management.

McLaren, K. (1986) *The colour science of dyer and pigments*, 2nd edn. (Institute of Physics) covers the physics, chemistry and applications of colouring matters.

Agoston, G. A. (1987) *Colour theory and its applications in art and design*, 2nd edn. (Springer).

Macadam, D. L. (1985) *Colour measurement: theme and variations*, 2nd edn. (Springer).

Healey, G., Shafer, S. A. and Wolff, L. B. (1992) *Colour* (Jones and Bartlett) is a comprehensive review covering colour image formation and segmentation; colour reflection models and inter-reflection.

Woo, G. C. (ed.) (1987) *Low vision. Principles and applications* (Springer) is the proceedings of an International Symposium on this subject held at University of Waterloo, 1986.

Fiorentini, A., Guyton, D. L. and Seigel, I. M. (1987) *Advances in diagnostic visual optics* is the proceedings of the Third International Symposium, Tirrenia, Italy, 1986 (Springer).

Garfield, B. and Rendell, J. (1991) *The 19th international conference on high speed photography and photonics* (SPIE vol. 1358) is the proceedings of a conference held at Cambridge, UK in 1990 and covers visualization and applications, opto-mechanical cameras and systems, holography and interferometry and gated intensifiers, etc.

Ragowitz, B. E., Brill, M. H. and Jan, P. (eds.) (1991) *Human vision. Visual processing and digital display II* (SPIE vol. 1453) is the proceedings of a conference held in San Jose, Ca., in 1991 covering quality of displayed information, processing of spatial and spatial-temporal images, modelling and biologically based machine vision.

PHOTOMETRY, ILLUMINATION AND LUMINESCENCE

Walsh, J. W. T. (1968) *Photometry*, 3rd edn. (Constable).

Keitz, H. A. E. (1971) *Light calculation and measurements*, 2nd edn. (Macmillan).

Lau, E. and Krug, W. (1968) *Equidensitometry* (Focal).

A contemporary publication is to be found in Egan, W. G. (1984) *Photometry and polarisation in remote sensing* (Elsevier).

The Illuminating Engineering Society of New York has published Kaufman, J. E. (1966) *The lighting handbook* which contains sections on light measurement and sources; the Society also publishes the *Journal of the Illuminating Engineering Society*.

Marshak, I. S. (1984) *Pulsed light sources* (Consultants) is concerned with the physical and technical characteristics of pulsed discharge in gases and their complementation in pulsed light sources.

Elmer, W. B. (1980) *The optical design of reflectors* (Wiley) discusses optimum illumination in different lighting tasks.

The subject of luminescence is reviewed in the *Journal of Luminescence* (North-Holland).

Parswater, R. A. *Guide to fluorescence literature* (Plenum, 1967, 1970) is an early source book.

Stepanov, B. I. and Gribovskii, V. P. (1968) *Theory of luminescence* (Iliffe) covers both classical and quantum theoretical aspects.

A review of photoemission at the theoretical level by M. Cardona and L. Ley is *Photoemission in solids* in the Springer Series on Applied Physics (1978) and other books are: Lumb, D. (ed.) (1978) *Spectroscopy* (Academic) and Schmillen, A. and Legler, R. (1967) *Luminescence of organic substances* (Springer). A book bridging nuclear and optical science by A. E. McKinlay is *Thermoluminescence dosimetry* (Hilger, 1981).

Shinoya V. and Kobayashi, H. (eds.) (1989) *Electroluminescence* (Springer) is the proceedings of the 4th International Workshop, Tottori, Japan, in 1988.

McKeever, S. W. S. (1985) *Thermoluminescence of solids* (CUP) covers the theoretical background, thermoluminescence analysis, dosimetry and dating, instrumentation and geological applications.

OPTICAL RADIATION MEASUREMENTS

Grun, F. C. and Becherer, J. (eds.) *Optical radiation measurements* (Academic) is a series of volumes covering *Radiometry*, vol. 1 (1979); *Colour measurement*, vol. 2 (1980); *Measurement of photoluminescence*, vol. 3 (1982), *Physical detectors and optical radiation*, vol. 4 (1983), and *Visual measurements*, vol. 5 (1984).

Applied Optics and Optical Engineering (Academic) is a series relating classical concepts in optics to modern detector use. R. Kingslake was editor from 1965 to 1969 and 1980; R. R. Shannon and J. C. Wyart were editors in 1979, 1980 and 1983.

Boyd, R. W. (1983) *Radiometry and the detection of optical radiation* (Wiley) examines the practical aspects and the principles of operation of detectors.

Kingston, R. H. (1978) *Detection of optical and infra-red radiation* (Springer) is published in the Springer Series of Optical Sciences (1978).

Keyes, R. J. (ed.) (1980) *Optical and infra-red detectors* (Springer) deals with photon detection, thermal and photoemissive detectors, non-linear heterodyne detection; charge transfer devices; and optical and infra-red technology.

Dereniak, E. L. and Crowe, D. G. (1984) *Optical radiation detectors* (Wiley) covers the physics of detection, photovoltaic and photoemissive detection theory, radiometry, photoconductors, thermal detectors, bolometers, pyroelectric detectors and charge-sensitive devices including mechanisms, read-out techniques, focal plane arrays and other imaging arrays.

Schwarz, K. K. (1991) *Physics of optical recording* (Springer).

Fox, N. P. and Nettleton, D. H. (eds.) (1992) *New developments and applications in optical radiometry* (IOP) is the proceedings of the 2nd International Conference at NPL London in April, 1988.

Tombesi, P. and Watts, D. F. (eds.) *Quantum measurements in optics* is the proceedings of the NATO Advanced Research Workshop held in Cortina, Italy, in 1991.

ELECTRON OPTICS

The subject of electron-optical techniques is not treated in this chapter, but two references on the fundamentals of electron optics are given as follows:

Hawkes, P. W. and Kasper, E. (1989) *Principles of electron optics*, vols. I and II (Academic) is a source book at advanced level with applications to instrumentation viz., electrostatic lenses, magnetic lenses, quadrupole lenses, electron mirrors and electron guns.

Vol. III is in the course of preparation and features image processing, electron holography and interference; analysis, description and pattern recognition and image enhancement.
De Wolf, D. A. (1990) *Basics of electron optics* (Wiley) provides a basic introduction at a practical level covering trajectory equations, electromagnetic fields and forces; basics of analytical and numerical analysis and phase-space methods.

Optical systems

Books and Proceedings

LENS DESIGN

The classical source book on lens design is Conrady, A. E. (1929) *Applied optics and optical design* (OUP). A modern equivalent is by R. Kingslake (1978) *Lens design fundamentals* (Academic), which covers the detailed design of lens and mirror systems at the fundamental level with worked examples to illustrate the practical aspects.
Several books specifically on lenses published by Hilger in the Monographs on Applied Optics series are:

Barnes, K. R. (1971) *The optical transfer function.*
Jamieson, T. H. (1971) *Optimization techniques in lens design.*
Palmer, J. M. (1971) *Lens aberration data.*
Horne, D. F. (1975) *Lens mechanism technology.*
Horne, D. F. and Moore, A. A. S. (1978) *Spectacle lens technology.*
Baum, C. E. and Stone, A. P. (1991) *Transient lens synthesis* (Hemisphere Publishing) uses a mathematical electromagnetic approach to lens design utilizing Maxwell's equations in differential geometry which combine physical components with scale factors of the co-ordinate transformation.
Laikin, M. *Lens design* (Marcel Dekker, Opt. Eng. Series, vol. 27).

SYSTEMS DESIGN

Books on system design are *A system of optical design* by A. Cox (Focal, 1964) which includes a computerized application of over 300 lens designs; Kingslake, R. (1983) *Optical system design* (Academic) which gives a comprehensive treatment of optical system fundamentals including ray-tracing procedures, photometry, physiological and photographic optics; Boutry, G. A. (1961) *Instrumentation optics* (Hilger, 1961); and *Modern optical engineering* by W. J. Smith (McGraw-Hill, 1966), which discusses image formation and evaluation with reference to the design of optical systems.

Heddle, D. W. O. (1991) *Electrostatic lens systems* (Hilger) is an introductory text containing an IBM program which is designed to calculate the focal and aberration properties of a range of optical lenses and imaging behaviour.

Hutley, M. C. (1991) *Microlens arrays* (IOP) is the result of a seminar and focuses on techniques for manufacturing lens arrays and their applications.

INSTRUMENTS AND TECHNIQUES

Books on optical instruments are:

Habell, K. J. (ed.) (1962) *Optical instruments and techniques* (Chapman and Hall), which gives an account of the proceedings of a conference on this subject in London, 1961.

Dixon, J. H. (1970) *Optical instruments and techniques* (Oriel), which is a Report on the Proceedings of the International Commission for Optical Instruments and Techniques, Reading, 1969.

Horne, D. F. (1980) *Optical instruments and their applications* (Hilger).

Van Heel, A. S. C. (1967) *Advanced optical techniques* (North-Holland).

Flügge, S. (ed.) (1967) *Handbuch der physik*, vol. 29, *Optische instrumente* (Springer).

Cornbleet, S. (1976) *Microwave optics* (Academic) covers the optics of microwave antenna design.

Books on specific instruments include B. V. Barlow and A. S. Everest, *The astronomical telescope*, and H. N. Southworth and R. A. Hull *Introduction to modern microscopy*, both published in 1975 in the Wykeham Science Series by Lewis.

Tarasov, K. I. *The spectroscope* is a monograph in the Hilger Applied Optics series and was published in 1974.

More recent publications are:

Kafri, O. and Glatt, I. (1990) *The physics of Moiré metrology* (Wiley) is an introduction to the subject covering the underlying physics, applications to optics and comparison with conventional methods.

Reidling, K. (1988) *Ellipsometry for industrial applications* (Springer).

Azzam, R. M. A. (ed.) (1991) *Ellipsometry* (SPIE vol. MS 27) has selected papers giving a comprehensive and definitive view of the subject.

Bradbury, S. (1991) *Basic measurement techniques for light micro-scopy* (OUP) provides a practical guide to the experimental and theoretical aspects and use of available accessories.

Cherry, R. J. (1990) *New techniques of optical microscopy and microspectroscopy* (Macmillan) is a collection of papers on the latest techniques, including video and optodigital image micro-scopy, quantitative interference and laser light scattering micro-scopy.

Gouesbet, G. and Grehan, G. (1988) *Optical particle sizing* (Plenum).

Harding, K. (ed.) (1989) *ICALEO '89: Optical sensing and measurement* (SPIE, vol. 1375, 1989) is the proceedings of a conference held at Orlando, Florida, in 1989 covering three-dimensional contouring, the techniques of optical metrology, holography and speckle, and applications of laser diagnostics (to criminalistics and to coal combustion).

Long, M. B. (ed.) (1989) *ICALEO '89: Optical methods in flow and particle diagnostics* (SPIE, vol. 1404) is the proceedings of a conference held at Orlando, Florida, and covers laser doppler velocimetry, particle velocimetry, field measurements and volu-metric imaging.

Fongshu, L. (1991) *Theory of conjugation for reflecting prisms* (Pergamon – CNPIEC) deals with the optical characteristics of reflecting prisms in optical systems.

Sirohi, S. and Kothiyal, M. P. (eds.) *Optical components, systems and measurement techniques* (Marcel Dekker, Opt. Eng. Series, vol. 28).

Mentzer, M. A. *Principles of Optical Circuit Engineering* (Marcel Dekker, Opt. Eng. series, vol. 26).

WORKSHOP PRACTICE

Books of a practical nature are:

Horne, D. F. (1972) *Optical production technology* (Hilger).

Elmer, W. B. (1980) *The optical design of reflectors* (Wiley), which gives requirements for optimum illumination, with bibliography.

Slyusarev, G. G. (1983) *Methods of designing optical systems* (Hilger) is an English translation of the second edition of a classic Russian text on the detailed procedures of optical system design.

Shannon, R. R. and Wyant, J. C. (eds) (1983) *Applied optics and optical engineering*, vol. 9 (Academic) deals with BASIC algo-rithms for optical engineering, diffraction gratings and optical recording. A sourcebook dealing with workshop practice of semi-historical, as well as having current, value is C. Deve (1949) *Optical workshop principles* (translated) 2nd edn. (Hilger).

Boardman, A. D. (ed.) (1980) *Physics programs. Vol. 1 Optics* (Wiley) is one of a series of manuals based upon computer exercises in selected topics.

Malacara, D. (ed.) (1990) *Optical shop metrology* (SPIE, vol. MS 15) covers selected papers on interferometric, holographic and other optical techniques at the workshop level.

Lorensen, M., Campbell, D. R. and Johnson, C. W. (1991) *Optical fabrication and testing* (SPIE, vol. 1400) is the proceedings of a conference held in Singapore, 1990, covering interferometry, metrology and associated fabrication and manufacturing processes.

Taylor, H. D. (1983) *The adjustment and testing of telescope objectives*, 5th edn. (IOP) covers the appropriate techniques for carrying out these operations.

Twyman, F. (1988) *Prism and lens making* (Hilger) demonstrates hands-on experience and covers all the optical workshop techniques likely to be needed by all groups of optical workers and specialists.

Malacara, D. (1992) *Optical shop testing*, 2nd edn. (Wiley) develops the theme of the first edition to cover fringe scanning techniques, wave front fitting and holographic and Moiré methods, etc.

OPTICAL PROPERTIES OF MATERIALS

Books concerned with optical properties of materials are:

Levi, L. (1968) *Applied optics*, volume 1 (Wiley), which gives much engineering data.

Abels, F. (ed.) (1972) *Optical properties of solids* (North-Holland) and a companion volume by B. O. Seraphin (ed.) (1976) *Optical properties of solids new developments* (North-Holland) are reference works covering a wide range of topics concerned with the theoretical and experimental aspects of optical properties of solids.

The handbook on semiconductors is a set of four volumes of which vol. 2 by M. Balkanski (ed.) (1980) *Optical properties of solids* (North-Holland) reviews the optical properties of semiconductors on the basis of interaction of light, dealing with properties of bulk material defects and impurities, phonon interaction and vibrational behaviour including spectroscopy.

Hodgson, J. N. (1970) *Optical absorption and dispersion in solids* (Chapman and Hall) and Lee, P. A. (ed.) (1976) *Optical and electrical properties* (Reidel) are two other books on the subject.

The following books on optical materials have been published since the listing in the second edition of this work:

124   *Classical optics*

Ward, L. (1988) *The optical constants of bulk materials and films* (Hilger) presents a broad picture of the field of optical constants covering theoretical background, experimental techniques and typical results, with references.

Sotomayor Torres, C. M., Portal, J. C., Maan, J. C. and Stradling, R. A. (1988) *Optical properties of narrow-gap low-dimensional structures* (Plenum).

Hutchins, M. G. (1988) *Solar optical materials* (Elsevier).

Lakhtakia, A. (ed.) (1990) *Natural optical activity* (SPIE) covers selected papers on this subject.

Shifrin, K. S. (1988) *Physical optics of ocean water* (AIP/Hilger) brings together widely dispersed information on ocean optics.

Palik, E. D. (1991) *Handbook of optical constants of solids*, 2nd edn. (Academic) is a comprehensive data source and is a sequel to *Handbook of optical constants*.

Elliott, R. J. and Ipatova, I.P. (eds.) (1988) *Optical properties of mixed crystals* (North Holland) is a compendium of specialized selected topics within the Modern Problems in Condensed Matter Sciences series of this publisher.

Fuxi, G. (1992) *Optical and spectroscopic properties of glass* (Springer) relates the composition and structure of glass-forming materials to their optical and spectroscopic properties.

Izumitani, T. S. (1986) *Optical glass* (AIP/Hilger) features the history of optical glass, current research on surface characteristics, glass fabrication and engineering and glass fabrication.

Rashkovich, L. N. (1991) *KDP-family single crystals* (Hilger) provides a comprehensive investigation of the family of crystals which are used to control parameters of laser emission.

Savage, J. A. (1985) *Infra-red optical materials and their anti-reflection coatings* (IOP) covers surface and bulk optical properties, sample preparation and testing, optical fibres and coatings.

Günther, P. and Haignard, J. P. (eds.) (1988) *Photorefractive materials and applications. I. Fundamental phenomena* and *II. Survey of applications* (Springer).

Günther, P. (ed.) (1987) *Electro-optic and photorefractive materials* (Springer) is the proceedings of the International School on Materials, Science and Technology held at Erice, Sicily, in 1986.

Kaminskii, A. A. (1990) *Laser crystals: their physics and properties*, 2nd edn., (Springer) analyses the spectrophysical properties of activated insulating laser crystals and systemises mechanisms for obtaining stimulated emission based on those properties.

Birman, J. L., Cummins, H. Z. and Kaplyanskii, A. A. (1988) *Laser optics and condensed matter* (Plenum).

Garmire, E., Maradudin, A. A. and Rebane, K. K. (1991) *The physics of optical phenomena and their use as probes of matter* (Plenum).

Lin, J. T. (ed.) (1989) *Growth characterisation and applications of laser host and nonlinear crystals* (SPIE vol. 1104) is the proceedings of a conference held in Orlando, Florida, in 1989 and addresses new optical crystals, crystal growth and properties.

Klocek, P. (ed.) *Handbook of infra-red optical materials* (Marcel Dekker, Opt. Eng. Series, vol. 30).

Joshi, N. V. *Photoconductivity:art science and technology* (Marcel Dekker, Opt. Eng. Series, vol. 25).

Petrov, M. P., Stepanov, S. I. and Khomenko, A. V. (1991) *Crystals in coherent optical systems* (Springer).

OPTICAL IMAGE PROCESSING

Taylor, V. A. (1978) *Images* (Wykeham) is a unified view of diffraction and image formation for different regions of the electromagnetic spectrum using Fourier transforms.

Marathay, A. S. (1982) *Elements of optical coherence theory* in the Wiley series of Pure and Applied Optics covers image formation and resolution criteria with worked examples.

Another book on the same theme is by M. Schlenker, (ed.) (1980) *et al. Image processing and coherence in physics* (Springer),which is the proceedings of a Workshop at Les Houches, France, 1979.

Rogers, G. L. (1978) *Noncoherent optical processing* is another volume in the Wiley series of Pure and Applied Optics dealing with temporal and spatial modulation and applications to optical systems.

Optical data processing by A. R. Shulman is also a Wiley publication (1970) and Casasent, D. (ed.) (1978) *Optical data processing*, is published by Springer, as also is Lee, S. H. (ed.) (1981) *Optical information processing fundamentals* (Topics in Applied Physics, 48) (Springer). *Optics and information theory* and *Optical information processing: signal processing and Fourier optics* (1982), both by F. T. S. Yu and *Optical information processing and holography*, by W. T. Cathey (1974) are published by Wiley.

Françon, M. (1979) *Optical image formation and processing* is published by Academic.

Books on image display are:

Kazan, B. (1983) *Advances in image pick-up and display* (Academic), dealing with image sensors and gas discharge panels.

Faugeras, O. D. (1983) *Fundamentals in computer vision* (CUP) is the proceedings of an advanced course concerned with the theoretical aspects of vision system design and *Techniques for image processing* and *Classification in remote sensing* are two books by R. A. Schowengerdt (Academic, 1983) which are concerned with fundamental mathematical concepts of image processing at the engineering level.

Kasturi, R. and Trivedi, M. M. (eds.) *Image analysis applications* (Marcel Dekker, Opt. Eng. Series, vol. 24).

New publications in this field are:

Korsch, D. (1991) *Reflective optics* (Academic) which is an advanced mathematical treatment concerning the design of imaging systems from near-normal to grazing incidence.

Aminzadeh, F. (1988) *Pattern recognition and image processing* (Elsevier).

Marshall, G. E. (1991) *Optical scanning* (Marcel Dekker) addresses information scanning by optical means and is part of a Dekker Series on Optical Engineering.

Daz, P. K. (1991) *Optical signal processing* (Springer) focuses on the fundamentals and discusses adaptive, Kalman and lattice filters; and two-dimensional signal processing.

Fauqeras, O. (ed.) (1990) *Computer vision* (Springer) is the proceedings of the First European Conference on Computer Vision held at Antibes, France, in 1990.

Nasr, H. N. (1991) *Image understanding for aerospace applications* (SPIE vol. 1521) is the proceedings of a conference held at Munich in 1991, covering motion analysis in navigation, image interpretation, systems and sensors, phenomenology and modelling.

Mahajan, V. N. *Aberration theory made simple* (SPIE vol. TT 6) is a basic book identifying the characteristics of aberrations in optical imaging systems and is intended for engineers and scientists who have a practical interest.

Mitchell, B. T. (ed.) (1991) *Image understanding in the '90s: building systems that work* (SPIE vol. 1406) is the proceedings of a conference held at McLean, Virginia, in 1990, covering techniques and technologies, systems and applications, algorithms in architectures.

Rabbani, M. and Jones, P. W. (eds.) (1991) *Digital image compression techniques* (SPIE vol. TT 7) is a tutorial text and gives the groundwork for understanding these techniques and processing characterization.

Trivedi, M. (ed.) (1990) *Digital image processing* (SPIE vol. MS 17) are selected papers giving a comprehensive and definitive review of the subject.

Neimann, H. (1990) *Pattern analysis and understanding*, 2nd edn. (Springer).

Nasr, H. N. (ed.) (1991) *Automatic object recognition* (SPIE vol. IS 7) is a review of techniques, modelling and evaluation.

Civanlar, M. R. and Mitra, S. K. (eds.) (1991) *Image processing algorithms and techniques II* (SPIE vol. 1452) is the proceedings of a Conference held at San Jose, California, in 1991 and covers colour imagery, image processing and filtering techniques, image reconstruction, image analysis and pattern recognition: focussing primarily on military applications.

Rogers, S. K. and Kabrisky, M. (1991) *An introduction to biological and artificial neural networks for pattern recognition* (SPIE vol. TT 4) is a tutorial text giving the underlying theory.

Journals

Journals specializing in optical systems and technology include:

Engineering Optics (IOP)
Optical Engineering (SPIE)
Review of Scientific Instruments (AIP)
Optics and Laser Technology (Butterworths and Heinemann)
Soviet Journal of Optical Technology (OSA/SPIE)
Microwave and Optical Technology Letters (Wiley)
Jena Review (VEB Carl Zeiss Jena)
Leitz-Mitteilungen für Wissenschaft und Technik (Leitz)
Zeiss Information (Carl Zeiss)
Journal of Optical Society of America, Part B: Optics and Image Science (OSA)
Journal of European Optical Society, Part A: Pure and Applied Optics (IOP)

Journals specializing on imaging science and technology are:

Journal of Pattern Recognition (Pergamon)
Journal of Display and Imaging Technology (Gordon and Breach)
Imaging Science (Crane Russak)
Physics-Based Vision (Jones and Bartlett)

Ultra-violet, infra-red and millimetre waves

Books and proceedings

Ultra-violet radiation by L. R. Koller (Wiley, 2nd edn., 1965) and Green, A. E. S. (1966) *The middle ultra-violet* (Wiley) are introductory sourcebooks.

Phillips, R. (1983) *Sources and applications of ultra-violet radiation*

(Academic) reviews the production, detection, application and hazards involved in using ultra-violet radiation and *Ultra-violet radiation in medicine* by B. L. Diffey (Hilger, 1982) discusses medical applications.

Chantry, G. W. (1984) *Long wave optics* (Academic) is produced in two volumes and covers the generation, propagation and detection of long-wave optical radiation from the infra-red to the radio-frequency region. It deals with general principles, adopting an integrated approach based upon coherence and guided waves and there is an extensive bibliography. Academic Press, under the general editorship of K. J. Button publish a series of books entitled *Infra-red and Millimetre Waves.* vol. 1, 1979 covers *Sources of radiation*; vol. 2, 1979 *Sub-millimetre techniques I*; vol. 3, 1980 *Sub-millimetre techniques II*; vol. 4, 1981 *Millimetre systems*; vol. 5, 1982 *Coherent sources and applications I*; vol. 6, 1982 *Systems and components*; vol. 7, 1983 *Coherent sources and applications II*; vol. 8, 1983 *Electromagnetic waves in matter*; vol. 9, 1983 *Millimetre components and techniques I*; and vol. 10, 1983 *Millimetre components and techniques II*. As can be appreciated, the series is a very comprehensive review of the field, covering all aspects from principles and techniques to applications.

Button, K. J. (1983) *Review of infra-red and millimetre waves*, vol. I, (Plenum) is a state-of-the-art review covering detectors, mixers and receivers.

Button, K. G., Inguisco, M. and Strumia, F. (1984) *Review of infra-red and millimetre waves* vol. II (Plenum) is a compendium of research reports on 20 types of laser in this spectral region.

Robinson, L. C. (1973) *Physical properties of far infra-red radiation* (Academic) covers the whole spectrum of infra-red physics, dealing with the principles of wave generation, detection, transmission and their sensitivity limits.

Wolfe, W. L. and Zissus, G. J. (1978) *The infra-red handbook* (ONR) is both a comprehensive review of the field and a detailed reference work. The same organization has published, in conjunction with the University of Michigan, *Handbook of military infra-red technology*, by W. L. Wolfe (ed.) (1965) which is an early review of the scientific basis of surveillance instruments.

Lesurf, J. C. G. (1990) *Millimetre wave optics, devices and systems* (IOP) aims to describe the fundamental physics of the techniques, devices and system design for processing millimetre wave signals.

Baden-Fuller, A. J. (1990) *Microwaves*, 3rd edn. (Pergamon) is an introduction to microwave theory and techniques.

Books dealing with techniques and technology are:

Kruse, P. W., McGlauchlin, L. D. and McQuiston, R. B. (1963) *Elements of infra-red technology* (Wiley).

Colliver, D. J. (ed.) (1982) *Advanced millimetre wave technology* (Microwaves Exhibitions and Publications).

Vanzetti, R. (1972) *Practical applications of infra-red techniques* (Wiley).

Aramos, F. R. (1973) *Infra-red to millimetre wavelength detectors* (Artech) and *Applications of Infra-Red Detectors* (Mullard, 1971), which is systems-orientated.

Moss, J. S. (1976) *Infra-red detectors* (Pergamon) is a collection of papers presented to a meeting of a US Speciality Group on infra-red detectors. Aspects of infra-red detection are covered in *Optical radiation measurements* by F. C. Grum and R. J. Becherer (eds.), vols 1–4 (Academic, 1979–1983).

Kimmitt, M. E. (1970) *Far infra-red techniques* (Pion) deals with optical components, sources, detectors, methods including Fourier transform and interferometric spectrometers, and selected infra-red experiments.

Lloyd, J. M. (1979) *Thermal imaging system*, in the Optical Physics and Engineering series published by Plenum gives a comprehensive introduction to the basis of the technology of thermal imaging with treatment at a fundamental level and covers line filter theory of imaging, optical processes and aberrations, systems, acquisition, detection and recognition of targets. It is intended to complement other books which deal with detectors, cryogenic coolers and circuit design. IEE Conference Publication no. 173 entitled *Low light and thermal imaging* is published by Peregrinus.

Spiro, I. J. and Schlessinger, M. *Infra-red technology fundamentals* (Marcel Dekker, Opt. Eng. Series, vol. 22).

Journals

Infra-red Physics (Pergamon)
Advanced Infra-red Detectors and Systems (IEE)
International Journal of Infra-red and Millimetre Waves (Plenum)
IEE Proceedings on Microwave Optics and Antennae (Peregrinus)

CHAPTER SEVEN

Computational physics*

MANFRED KREMER and STEVE O'NEALE

Computers have an ever-growing impact in all aspects of everyday life and professional activity. Nowadays nearly every scientist must use them for his experiments, calculations, data processing or information retrieval. There has been a tremendous increase in computing power during recent years. Today, desktop workstations, usually running a Unix based operating system, are able to perform many tasks which would have been considered a super-computer application three years earlier.

The increased processing speed of today's computers has been achieved by a number of structural developments in computer architecture (there are now superscalar workstations, vector supercomputers, and SIMD and MIMD parallel computers). In order to use the power of today's computers efficiently it is important to have a basic knowledge of their architecture. The increased computing power manifests itself also in an increased volume of data to be handled. The requirement to handle and eventually interpret these large amounts of data has promoted the development of a number of visualization tools, which are available for a large variety of computer platforms.

For these reasons we do not confine the scope of this chapter to computational physics in its classical sense, but try to provide information sources to all computer related topics which could be useful to a physicist. It includes both introductory texts and specialized books and journals on the one hand and presents literature on applications in different branches of physics in addition to the monographs or reports on some branches of computer science on

* based on the second edition chapter written by J. Nadrchal

the other. The books and proceedings concerning special fields of physics (fluid dynamics, plasma physics, etc.) are presented only occasionally due to limitations on space.

The rapid development of computers and their applications makes many books obsolete in a few years after their publication. Therefore we had to apply major changes to this chapter for this third edition of *Information sources in physics*.

In the following two paragraphs we list a small number of information sources which could be of general interest in the context of computing.

There are five sections in *Encyclopedia of applied physics* edited by G. L. Trigg (VCH, 1992–97) volume 4 concerning computers, hardware, programming languages, graphics and databases which present concise current views of these topics, with glossaries and recommendations for further reading. The first orientation in the mass of computer terms may be found in *Dictionary of computing* edited by V. Illingworth (OUP, 1991) and *Encyclopedia of computer science and engineering* edited by A. Ralston and E. Reilly (Van Nostrand Reinhold, 2nd edn., 1983). The *Software engineer's reference book* (SERB) edited by J. A. McDermid (Butterworth-Heinemann, 1991) has 70 contributors covering the theory, techniques and application of software engineering. Each section has a bibliography or references to further work.

Computers in Physics (AIP) is a bi-monthly journal which carries peer-reviewed papers and 'departments' on numerical techniques, scientific programming (since 1993), simulations, education and visualization in each issue, as well as reviews of books, software, computers and other products. Perhaps because it is unique in its selection of material, it attracts advertisements for many data manipulation packages and program libraries for which it is impossible to sustain an up-to-date comprehensive review.

Scientific programming (Wiley) promises to have an even tighter focus and is promoted as a forum for papers and product reviews of software engineering environments, tools, languages and models of computation. The implementation of scientific applications on parallel computers and novel architectures are covered.

This chapter is divided into the following sections:

- Programming and computer architecture
- Numerical methods and algorithms
- Computational physics
- Non-numerical methods
- Computer graphics and image processing
- Databases and data handling

- Software for process control
- Publishing and retrieving information

Programming and computer architecture

Books

PROGRAMMING LANGUAGES

Fortran has been the language of choice for scientific programming for over 30 years. The new Fortran 90 language adds many facilities for numerical computation and data abstraction to the 77 standard. *Fortran 90 explained* by M. Metcalf and J. Reid (OUP, 1990) gives a concise description of the whole of Fortran which is suitable for learning, or teaching, the language. This book may be complemented with the much larger volume *Fortran 90 handbook* by J. C. Adams *et al.* (McGraw-Hill, 1992) which follows the standard, explaining the use of, and often the motivation behind each feature. Many problems in Fortran programming are explained and solved in *Fortran 77: featuring structured programming* by E. I. Organick and L. P. Meissner (Addison-Wesley, 3rd edn., 1980), and *A guidebook to Fortran on supercomputers* by J. M. Levesque and J. W. Williamson explains the restructuring needed to approach the potential benefits from vector and parallel processors. The 'High Performance Fortran Forum' is aiming for extensions to Fortran 90 for high performance programming on a variety of massively parallel SIMD and MIMD systems and vector processors to be released in 1993. Their progress is reported in the ACM SIGPLAN's *Fortran Forum*. Users of Watfor-77 extensions to the Fortran 77 language will find a comparison with the Fortran 90 standard in *Fortran Forum* (1991) 10, (2).

Pascal is a subject of numerous books from which *Programming in Pascal: computer science* by P. Grogono (Addison-Wesley, 2nd rev. edn., 1984) and *Computing* by J. D. Lawson *et al.* (Prentice-Hall, 1982) may be recommended for their clear language presentation. The central Pascal document is the *Pascal user's manual and report* by K. Jensen *et al.* (Springer, 4th edn., 1991). Pascal was very popular in the 1980s not only because of its clear data structures, but also because Borland's Turbo Pascal Compiler offered a cheap and comfortable programming environment for the IBM PC. Today one can say that Pascal was not able to replace Fortran in the field of programming of numerical problems. It still plays a significant role, however, and it should be mentioned that D. Knuth's typesetting system TeX is written in Pascal.

C described in *The C programming language* by B. W. Kernighan and D. M. Ritchie (Prentice Hall, ANSI 2nd edn., 1989) originated as a system programming language on minicomputers. As it is available on every Unix workstation, C is now popular in a wide range of scientific applications. It should be noted, however, that at the time of this revision there is still very limited support for C (and essentially any language other than Fortran) in supercomputing environments. Note also, that good vectorization of numerical code for Cray supercomputers can most easily be achieved in Fortran programs. Also the libraries with optimized subroutines are larger and more numerous for Fortran than for any other language.

A book that presents a survey of both classical and modern languages including applicative data-abstraction languages and languages with concurrency features is *Programming languages – a grand tour*, edited by E. Horowitz (Springer, 3rd edn., 1987). *Object-oriented languages* by G. Masini *et al.* (Academic, 1992) is a comprehensive introduction to object-oriented programming with comparisons of Smalltalk Objective-C, CLOS, Simula, C++, Eiffel, and many other case studies, examples, a glossary, and a list of vendors.

C++ is an object-oriented extension to the C language. *The annotated C++ reference manual* by M. A. Ellis and B. Stroustrup (Addison-Wesley, 1990) is the ANSI base document for this language which has developed considerably since Stroustrup's early work, and *An introduction to object-oriented programming and C++* by R. Wiener and L. J. Pinson (Addison Wesley, 1988) and *Algorithms in C++* by R. Sedgewick (Addison Wesley, 1992) are recommended.

Eiffel 3.0 is another choice for object-oriented programming, and *Object-oriented software construction* by B. Meyer (Prentice-Hall, 1988) and *Eiffel: the language* (Prentice-Hall, 1992) are recommended reading.

Lisp has been the language of artificial intelligence for a long time. The advent of object-oriented languages may challenge the position of Lisp. It is, however, important as the basis of several computer algebra systems, notably Reduce and Macsyma. The most important Lisp dialect is described in *Common Lisp: the language* by G. L. Steele (Digital Press, 2nd edn., 1990).

The *GNU* C and C++ compiler as well as many additional programming tools, mainly for Unix environments, are freely available form the Free Software Foundation. The *GNU project* of the Free Software Foundation may be contacted at *gnu@prep.ai.mit.edu*. The software is available from many anonymous ftp-servers in the USA and Europe.

SOFTWARE ENGINEERING

To solve the complicated problems in programming, knowledge of the syntax of a programming language is not sufficient. There are many fundamental topics that must be studied, if one wishes to obtain a more detailed theoretical understanding of programming. This subject is computer science. Also, one should study the effective methods of writing programming systems – especially of those that require many years of work. These methods are covered by the topic of software engineering, an application of computer science. An overview about these subjects, their development and the relations between them, may be gained from collections of classical papers *Classics in software engineering* (1979) and *Writing of the revolution* (1982) both edited by E. Nash Yourdon and published by Yourdon Press.

Software tools by B. W. Kernighan and P. J. Plauger (Addison-Wesley, 1976), and *The elements of programming style* by the same authors (McGraw-Hill, 2nd edn., 1978), present pragmatically the analysis of real programs and their style of programming. *Structured programming* by O. J. Dahl *et al.* (Academic, 1972) explains basic concepts of structured programming, and *Systematic programming* by N. Wirth (Prentice-Hall, 1973) 'introduces programming as the art and technique of constructing and formulating algorithms in a systematic manner'.

A modern view of computer science is given in the *Handbook of theoretical computer science* edited by J. van Leeuwen (MIT, 2 vols, 1990). *Formal methods of program verification and specification* by W. R. Frantu *et al.* (Prentice-Hall, 1982) gives an introduction to the fundamental techniques of program specification and verification. A general outlook on programming theory and practice with practical implications is presented in *The science of programming* by D. Gries (Springer, 1981, repr. 1991). The book concerns the basic principles of programming logic, semantics of languages, and the development of programs. Similar coverage is given in *Computer science* by D. Woodhouse *et al.* (Wiley, 1992). The text is illustrated by examples and exercises.

The book of essays *The mythical man-month* by F. P. Brooks Jr. (Addison-Wesley, 1975) is a sparkling classic work which exposed the problems of managing a large software project. *Decline & fall of the American programmer* by E. Yourdon (Prentice Hall, 1993) shows the 'software crisis' is still with us and gives a clear non-technical exposition of methods for better programming.

Software engineering by I. Sommerville (Addison-Wesley, 4th edn., 1992) gives a pragmatic view of state-of-the-art software engineering techniques with the standard *Instructor's guide, source code of examples (in Ada)* on PC disc and *Transparency masters* available separately.

Although Sommerville and Yourdon (above) emphasize the need to adopt new ideas and technologies they do not endorse or recommend any specific methodology or product; perhaps none are yet sufficiently mature for a large scale physics project. Nevertheless *Case outlook: guide to products and services* (CASE Consulting Group, November 1991) catalogues many computer assisted software engineering tools, and is maintained electronically as TOOL FINDER/PLUS. Yourdon Press specializes in education and consulting in computer technology.

The object-oriented paradigm (OO) is in vogue as we go to press. *Object-oriented modeling and design* by J. Rumbaugh *et al.* (Prentice Hall, 1991) describes the object modeling technique (OMT) applying OO technology through the entire software life-cycle. OMT is presented and compared with other OO work; *Object-oriented design with applications* by G. Booch (Benjamin/Cummings, 1991) and *Object-oriented analysis* by P. Coad and E. Yourdon (Yourdon Press, 2nd edn., 1990); with ER models arising from *The entity-relationship model – towards a unified view of data* by P. Chen (*ACM Transactions on Database Systems*, 1 March 1976); with Structured Analysis/Structured Design (SA/SD) as in *Modern structured analysis* by E. Yourdon (Yourdon, 1989), *The practical guide to structured systems design* by M. Page-Jones (Yourdon, 2nd edn., 1988), *Structured analysis and system specification* by T. De Marco (Prentice Hall, 1983), and *Structured design* by E. Yourdon and L. Constantine (Yourdon, 1979); and with JSD Jackson Structured Design in *System development* by M. A. Jackson (Prentice-Hall, 1983). Rumbaugh *et al.* also present a survey of OO languages and discuss the implementation of OMT designs with non-OO languages (Fortran, C, and Ada) and with relational databases. The books of Booch and Meyer (above) may be recommended to a physicist-programmer thinking of moving to an object-oriented language.

Many OO bibliography references are available via anonymous ftp from cui.unige.ch:PUBLIC/oscar/main.lib, and by following the link to 'Object Oriented Information Sources' through http://cui_unige.ch on the World Wide Web.

The *Software reliability guidebook* by R. L. Glass (Prentice-Hall, 1979) should be studied by anyone who is concerned about the reliability of his programs. The book contains a critical survey of the literature in the field.

COMPUTER ARCHITECTURE AND PROGRAMMING

The variety of computer architectures ranging from PCs and super-scalar workstations to high speed vector supercomputers and massively parallel systems requires special consideration. Even though the compilers for vector computers have been improved over the years and offer much support for automatic vectorization of the code, it is still necessary to have a basic understanding of the vector architecture in order to write efficient programs.

Parallel and massively parallel computers are still relatively new on the market, and programming these machines is still difficult. But it can be expected that the most powerful supercomputers of the future will be of massively parallel architecture.

All aspects of vector and parallel computers are covered in *Parallel computers 2 – architecture, programming, and algorithms* by R. W. Hockney and C. R. Jesshope (Hilger, 2nd edn., 1988). Parallel computer systems are also discussed in *Highly parallel computing* by G. S. Almasi and A. Gottlieb (Benjamin/Cummings, 2nd edn., 1989).

A survey of parallel computer systems is given in *Past, present, parallel: a survey of available parallel computer systems* edited by A. Trew and G. Wilson (Springer, 1991). The most recent results in computer architecture and programming can be found in the proceedings of the annual Supercomputing conferences of the ACM.

In particular in high energy physics massively parallel special purpose computers play an important role. Their architecture and recent new developements are described in the proceedings of the annual conference LATTICE, which are published in *Nuclear Physics* (see section below in 'Computational physics').

The report of the High Performance Computing and Networking Advisory Committee (CEC, Oct 1992, 2 vols) includes many references to major European computing and network projects and their scientific and industrial application.

Periodicals

The following list consists of a selection of academic journals concerning computers, but containing some papers of interest to physicists, mainly those who write more general and larger programs and software systems:

Journal of the ACM (ACM) specializes in theoretical computer science.

Communications of the ACM (ACM) publishes original contributions to the computing field. The 'Computing practices' section presents articles of current use to practitioners.

IEEE Transactions on Computers (IEEE).

Parallel Computing (Elsevier).

Computer Standards and Interfaces (North-Holland) concerns standards of programming languages, databases, and information

systems, computer graphics, I/O interfaces, media, data dictionary, documentation, etc.

Computing (Springer) includes book reviews.

Computing Surveys (ACM) is aimed at the broader interests of the
computational community and is oriented to readers with a
minimal background in the areas covered. Each issue is devoted to
a special topic.

Software-Practice and Experience (Wiley) contains papers on
systems and applications software implementation and use. The
contents of the journal will be of interest to both specialists and
non-specialists.

As a useful source of information about all computer literature the
annual compendium *ACM Guide to Computing Literature* (ACM)
is recommended. This, and a 10-year archive, are also available on
CD-ROM.

Numerical methods and algorithms

Numerical methods form the basis of computational physics. It is, of
course, not possible to review here the whole area of numerical
mathematics. Therefore we have selected a few handbooks and
books which are especially useful for a physicist.

Software packages like the IMSL (International Mathematical
and Statistical Libraries) or NAG libraries contain each about 1000
subroutines which offer solutions to standard problems of numerical
mathematics and statistics. These software packages are available
and optimized for a large variety of hardware platforms and thus
contribute to the portability of applications.

Books

HANDBOOKS

Handbook of applicable mathematics, chief editor W. Ledermann
(Wiley, 1980–1981) is published in six volumes: I *Algebra*; II *Probability*; III *Numerical methods*; IV *Analysis*; V *Geometry and
combinatorics*; VI *Statistics*. It covers many practical aspects of
mathematics and is especially addressed to users of mathematics
who are not professional mathematicians.

*Handbook of mathematical functions with formulas, graphs, and
mathematical tables*, edited by M. Abramowitz and I. A. Stegun
(Dover Publications N.Y., 1972) and *Tables of integrals, series, and
products* by I. S. Gradshteyn and I. M. Ryzhik (Academic, 4th edn.,
corrected and enlarged, 1980), originally published in Russian and

translated also into German, are invaluable sources of information in their fields. The second book should be used with lists of errata and addenda published by H. Haeringen and L. P. Kok (Delft, 1981).

NUMERICAL ALGORITHMS AND SOFTWARE PACKAGES

Numerical methods for mathematics, science, and engineering by J. H. Mathews (Prentice Hall, 1992) can be used as an introduction to numerical mathematics. Other introductory books include *The numerical analysis problem solver* by M. Fogiel (Research and Education Association, Piscataway, NJ, 1988) and *Numerical analysis for applied mathematics, science, and engineering* by V. Casulli and D. Greenspan (Addison Wesley, 1988); *Introduction to numerical analysis* by F. B. Hildebrand (McGraw-Hill, 2nd edn., 1974); *Numerical methods for scientists and engineers* by R. W. Hamming (McGraw-Hill, 2nd edn., 1973, repr. Dover 1987); *Theory and applications of numerical analysis* by G. M. Phillips and P. J. Taylor (Academic, 1973).

The following books deal with specialized subjects in numerical analysis: *Matrix computations* by G. H. Golub and Ch. F. van Loan (Johns Hopkins U. Pr., 2nd edn., 1989); *Solving linear systems on vector and shared memory computers* by J. J. Dongarra *et al.* (SIAM, 1991); *An introduction to multigrid methods* by P. Wesseling (Wiley, 1992); *The finite element method: linear, static and dynamic finite element analysis* by T. J. R. Hughes (Prentice Hall, 1987); *Fitting equations to data: computer analysis of multifactor data* by C. Daniel and F. S. Wood (Wiley, 2nd edn., 1980). *Statistical methods in experimental physics* by W. T. Eadie *et al.* (North-Holland, 1971) is still a valuable source in the field of statistical methods. Other books are *Statistical and computational methods in data analysis* by S. Brandt (North-Holland, 2nd edn., 1983), *Spectral analysis and time series* by M. B. Priestley (Academic, 1981), and *Probability and statistics in particle physics* by A. G. Frodensen *et al.* (Universitetsforlaget, 1979). More recent books include *Statistics for nuclear and particle physicists* by L. Lyons (CUP, 1986), *Statistics: a guide to the use of statistical methods in the physical sciences* by R. J. Barlow (Wiley, 1989) and *Modern methods of data analysis* by J. Fox and S. J. Long (Sage, 1990).

In recent times neural networks (NN) have become increasingly important as a data analysis tool. Good general introductions to neural networks are given in *Neural computing: an introduction* by R. Beale and T. Jackson (IOP, 1990) and *Neural computing: algorithms, applications, and programming techniques* by J. Freeman and D. Skapura (Addison-Wesley, 1991) which also provides a software package in C. *Introduction to the theory of neural computation* by J. Hertz *et al.* (Addison-Wesley, 1991) has a strong physics content and a discussion of optimization problems.

As neural networks are a relatively new analysis tool we also note some sources for applications. Neural networks with a feed-forward topology are widely used in pattern recognition. 'Identification of b Jets using neural networks' by G. Bahan and R. Barlow in *Computer Physics Communications*, 74, (North Holland, 1993) is a typical example and refers to the work of the Lund group, notably C. Peterson, and to the 'JETNET program in Fortran 77' which is published in the same journal. *New computing techniques in physics research II* edited by D. Perret-Gallix (World Scientific, 1992) has 27 papers on neural networks, mostly applications, but also some overviews of NN algorithms and a paper on Intel's ETANN (Electronically Trainable Analog Neural Network). The CHEP 92 NN session includes *Finding the decay vertex of a charged track with neural networks* by G. Stimpfl-Abele (where comparison with the conventional Kalman filter applied to Aleph's TPC are favourable) and plans for ETANNs in the trigger for the CDF and D0 experiments at Fermilab.

The implementation of numerical algorithms on computers is the subject of the following books:

The art of computer programming by D. E. Knuth (Addison-Wesley) is one of the classical works in this field. The first two volumes will be of interest to computational physicists: vol. 1: *Fundamental algorithms* (1973) and vol. 2: *Seminumerical algorithms* (2nd edn., 1981).

Numerical recipes are available in a FORTRAN, a PASCAL, and a C version *Numerical recipes: The art of scientific computing in . . .* by W. H. Press *et al.* (CUP, 1988–1990) together with the example book and diskette.

Parallel algorithms in computational science by D. W. Heermann and A. J. Burkitt (Springer, 1991) explains MIMD parallelism with specific examples from statistical physics. Algorithms for parallel computers are also discussed in: *Numerical techniques and parallelism in physics* (Proceedings of the European Summer School on Computing Techniques in Physics) edited by J. Nadrchal (North Holland, 1992); *Numerical algorithms for modern parallel computer architectures*, edited by M. Schultz (Springer, 1988); *Numerical methods for ordinary differential equations*, edited by A. Bellen (North Holland, 1993); *Introduction to parallel and vector solution of linear systems* by J. M. Ortega (Plenum, 1988); *Solving problems on concurrent processors*, vol. 1: *General techniques and regular problems* by G. C. Fox *et al.* (Prentice Hall, 1988) and vol. 2:

Parallel and distributed computation – numerical methods by D. P. Bertsekas and J. N. Tsitsiklis (Prentice Hall, 1989).

The role of mathematical software libraries has become increasingly important in recent times. The IMSL and the NAG library are general purpose numerical software packages. Their manuals *IMSL MATH/LIBRARY user's manual* (IMSL, Houston 1992) and *NAG Fortran library manual, Mark 15* (NAG, Oxford 1991) are separately available.

LAPACK is a linear algebra package which is aimed at providing a universal standard for basic linear algebra algorithms on most hardware platforms. It is described in *LAPACK user's guide* by E. Anderson *et al.* (SIAM, 1992).

CERNLIB, an extensive scientific library of the European high energy physics laboratory, is available from CERN. (See the article by T. Lindelöf (1980) 'Maintenance of the CERN library in a multi-computer environment' in the book *Production and assessment of numerical software*, ed. M. A. Hennell and L. M. Delves. (Academic).)

The following software packages provide the tools for more specialized numerical problems: *Matrix eigensystem routines – EISPACK guide* by B. T. Smith *et al.* (Springer, 1976) and *Matrix eigensystem routines – EISPACK guide extension* by B. S. Garbow *et al.* (Springer, 1977) represent user guides to the EISPACK system for the computation of eigenvalues and/or eigenvectors of matrices.

ELLPACK is a system for the solution of elliptic partial differential equations. It is described in *Solving elliptic problems using ELLPACK* by J. R. Rice and R. F. Boisvert (Springer, 1985).

SAS is a powerful software system for statistical data analysis. With SAS/GRAPH it also offers tools for the graphical presentation of data. It is developed by SAS Institute Inc., Cary, NC.

NAG, who offer numerical, symbolic, statistical and visualization libraries for sale, also distribute some public domain products on a cost-recovery basis: Lapack, Eispac, Linpack and Minpack. For numerical software *netlib* is a useful source, described in 'Distribution of mathematical software via electronic mail', by J. Dongarra and E. Grosse, *Comm. ACM*, 30, 403–407 (ACM, 1987). A starting point for *netlib* is to mail *netlib@ornl.gov* or in Europe *netlib@nac.no* with the line *send index*. A similar collection of statistical software is available from *statlib@temper.stat.cmu.edu*.

Periodicals

Transactions on Mathematical Software (ACM) publishes papers concerned with fundamental algorithms.

Numerische Mathematik (Springer). Most papers are published in English.

Journal of Mathematical Physics (AIP).

Computers & Mathematics with Application 1 (Pergamon).

International Journal of High Speed Computing (World Scientific).

USSR Computational Mathematics and Mathematical Physics (Pergamon) contains translations from *Zhurnal vychislitel'noi matematiki i matematiceskoi fiziki*.

SIGNUM Newsletter (ACM).

Applied Mathematics and Computation (North-Holland).

SIAM Journal on Applied Mathematics (SIAM).

SIAM Journal on Numerical Analysis (SIAM).

Collected Algorithms from ACM (ACM) is a journal that includes algorithms from *Communications of the ACM* (1960–1975) and supplementary collected algorithms.

A most important collection of algorithms in computational physics is in the Queen's University Belfast Library described in *Computer Physics Communications* (see below in 'Computational physics: periodicals').

Concurrency-Practice and Experience (Wiley) contains papers on algorithms for parallel computer systems.

Computational physics

Books

There exists a large number of introductory textbooks for computational physics. *Computational physics* by S. E. Koonin (Benjamin/ Cummings, 1986) provides in eight chapters introductions to many topics, among others ordinary differential equations, boundary value and eigenvalue problems, matrix operations, elliptic and parabolic differential equations, and Monte Carlo methods. Example programs (in BASIC) are included on a diskette. Similar in scope is the book *Computational techniques in physics* by P. K. MacKeown and D. J. Newman (Hilger, 1987).

Computer simulation and computer algebra – lectures for beginners by D. Stauffer *et al.* (Springer, 1988) gives an introduction to computational methods in classical and statistical physics as well as an introduction to the computer algebra system REDUCE (see the section on non-numerical applications).

Computer simulation methods in theoretical physics by D. W. Heermann (Springer, 1986) contains introductions to Monte Carlo methods and to molecular dynamics simulations.

Further introductory books are: *Monte Carlo methods* by J. M. Hammersley and D. C. Handscomb (Methuen, 1975) and *Probability and random processes: a first course with applications* by A. B. Clarke and R. L. Disney (Wiley, 1985). The following book presents the theory and a thorough survey of Monte Carlo methods: *Monte Carlo methods in statistical physics* edited by K. Binder (Springer, 1986). It contains examples of applications in various branches of physics.

Further books on Monte Carlo methods include: *Monte Carlo simulation in statistical physics* by K. Binder and D. W. Heermann (Springer, 1988); *Applications of the Monte Carlo method in statistical physics* edited by K. Binder (Springer, 1987); *Computer simulation studies in condensed matter physics I–III* edited by D. P. Landau *et al.* (Springer, 1988–1991); *Monte Carlo methods, Vol. 1* by M. H. Kalos and P. A. Whitlock (Wiley, 1986); *An introduction to computer simulation methods – application to physical systems* by H. Gould and J. Tobochnik (Addison-Wesley, 1988); *Monte Carlo methods in quantum problems* edited by M. H. Kalos (Reidel, 1984); *Quantum Monte Carlo methods* edited by M. Suzuki (Springer, 1987); *Quantum simulations of condensed matter phenomena* edited by J. E. Gubernatis and J. D. Doll (World Scientific, 1990); *The Monte Carlo method in condensed matter physics* edited by K. Binder (Springer, 1992).

Molecular dynamics simulations are discussed in the following books: *Molecular dynamics* by W. G. Hoover (Springer, 1986); *Molecular dynamics simulation of statistical mechanical systems* edited by G. Ciccotti and W. G. Hoover (Elsevier, 1987); *Computer simulations using particles* by R. W. Hockney and J. W. Eastwood (Hilger, 1988); *Computer simulation of liquids* by M. P. Allen and D. J. Tildesley (Clarendon Press, 1987).

A description of Monte Carlo methods used to simulate fermionic systems (Langevin, Molecular Dynamics, Hybrid Monte Carlo, Pseudofermion Method) as well as a guide to related high energy physics references can be found in: H. J. Rothe's *Lattice gauge theory: an introduction* (World Scientific, 1992). *Workshop on fermion algorithms* edited by H. J. Herrmann and F. Karsch (World Scientific, 1991) presents the current state of Monte Carlo simulations for fermionic systems.

There are two very valuable series devoted to computational physics that include works on specialized branches of physics. The first is Methods in Computational Physics – Advances in Research and Applications (Academic). So far the series has included volumes on statistical physics (1963), quantum mechanics (1963) hydrodynamics (1964, 1965), nuclear particles and physics (1966), energy bands in solids (1968), plasma physics (1970), and controlled fusion

(1976). Unfortunately there have been no futher additions to this series. The other is the Springer Series in Computational Physics, volumes of which include topics on fluid dynamics (1977, 1981, 1988, 1990), plasma physics (1978, 1986, 1991), unsteady viscous flows (1981), fluid flow (1983), Monte Carlo methods (1991), non-linear variational problems (1984), orthogonal polynomial of discrete variables (1991), and finite elements (1981). There are also a few volumes of another series Topics in Applied Physics (Springer) that concern computational physics e.g. *Numerical and asymptomatic techniques in electromagnetics* (1975), *The computer in optical research* (1980), *Two-dimensional digital signal processing* (1981), *Optical information processing* (1981), and *The Monte Carlo method in condensed matter physics* (1992).

A series of conferences, International Conference on Computational Physics, was started in 1989 in Boston. Proceedings of the, 1990 conference were published as *Computational physics proceedings of the CP90, Amsterdam, 10–13 Sep., 1990* edited by A. Tenner (World Scientific, 1991). CP92 in Prague has abstracts in *Computers in Physics* (AIP), the proceedings (edited by J. Nadrchal) will be published in *International Journal of Modern Physics C* (World Scientific).

Periodicals

Computer Physics Communications (North-Holland) is the leading international journal publishing descriptions of computer programs in physics papers on computational methods and the application of computers in physics. The programs described are available in the Program Library from Queen's University in Belfast.

Journal of Computational Physics (Academic) is the main American journal in the field publishing not only papers concerning computer applications, but also descriptions of data-handling methods that can be used without a computer.

Computer Physics Reports (North-Holland) was started in, 1984 and aimed to publish primarily reviews devoted to distinct areas of computational physics. It has been integrated into *Physics Reports* in 1990.

International Journal of Modern Physics C started in 1990 and is today one of the the major journals in the areas of computational high energy physics and computational statistical physics. It publishes reviews also, and has put emphasis on algorithms for vector and parallel computers. Further journals are *Computers in Physics* (AIP); *Computational Polymer Science* (PRA Press, since 1991); *Computational Materials Science* (IDP, since 1992).

A special section in the proceedings of the annual conference

LATTICE is devoted to new developments of simulation techniques in high energy physics. These proceedings are published by North Holland as supplement to *Nuclear Physics B*. Recent issues are: *Lattice '91: Nuclear Physics B (Proc. Suppl.)* 26 (1992); *Lattice '90: Nuclear Physics B (Proc. Suppl.)* 20 (1991); *Lattice'89: Nuclear Physics B (Proc. Suppl.)* 17 (1990); *Lattice'88: Nuclear Physics B (Proc. Suppl.)* 9 (1989). These proceedings contain also a special section on commercial and special purpose computers.

Many details on computing, data handling, and networking in big experimental environments are reported in the series: *Proceedings of the International Conference on Computing in High Energy Physics*. Recent proceedings are: *CHEP 1992* edited by C. Verkerk and W. Wojcik (Geneva: CERN 92–07); *CHEP 1991* edited by Y. Watase and F. Abe (Academic, 1991); *CHEP 1990* edited by J. Lillberg and M. Oothoudt (AIP Conf. Proc. 209, 1990).

The proceedings of the series *International Workshop on Software Engineering, Artificial Intelligence and Expert Systems in High Energy and Nuclear Physics*, entitled *New Computing Techniques in Physics Research* (AI-HENP) include major sections on neural nets and symbolic manipulation techniques. They are published as: *AI-HENP* edited by D. Perret-Gallix and W. Wojcik (Editions du CNRS, 1990) and *AI-HENP II* edited by D. Perret-Gallix (World Scientific, 1992)

Non-numerical methods

Symbolic computation or computer algebra is the performance of non-numerical mathematical computations on a computer. Since the early days of SCHOONSHIP and FORMAC much has changed in the field of computer algebra. Several computer algebra packages are available today, and in most cases they are not restricted to mainframes, but perform very well on workstations. Possible applications of these systems include manipulation of polynomials, matrix computations, series expansions, differentiation, integration, differential equations, differential forms, traces of Dirac's gamma matrices, arbitrary precision arithmetics, etc. Most of the modern packages include a graphics output module, interfaces to generate output, which can be further processed by Fortran or C programs, and a TeX interface.

Books

A guide to computer algebra systems by D. Harper, D. Hodgkinson and C. Wooff (Wiley, 1991) reviews Reduce, Macsyma, Maple,

Mathematica, and Derive. Chapter 9 of this book is an information source for newsletters, electronic mail discussion groups, and bibliographies. An easy introduction to computer algebra for beginners is contained in *Computer simulation and computer algebra – lectures for beginners* by D. Stauffer *et al.* (Springer, 1989). More specialized books are *Computer algebra: systems and algorithms for algebraic computations* by J. Davenport *et al.* (Academic, 2nd edn., 1993) and *Perturbation methods, bifurcation theory, and computer algebra* by R. H. Rand and D. Armbruster (Springer Applied Mathematics Series, 1987). The following handbooks for specific computer algebra systems are separately available: *Mathematica – a system for doing mathematics by computer* by S. Wolfram ((Addison-Wesley, 2nd edn., 1991); *Maple V – language reference manual, Maple V – library reference manual, First leaves: a tutorial introduction to Maple V*, all by B. W. Char *et al.* (Springer, 1991); *REDUCE – software for algebraic computation* by G. Rayna (Springer, 1987).

An introduction to artificial intelligence is contained in *Artificial intelligence* by P. H. Winston (Addison-Wesley, 1992). The *catalogue of artificial intelligence techniques* edited by A. Bundy *et al.* (Springer, 3rd edn., 1990) is a comprehensive work on artificial intelligence.

Periodicals

Journal of Symbolic Computation (Academic) contains descriptions of systems, presentations of new applications, and research articles. ACM publish the *SIGSAM Bulletin*.

Conference proceedings

ACM publish the *SIMSAC Proceedings*; the proceedings of the European conferences EUROSAM, EUROCAM and EUROCAL are all published in the series: Lecture Notes in Computer Science (Springer).

Electronic news and libraries

- Maple	via anonymous ftp from
	- neptune.inf.ethz.ch
	- daisy.waterloo.edu
	- daisy.uwaterloo.ca
- Mathematica	via anonymous ftp from
	- ftp.ncsa.uiuc.edu
	- otter.stanford.edu
	via electronic mail from
	- mathresource@wri.com

-Reduce   via anonymous ftp from
                 -elib.zib-berlin.de
                 via electronic mail from
                 -reduce-netlib@rand.org
                 -reduce-netlib@can.nl
                 -redlib@elib.zib-berlin.de
                 -reduce-netlib@pi.cc.u-tokyo.ac.jp

Computer graphics and image processing

The capabilities of supercomputers and modern experimental facilities to produce enormous amounts of scientific data require new ways of data evaluation. The visualization of scientific data is of central importance in the evaluation of these data.

Books

From a deluge of books the following are of special interest for physicists.

The mathematical structure of raster graphics by E. L. Fiume (Academic, 1989) is an introduction to the theory of computer graphics.

Visualization in scientific computing edited by G. M. Nielson and B. Shriver (IEEE Computer Society Press, 1990) with its companion (VHS) videotape shows the usefulness of visualization techniques.

Scientific visualization techniques and applications edited by K. Brodie *et al.* (Springer, 1991) summarizes current techniques and applications and contains an extensive bibliography and product review. It is particularly recommended to anybody responsible for interactive graphics facilities in Europe.

The following books may be used for introductory reading and as reference books: *Computer graphics: principles & practice* by J. D. Foley *et al.* (Addison-Wesley, 2nd edn., 1990); *Principles of interactive computer graphics* by W. M. Newman and R. F. Sproull (McGraw-Hill, 2nd edn., 1979); *Procedural elements for computer graphics* by D. F. Rogers (McGraw-Hill, 1985); *Volume visualization* by A. Kaufman (IEEE Computer Society Press, 1990). *Computer graphics* by E. Angel (Addison-Wesley, 1990) gives an easy introduction to computer graphics with many practical examples of C-code.

Computer graphics algorithms are published in the following books: *State of the art in computer graphics and graphics programming* by D. F. Rogers and R. A. Earnshaw (Springer, 1991);

An introduction to ray tracing by A. S. Glassner (Academic, 1989); *Graphic gems* by A. S. Glassner (Academic, 1989); *Graphic gems II* by J. Artvo (Academic, 1991); *Graphic gems III* by D. Kirk (Academic, 1992); *Visual cues: practical data visualization* by P. R. Keller and M. M. Keller (IEEE Comp. Soc. Pr., 1992).

The following books can be used as introductory reading to image processing: *An introduction to digital image processing* by R. J. Schalkoff (Wiley, 1989); *Algorithms for graphics and image processing* by T. Pavlidis (Computer Science Press, 1982); *Digital picture processing* by A. Rosenfeld and A. C. Kac (Academic Press, 1982); *An introduction to image processing* by A. Marion (Van Nostrand, 1991).

Periodicals

IEEE Computer Graphics and Applications (IEEE) deals with methods of the design and the applications of computer graphics systems.

ACM Transactions on Graphics (ACM) also deals with all aspects of the development and use of computer graphics.

SIGGRAPH (ACM) publishes mainly short methodical notes. Further journals include *The Visual Computer* (Springer) and *Computer Vision Graphics and Image Processing* (Academic). Since 1991 the latter journal has been published in two parts: *CVGIP: Graphical Models and Image Processing* and *CVGIP: Image Understanding*.

Conferences

Among many conferences on computer graphics the following may be especially recommended: *EUROGRAPHICS* (proceedings of this conference are published by North-Holland every year); *IEEE Visualization* (IEEE); and *ACM SIGGRAPH* (ACM).

Software packages

The Graphical Kernel System (GKS) provides a functional interface between an application program and a configuration of graphical input and output devices. The two following books provide all necessary information: *Computer graphics programming: GKS – the graphics standard* by G. Enderle, K. Kansy and G. Pfaff, (Springer, 2nd edn., 1987) which also covers the GKS-3D standard, the FORTRAN and Pascal language bindings, and mentions the Programmers Hierarchical Interactive Graphics Standard (PHIGS); and *Introduction to the graphic kernel system (GKS)* by F. R. A. Hopgood (Academic, 1983).

Apart from the official ISO standard papers, the following is a rather good introduction to PHIGS: *A primer for PHIGS* by F.R.A. Hopgood and D. A. Duce (Wiley, 1991). *A practical introduction to PHIGS and PHIGS PLUS* by T. Howard *et al.* (Addison-Wesley, 1991) gives a comprehensive coverage of interactive modelling. There are example programs in C and Fortran and an extensive annotated bibliography. The PHIGS extensions to X are described in *The PEXlib programming manual* by T. Gaskins (O'Reilly, 1992).

There is also a number of high-end ready-to-use visualization systems available. Among these are systems which were developed at some supercomputer centres in the USA which are freely available via anonymous ftp.

AVS by Advanced Visual Systems Inc. Concord MA is available for many platforms from Cray and Convex supercomputers to Unix workstations of almost every vendor. *apE III* by TaraVisual Corporation, Columbus, Ohio, offers comparable capabilities. Both systems come with an easy-to-use visual programming interface which allows rapid application building. *KHOROS* is a similar system for image processing and data visualization which is freely available from the University of New Mexico. It can be obtained via anonymous ftp from *pprg.eece.unm.edu*. A number of visualization programs for two-dimensional data is available via anonymous ftp from the National Center for Supercomputer Applications (NCSA) at Urbana-Champaign from *ftp.ncsa.uiuc.edu*. Some of these are for the Apple Macintosh the others for use with Unix workstations and the X-Window system. The San Diego Supercomputer Center's *Image Tools* for the processing of raster images are available via anonymous ftp from *sdsc.edu*.

Databases and data handling

Books

Physicists are not alone in their need to store and manage data, so for many applications there are excellent commercial products available at reasonable prices with adequate information in their user guides and reference manuals.

PC databases such as *dBase* (Borland) are very common. The dBase language is interpreted, which is useful for development, and compilers like *Clipper* (Nantucket Corporation) give performance improvements. There is an ANSI standards workgroup to define the *Xbase* language: a consistent version derived from *dBase*, its clones and extensions. Other popular PC database products are *Hyper-Card* with the *HyperTalk* programmimg language (Macintosh) and *FileMaker Pro* (Claris). Spreadsheet packages like *1-2-3* (Lotus) and *Excel* (Microsoft) have excellent graphics and database features. Data interchange between different spreadsheet packages and with word-processing systems is well supported.

Central computing facilities, often mainframes, tend to offer database systems like *Oracle, Sybase, Ingres* and *Informix*. These products offer local and remote access for many concurrent users to distributed databases. The structured query language, SQL, is the standard for communicating with these systems. *A visual introduction to SQL* by J. H. Trimble and D. Chappell (Wiley, 1989) is recommended to the end user for its clarity. There is a considerable literature for database designers which is not aimed explicitly at physicist users. *An introduction to database systems* by C. J. Date (Addison-Wesley, 2nd edn., 1990), is a textbook covering the fundamental concepts of database systems. The architecture of database systems, the relational, hierarchical and network approach are explained in a clear manner. For SQL, *A guide to the SQL standard: a user's guide to the standard relational language SQL* (Addison-Wesley, 3rd edn., 1992) by C. J. Date may be consulted. Date's 1986 tome *Relational database: selected writings* has six chapters on SQL and one long one on relational database design methodology. This has been succeeded by two further volumes: *Relational database writings: 1985–1989* (illus.) and *Relational database writings: 1989–1991* (illus.), which are both published by Addison-Wesley.

The 'Computer databases' article in the *Encyclopedia of applied physics*, cited above, gives a general view of databases and online services used by physicists. The first conference covering the whole subject was the 5th Summer School on Computing Techniques in Physics, Bechyne, 1983. Proceedings of the school – *Data bases and data structures in physics* edited by J. Nadrchal are published in *Computer Physics Communications*, (1984) 33, 1–297. Databases and data management are represented in most CHEP conferences, for example, research towards very large relational and object databases (many terabytes) on massively parallel systems by companies such as Oracle and by physicists hoping to acquire such data are reported in *CHEP92* (see the section on 'Computational physics').

Database computing is currently prohibitively expensive, both in computer access time and in license fees to vendors, to be acceptable for many physics analyses. The Fortran and C interfaces to access the data are not notably easy or efficient to use. Most data is therefore stored on tape or disc using a machine independent data representation and accessed through the operating system's file access methods. *Data structures for particle physics experiments: evolution or revolution?* edited by R. Brun, P. Kunz and P. Palazzi (World Scientific, 1991) contains reviews of the packages in current use to read and write, in a machine independent way, complex data structures in C and Fortran programs. Some insight into future data handling facilities may be obtained from the *Mass storage reference model version 5* edited by S. Coleman and D. Isaac (IEEE Computer Society, 1993) and into the state of progress from *Eleventh IEEE symposium on mass storage systems* (IEEE, 1991) and subsequent meetings in the same series.

Periodicals

A journal specializing in databases is *ACM Transactions on Database Systems* (ACM). This includes papers on data structures and models, data security and protection, database architecture and related topics.

Software for process control

Books

Real time software for small systems by A. Leigh (Wiley, 1988) covers the design and practical implementation of real-time programs on small microprocessor based systems. Structured design and high level languages are reviewed for applications and example systems are written in Z80. *Operating system principles* by P. Brinch Hansen (Prentice Hall, 1973), *The architecture of concurrent programs* by P. B. Hansen (Prentice Hall, 1977) and *Principles of concurrent programming* by M. Ben-Ari (Prentice-Hall, 1982) are recommended further reading. There are many industrial applications of real-time systems which provide tools and examples for physics use. SERB provides a concise overview and extensive bibliographies for applications like *Real time:section 56, Interactive (human interface):56, Safety:60, Fault tolerance:61* and *AI/IKBS artificial intelligence / intelligent knowledge based systems:38*.

There are several books on experimental physics such as *Data analysis techniques for high energy physics experiments* edited by M. Regler (CUP, 1990) which include the principles of an online data acquisition (DAQ) system, but none giving any depth of coverage to a complete system. The periodicals and summer schools cited below are useful information sources. 'The data acquisition system of the OPAL detector at LEP' by J. Baines *et al.* in *Nuclear Instruments and Methods in Physics Research* A325, pp. 271–293 (North-Holland, 1993) is a paper with extensive references and appendices for the *VMEbus family*, the *OS-9 environment* and the online *Infrastructure*. 'DAQ software architecture Aleph, a large HEP experiment' by A. Belk *et al.* and 'The ALEPH Data Acquisition System' by W. von Ruden in *IEEE Transactions on Nuclear Science* 36 (IEEE, 1989) have more emphasis on *FASTBUS*.

SASD/RT and Petri nets are the formal methods suitable for the design of real-time systems. *Structured development for real time system* by P. Ward and S. Mellor (Yourdon, 1985) supported by a CASE tool such as StP (Software through Pictures) by IDE (Interactive Development Environments) adds state transition to dataflow as a real-time feature.

A better treatment (than SASD/RT) is achieved using Petri net theory, which is introduced in *Petri net theory and the modelling of systems* by J. Peterson (Prentice Hall, 1981). The CASE tool *Artifex*, described in *Operational development of discrete event dynamic systems with PROTOB, an object-oriented CASE tool* by G. Bruno *et al.* (Artis, Turin, Italy) covers most of the software development life cycle and generates code in ANSI-C or Ada. Artifex is evaluated for use by the DAQ R&D projects at LHC in CHEP92. There are several titles, for example *Coloured Petri nets* by K. Jensen (Springer, 3 vols, 1992), in Springer's series of monographs on theoretical computer science for current development of Petri net theory.

At the local process control level and below, where fast real-time response is essential, the proprietary OS-9 operating system is often used with Motorola microprocessors in VME systems. *Basics of real-time operating systems* by D. Kemp (CERN Computing School, 1991) describes systems in use today, exemplified by OS-9. *OS-9 insights: an advanced programmers guide to OS-9* by P. Dibble (Microware, 1992) enhances the user's manual. There is also a niche for transputers at this level, particularly where a fast LAN (local area network) would not handle the dataflow. *The Helios operating system* by Perihelion Software (Prentice Hall, 2nd edn., 1991) helps with transputers near a Unix environment. *Occam2* by J. Galletly (Pitman, 1990) and *Parallel programs for the transputer* by R. Cok (Prentice Hall, 1991) are recommended.

There is a widespread choice of Unix in upper layers, where cheap processing power is more important than response. *A practical guide to the Unix system* by M. Sobell (Benjamin/Cummings, 2nd edn., 1989) is a good introduction to this operating system. *Writing a UNIX device driver* by J. Egan and T. Teixeira (Wiley, 1992) is useful for a physicist attaching devices which are not given full support by the computer vendor.

Real-time extensions to Unix are proposed in the standard POSIX 1003.4. *A portable uniform interface to real-time operating systems* by R. Russell (CERN Computing School, 1991) describes a virtual operating system (VOS) which includes the implementation of real-time extensions proposed for POSIX and allows identical software to run on Unix, OS-9 and VMS.

The C programming language is in common use in online Unix

systems and Fortran code is also run at high levels. Database products like *Oracle* and *HyperCard* are used for logging and presenting information to a human operator. The operator responsible for monitoring an apparatus is helped by the use of graphic tools to present information. Information sources for these tools are given in other sections of this chapter. The NCTPR conference series, cited below, contains papers describing expert systems used in the operation of accelerators, detector readout and experimental control. In general, AI languages like *LISP* and *Prolog* have been discarded in favour of C++.

A third major information source is the proceedings of conferences like the CHEP series. These attract short papers giving full details of components of complex systems such as the human interface, the event builder and data recording devices and papers proposing new methodologies, hardware and software standards.

Periodicals

Descriptions of complete systems are often published in NIM and IEEE journals. Lectures in the annual *CERN School of Computing* cover topical DAQ subjects at an advanced level. The CHEP series emphasize hardware and software issues in alternate conferences and is a forum in which DAQ subsystems, such as an expert system to control an experiment, may be found in parallel and poster sessions. *New Computing Techniques in Physics Research II* (NCTPR) edited by D. Perret-Gallix (World Scientific, 1992), the *Proceedings of the Second International Workshop on Software Engineering, Artificial Intelligence and Expert Systems* has papers on real-time work in astronomy and fusion as well as high energy and nuclear physics.

The ACM *SIGMICRO* and *IEEE Micro* cover microprocessor architectures.

Publishing and retrieving information

The classic English reference works for authors are *The complete plain words* by E. Gowers (HMSO, 3rd edn., 1986) and *A dictionary of modern English usage* by H. Fowler (OUP, 2nd revised edn., 1987). Americans may prefer *The Chicago manual of style: for authors, editors, & copywriters* (University of Chicago, 14th edn., 1993) or the more concise *The elements of style* by W. Strunk Jr. and E. B. White (Macmillan, 1979). *A practical English grammar* by A. Thomson and A. Marbinet (OUP, 4th edn., 1986) may be useful for authors whose first language is not English and *Words into type* (Prentice-Hall, 1974) is useful for both grammar and layout. *The visual display of quantitative information* and *Envisioning information* by E. Tufte (Graphics, 1989) and *The AIP style manual* (AIP, 4th edn., 1990) are particularly useful for figures and tables. The *Oxford science shelf* is available for PC and Mac giving interactive access to *The Oxford dictionary of computing, The Oxford concise science dictionary* and *The Oxford dictionary for scientific writers and editors*.

The TeX typography system is very popular for the production of structured scientific documents. *LaTeX: a document preparation system* by L. Lamport (Addison-Wesley, 1986) shows, mainly by examples, how to use a collection of commands which many physicists find adequate for their needs. *The TeXbook* (AMS, 1988) and *The METAFONTbook* by D. Knuth (Addison-Wesley, 1986) are needed if typesetting details are important. *TeX by topic* by V. Eijkhout (Addison-Wesley, 1991) is recommended for expert users, its references include macros for physicists such as PHYZZX and TeXsis and pointers to articles in *TUGboat*, the quarterly journal of the TeX Users Group (TUG@Math.AMS.org). TeX may be used semi-interactively by converting its device independent output into *PostScript* and then using a previewer such as *Ghostview* (FSF) or *Textures* (Blue Sky Research) for Macintosh. *PostScript* has become a *de facto* standard for which the *PostScript language reference manual* and *PostScript language tutorial and cookbook* by Adobe Systems (McGraw Hill, 1987) are available. Desktop publishing yields a more colourful finished product than the pages of scientific journals, *DTP the complete guide to corporate desktop publishing* by R. Jones (CUP, 1988) provides good coverage of the field, references for further reading, products and manufacturers. An editor might use *XPress* (Quark), *PageMaker* (Aldus) or *FrameMaker* (Frame Technology) to control the finished product. *Canvas* (Deneba Systems), *Illustrator* (Adobe) and *MacDraw* (Claris) are suitable for art work such as charts and schematics of apparatus. *Expressionist* (Prescience) and *MathType* (Design Science) are easy to use and give excellent results for numbered mathematical expressions and *Word* (Microsoft), *WordPerfect* (WordPerfect Corporation) and *MacWrite* (Claris) are some of the most popular text editors with considerable graphics facilities.

'As we may think' by V. Bush (*The Atlantic Monthly*, July 1945, p. 101–8) reprinted in *A history of personal workstations* by A. Goldberg (Addison-Wesley, 1988) considers the problems of physicists returning to academic pursuits after the projects of the Second World War. Particular attention is given to information storage and retrieval and 'new forms of encyclopedias, ready made with a mesh of associative trails running through them': the essential feature of hypertext systems. The proceedings of the 3rd ACM conference on hypertext are published as *Hypertext '91* (ACM, 1991). *Hypertext/Hypermedia handbook* edited by E. Berk and J. Devlin (McGraw-Hill, 1991) has some articles addressing the problems of the authors of documents. *Hypertext hands-on! An introduction to a new way of organizing & accessing information* by B. Shneiderman and G. Kearsley (Addison-Wesley, 1989) includes an IBM PC software package. A physicist browsing documentation distributed on CD-ROM and viewed through a hypertext system has no need to understand the underlying technology, but the person responsible for maintaining the documentation for a collaboration of physicists may find *The SGML handbook* by C. F. Goldfarb (OUP, 1991) useful. SGML is emerging as an exchange standard and is supported by Interleaf (Interleaf) and FrameBuilder (Frame). IEEE now manage journals through *The Publisher* (ArborText).

The Internet is a collection of networks around the world that link military, university and research sites using TCP/IP standards. There are terabytes of public domain software and other documents stored at these sites. *The whole Internet user's guide and catalog* edited by E. Krol (O'Reilly, 1992) is a guide to the services provided and *The Internet system handbook* by D. Lynch and M. Rose (Addison-Wesley, 1993) is suitable for the local expert.

Archie is a server supporting a software directory described in 'Archie – an electronic directory service for the Internet' in *Proceedings of the Winter 1992 USENIX Conference Berkeley, CA, USA* (USENIX Association, 1992). An article in *UnixWorld* (McGraw-Hill, Sept. 1992) may be more accessible. One European server is *archie.doc.ic.ac.uk* (146.169.11.3) – login as *archie*.

The *World-Wide Web* (W³) is a hypertext browser which may be obtained by anonymous FTP from *info.cern.ch*. Mail should be sent to *www-bug@info.cern.ch* (or from JANET, *www-bug@ch. cern.info*). There is a simple description of W³, 'Electronic publishing and visions of hypertext', in *Physics World* (IOP, June, 1992).

The American Institute of Physics promotes *PINET*: The Physics Information Network from which research papers and software may be downloaded and, in co-operation with FIZ, PHYS an online bibliographic database for physicists and astronomers: mail *elecpub* or *ellen@pinet.aip.org*.

For computing and graphics tools the *GNU Project* of the Free Software Foundation may be contacted at *gnu@prep.ai.mit.edu*.

Preprints and photocopies of preliminary results in conference papers are popular sources of information as there can be delays of several months before a paper, accepted by a referee, arrives on the library shelves. Computer and network tools described above have the potential to make radical improvements to the storage and selection of papers, software, pictures, speech, indeed any digital data. Many journals now accept computer files as well as camera ready copy and it is not unusual for conference organizers to issue a LaTeX style macro. There is speculation in *Europhysics News* (1992, vol. 23) that physics publishers are now open to the suggestion that papers may be stored and viewed from document servers. 'E-mail or perish' in *Physics World* (IOP, Feb. 1993) is just one of many articles from physicists crying out for such services to be implemented and responsibly managed.

Acknowledgements

Finally we would like to thank all those people who assisted us in collecting the material for this chapter. Special thanks for their contributions are due to K. Binder, H. Burckhard, R. Esser, J. Grotendorst, J. F. Hake, H. Herrmann, E. van Herwijnen, F. James, F. Karsch, W. Krüger, P. Kunz, S. Leech O'Neale, M. Metcalf, P. Palazzi, and D. Stauffer.

CHAPTER EIGHT

Crystallography*

E. MARSH

Few fields of science are as well organized as is crystallography, although, with interest in the solution of crystal structures as a problem in itself now past its apogee, there are signs of diversification in a hitherto monolithic edifice. It is appearing that the subject matter of crystallography is only coincidentally crystals. The real subject is the structure of matter at a level above that of the structure of the individual atom and below that of the reaction system. The techniques of microscopy, the use of waves of light, X-rays, electrons and neutrons, and also of protons, to see these structures, of course forms part of the subject, as does the technology of the computing which transforms the data and acts as the lens of the generalized microscope. The techniques form almost as large a body of published information as their target.

A great part of present-day crystallography supplies results which are essential to other branches of science, such as chemistry, biology and solid state physics, and consequently the production of compilations of data is especially well developed, although there is a tendency, facilitated by the computer, towards the provision by data services of answers to individual questions. The book is being overtaken by computer technology, networks and by CD-ROMs.

The International Union of Crystallography (now IUCr previously IUC), affiliated to the International Council of Scientific Unions, was founded in 1946, but before that there were co-operative projects for the rational publication of crystallographic data. Adherence to the IUCr is through national crystallographic organizations (in the case of the UK, the British National Committee

---

* based on the second edition chapter written by A. L. Mackay

for Crystallography, which is operated by the Royal Society). The British Crystallographic Association was founded in 1982 by the coalescence of a number of other groups.

The IUCr itself publishes a considerable number of books and periodicals. In this survey questions of mineralogy, other than the structures of crystals, are largely excluded, although crystallography is fundamentally linked with mineralogy (as well as with a number of other subjects).

Historical

The half-century from the discovery of the diffraction of X-rays by crystals (von Laue *et al.*, 1912) has been well documented by P. P. Ewald in *50 years of X-ray diffraction* (IUCr, 1962) which is a book produced to mark this anniversary. Ewald was an active participant in the events leading to this discovery and his book has become a classic of its type, furnishing material for the historians of contemporary science. Ewald's work has been commemorated in a memorial volume edited by D. W. J. Cruickshank *et al. P. P. Ewald and his dynamical theory of X-ray diffraction* (OUP, 1992). The noted neutron physisicist G. E. Bacon has emulated Ewald's work with a book, *50 years of neutron diffraction: the advent of neutron scattering* (Hilger for IUCr, 1987).

The earliest stages of the studies of the structures of crystals – from Pliny and Theophrastus to von Laue – are covered by J. G. Burke in *Origins of the science of crystals* (U. of Cal. Pr., 1966), who furnishes many references facilitating further investigation. The work of the first historian of crystallography, Carl Marx, *Geschichte der kristallkinde* (Karlsruhe u. Baden, 1825) is rare. There are also now many collections covering key figures, such as Federov, Bravais (whose works can now be obtained in facsimile reprints) and Pauling. The last named was presented with a Festschrift for his 65th birthday: *Structural chemistry and molecular biology,* edited by A. Rich and N. Davidson (Freeman, 1968). The IUCr has produced two volumes of *Early papers on diffraction of X-rays by crystals*, edited by J. Bijvoet *et al.* (1972) which makes the history accessible.

A Festschrift for Professor Dorothy Hodgkin adds to the useful compendia covering the fields of work associated with particular schools: *Structural studies on molecules of biological interest*, edited by G. Dodson and J. P. Glusker (OUP, 1981).

The massive *Dictionary of scientific biography* includes, of course, information on many crystallographers. J. D. Bernal's biography by C. P. Snow is in a supplementary volume. For British crystallographers, the *Biographical memoirs of Fellows of the Royal Society* is a major source. W. I.. Bragg summed up his own life's work with *The development of X-ray analysis* (1975). Shafranovskii has followed his history of crystallography in Russia, *Historiya kristallografii v Rossii* (Nauka, Leningrad, 1962) with a general history, *The history of crystallography from the earliest times to the beginning of the 19th century,* (vol. I, Nauka, Leningrad, 1978) and vol. II, *The 19th century* (1980).

Recently there has been a realization that the first generation of crystallographers has almost disappeared and an effort has been made to collect their information. *Crystallography in North America* is a splendid compendium of accounts of work in the USA and Canada (edited by the late D. McLachlan and J. Glusker, 1983).

The history of crystallography is experiencing a revival and a number of projects are in progress. A most interesting historical work edited by J. Lima de Faria *Historical atlas of crystallography* (Kluwer for the IUCr, 1990) shows crystallography as a series of time maps charting the development of the science from the 16th century to the present day. The time maps are complemented by short histories of specific areas and there are illustrations of more than 100 prominent crystallographers.

Crystallographic book lists

The IUCr maintains a Commission on Crystallographic Teaching and in 1965, under the editorship of Dr Helen Megaw, a comprehensive crystallographic book list was published by IUC: H. D. Megaw, *Crystallographic book list* (1965). A supplement to the book list followed in 1966. This covered books in English, French, German, Spanish and Russian, classifying them by level and by topic and giving an account almost complete up to 1964. The only major omission is of books in Japanese, the majority of which are either translations of material which has appeared elsewhere, or texts similar to those available in English. The Japanese publish increasingly in English for their more original material. The *Crystallographic book list* is comprehensive in including serials and conference proceedings as well as books. Supplements to the work were produced by M. M. Woolfson (1972) in *Journal of Applied Crystallography*, 5, pp. 148–162 and for the period covering 1970–1981, by J. Robertson (1982) in *Journal of Applied Crystallography*, 15, pp. 640–676. This included references to reviews, so that the enquirer can consult an informed opinion about any book. To some extent, then, the present review must be a personal exegesis of this list and of later publications.

Polycrystal Book Service (PO Box 3439, Dayton, Ohio 45401,

USA), booksellers specializing in crystallographic material, produce a catalogue: *CATALOG 1990.*

Periodicals

As in almost all other subjects, the most active part of the literature consists of papers in regular periodicals. Bibliometric methods (Hawkins, 1980) show that there are ten journals which have crystallographic papers as more than half their contents, and the periodicals which account for half the total papers in crystallography are the following:

Acta Crystallographica (1948–) *Section A (Foundations of Crystallography), Section B (Structural Science), Section C (Crystal Structure Communications)* and *Section D (Biological Crystallography)*
Crystal Lattice Defect (1969–)
Fizika Tverdogo Tela (1953–) (Leningrad)
Inorganic Chemistry
Izvestiya Akad. Nauk SSSR. Neorg. Mat.
Journal of the American Chemical Society
Journal of Applied Crystallography (1968–)
The Journal of Chemical Physics
The Journal of the Chemical Society, Dalton Trans.
The Journal of Crystal and Molecular Structure (Plenum)
The Journal of Solid State Chemistry
Kristallografiya (1956–) (translated as: *Soviet Physics: Crystallography)* (Akad. Nauk SSSR)
Molecular Crystals and Liquid Crystals (1966–) (Gordon and Breach) (also there are *Letters* and *Supplement Series)*
Physical Review B (1893–)
Physica Status Solidi a and b (1961–)
Progress in Crystal Growth and Characteristics
Solid State Chemistry
Zeitschrift fuer Kristallographie (1877–) (Akad. Verlag. Wies.)
Zhurnal Strukturnoi Khimii (1959–) (Acad. Sci. USSR) and in English translation as: *Journal of Structural Chemistry* (Consultants).

The list does not include the significant journals:

The American Mineralogist (1916–)
Bulletin de la Société Française de Mineralogie et Cristallographie (1878–) (Masson)
Journal of Crystal Growth (1967–) (North-Holland)
Crystal Research and Technology (Akademie)

Doklady Akademii Nauk SSSR (1933–) (English translation: *Soviet Physics – Doklady) Fizika*
Metallov i Metallovedenie (1955–) (English translation: *Physics of Metals and Metallography USSR*)
Journal of Materials Science (1966–)
The Mineralogical Magazine (Min. Soc.).

Valuable analyses of the use of crystallographic journals have been made, on the basis of citations, by Hawkins (1980) and Garfield (1974).

Databooks

The World directory of crystallographers now in its 8th edition (1990) and published by Kluwer for the IUCr, is an invaluable guide to prosopography. The directory will in future be published as a by-product of the world database of crystallographers established by IUCr.

The *International tables for crystallography*, previously the *International tables for X-Ray crystallography* referred to below as *The international tables*, are, of course, indispensable. A completely new redaction is in progress and a second edition of the new version (vol. A, *Space group symmetry)*, was published by Kluwer for the IUCr in 1987 and reprinted with corrections in 1989. A third revised edition of volume A is noted as underway in 1992. Volumes II *Mathematical tables*, III, *Physical and chemical tables* and IV, *Revision and supplementary tables for vols. II and III* are still available but have been updated by volume C of the new series as *Mathematical, physical and chemical tables* (Kluwer for the IUCr, 1992). Volume B, *Reciprocal space* and volume D *Physical properties of crystals* were still noted as in preparation in 1992.

Data on *Molecular structure and dimensions*, edited by O. Kennard *et al.* ceased publication in book form with volume 16 (1984). It presents data on the shapes of organic molecules arising from the Cambridge Crystallographic Data Centre. It is now distributed in electronic format either over the Joint Academic Network from the Daresbury Laboratory or as a stand-alone system for a variety of computers.

Structure Reports has been in continuous production since 1913 and provides complete, though tardy, coverage of the literature, with an index volume covering 60 years (to 1975) and a further cumulative index covering 1971–1980 as volume 47B. The publication of volumes on organic compounds was discontinued after the publication of the volume 52B for 1985. The stated aim is to discontinue volumes for metals and inorganic materials and allow database systems to takeover. *Crystal data* (now 4th edn.) provides unit cell dimensions (and is now published by the Joint Centre for Powder Diffraction Services, JCPDS).

Pearson's handbook of crystallographic data for intermetallic phases edited by P. Villars and L. D. Calvert is a databook in four volumes which covers 75 years of the world literature through 1989 with 50 000 entries from 130 current journals (ASM, 2nd edn, 1991) and with a companion work, also in four volumes *Atlas of crystal structure types for intermetallic phases*. An earlier work is *The Barker index of crystals* edited by M. W. Porter and R. C. Spiller (published for the Barker Index Committee by W. Heffer, 1951–1964) which provides a method for the identification of crystalline substances.

There are a number of handbooks dealing with minerals which are valuable: *The manual of mineralogy*, by J. D. Dana (and later C. S. Hurlbut and others) (Wiley), is more an institution than a book, there being 17 editions between 1912 and 1949 and the current edition is the 20th (1985) edited by C. S. Hurlbut and C. Klein; *Mineralogische tabellen* by H. Strunz (1941–) (a new (6th) English-language edition is current) (Akad. Verl. Leip.); *Microscopic characters of artificial inorganic solid substances* and *Optical mineralogy* by A. N. and H. Winchell (Academic, 1951). *Rock-forming minerals* (6 vols), by W. A. Deer *et al.*, (Longmans, 1962–1963), is also a basic compendium with vols 1A, 1B and 2A now appearing as a second edition (1978–1986).

Services and databases

There has been a steady shift from the use of books as sources of digested crystallographic information towards the consultation of databases. Many standard reference books have not been able to keep up with the generation of data still rising exponentially. In a further use of technology an information server has been established at the IUCr Education Office which makes considerable use of the Joint Academic Network (JANET) and permits the automatic retrieval of programs and documentation useful in handling Crystallographic Information Files (CIF).

An alternative service has been established by the British Library Document Supply Centre (BLDSC) under its Supplementary Publication Scheme for the deposition of data. Data which would lengthen periodical articles to unreasonable lengths may be deposited for reference at BLDSC, Boston Spa, and this includes crystallographic structure factor tables, spectroscopic frequency assignments, bond lengths and angles.

Databases in general are described in other chapters. The *CODATA directory of data sources for science and technology* has a special section on crystallography (Bulletin 24 of June 1977) by D. G. Watson and a more up-to-date list can be found in *Crystallographic databases* edited by F. Allen, G. Bergerhoff and R. C. Sievers and issued by the IUCr Data Commission in 1987. The databases are mostly accessible through the DIALOG, Data-Star and DIANE hosts over the international and academic networks.

The Cambridge centre is probably the most important, covering organic and organo-metallic structures where the molecular shape is of prime concern. The CAMBRIDGE STRUCTURAL DATABASE contains evaluated numerical data for some 40 000 structures and is increasing by some 6000 per year. Access to the database and programs for its utilization are provided through 22 National Affiliated Data Centres throughout the world. An important characteristic of this database is that the information is checked for internal consistency and mistakes detected are verified with the authors.

Altogether the database is being used in increasingly sophisticated ways (Allen, Kennard and Taylor, 1983) and is an essential basis of the emerging practical sciences of drug and protein design. Similar advances may be expected when the corresponding inorganic file is widely used.

A Protein Data Bank is also available, organized and maintained from Brookhaven National Laboratory, USA, and includes essentially all solved protein structures. The service is available on a CD-ROM as images for 600 crystallographic structures of proteins, polynucleotides and polysaccharides shown in 3–D to demonstrate the backbone, sidechains and secondary structure of each molecule. The new technology allows the molecules to be rotated at any angle. The late M. O. Dayhoff produced *An atlas of protein sequence and structure*, which is a valuable compendium which went through several editions before the volume of material became too great, but it includes invaluable material on protein sequences, phylogenies, etc. The data are available in machine-readable form as PROTEIN SEQUENCE DATA TAPES.

NBS Crystal Data Identification File (XTAL) has been replaced by the National Institute of Science and Technology (NIST) CRYSTAL DATA FILE which contains published information from the NIST Crystal Data Centre issued by the International Centre for Diffraction Data in CD-ROM format. The CD-ROM contains chemical, physical and crystallographic data on 150 000 crystalline materials including lattice parameters, crystal systems, space group symbol and number, chemical name and literature references. The material is updated twice a year and the CD-ROM can be used on most standard players.

The newer technology is also used for the NIST/Sandia/ICDD ELECTRON DIFFRACTION DATABASE which contains crystallographic and chemical information on more than 71 000 crystalline materials giving interplanar spacings, space group and unit cell data, compound name and formula. The database is available for any ISO 9660 drive and there are irregular updates.

JCPDS (Joint Centre for Powder Diffraction Services), now ICDD (International Centre for Diffraction Data), is a centre for diffraction data. As the successor of ASTM, it provides access to 139 collected data on powder diffraction with information on magnetic tape, microfiche, books and cards, handling the NBS files and other sources. Subsets of the main file are available in various forms. A book form version of data is available as *Mineral powder diffraction file*, (1JCPDS, 1980–83) with three volumes of which the first is the *Data book*, the second is the *Search manual* and the third is the *Group data book* which revises the mineral groups in Strunz.

The Fachinformationszentrum, Karlsruhe, is now offering the INORGANIC CRYSTAL STRUCTURE DATA BANK (ICSD) commercially. The file may be consulted online or a tape of structures may be leased. Crystallographic data are, of course, inseparable from other data and all the regular chemical information may have to be obtained. The Royal Society of Chemistry, for example, provides a guide, *Information tools*, to Chemical Abstracts Service products, Royal Society of Chemistry products, Mass Spectrometry Data Centre products and Analytical Abstracts.

CHEMICAL ABSTRACTS is, of course, in this and other fields, quite invaluable, and the French INIST DATA BASE (formerly BULLETIN SIGNALÉTIQUE) has a section 'Structure de la Matière: Cristallographie'.

The Chemical Structure Association (formerly the Chemical Notation Association) caters for the interests of those who have to catalogue and retrieve information about chemical structures, especially for patent work. It produces occasional publications and runs training courses in, for example, the Wiswesser Line Notation. There are commercial services for converting one notation for an organic structure to another, since different organizations use different systems.

Serials

Of the serially produced books, the principal is the series on *Structure Reports*, published for the IUCr by Kluwer, which has been appearing continuously since the first X-ray structure determinations were reported in 1913. Although somewhat behind the literature, it has now reached 57 volumes. It provides critical reports on virtually all determined crystal structures. From 1965 it is subdivided into two sections: A. Metals and inorganic compounds and B. Organic compounds. Each volume covers a year's literature and there are cumulative indices. Other source books on structures, particularly that produced by R. W. G. Wyckoff, *Crystal structures*, 6 vols. (Interscience), were reprinted during 1981–1983, and are useful in some areas, but have not been able to keep up with the rate of growth. Russian work may be found in *Growth of Crystals*, 1958– edited by A. V. Shubnikov and N. N. Sheftal' which contain primarily review articles with some original material of the growth of crystals and films. The Russian text is published by the Academy of Sciences of the former USSR as *Soveshchanie po Rosta Kristallov* and is published in English by Consultants Bureau.

There are also numerous serials on the borders of crystallography which appear at roughly annual intervals. They include: *Progress in Stereochemistry*, edited by W. Klyne and P. B. D. de la Mare (Butterworths); *Perspectives in Structural Chemistry*, edited by J. D. Dunitz and J. A. Ibers (Wiley); *Advances in X-ray Analysis*, edited by W. M. Mueller and M. Fay which are the *Proceedings of the Annual Conference on Applications of X-Ray Analysis* (Plenum); *Advances in Structure Research by Diffraction Methods*, edited by R. Brill and R. Mason (later Hoppe and Mason) (Interscience); *Progress in Biophysics and Molecular Biology* (Pergamon); and *Transactions of the American Crystallographic Association* (ACA).

Geometry, symmetry and mathematics

A sound knowledge of geometry is the basis of any study of structure in space. 'Euclid's Elements', indeed, is still essential but it is best approached through H. S. M. Coxeter's masterly account, *Introduction to geometry* (2nd edn., 1969 repr. in 1989, Wiley). The high reputation of the *Mathematical handbook* of G. A. Korn and T. M. Korn (McGraw-Hill, 1968) is justified and supplies most of the mathematics needed. The *International tables for X-ray crystallography* include in volume 2 a summary of much of the mathematics commonly encountered in crystallographic work and this has been updated in the new version as volume C *Mathematical, physical and chemical tables*. A standard text in a particular area of mathematics is provided by W. A. Wooster *Tensors and group theory for the physical properties of crystals* (Clarendon, 1973).

As almost every book on crystallography has some account of symmetry, and most are necessarily explanations of the *International tables for X-ray crystallography*, there is little need to mention any particular book, although the texts by M. J. Buerger: *X-ray crystallography* (Wiley, 1942); *Elementary crystallography* (Wiley, rev. edn., 1963); *Vector space* (Wiley, 1959); *Crystal structure analysis* (Wiley, 1960); and *Contemporary crystallography* (McGraw-Hill, 1970) are the most authoritative accounts. For the formal geometrical crystallography, *An introducton to crystallography* by F. C. Phillips (Longmans, 4th edn., 1971) remains unsurpassed and mention should also be made of P. B. Yale *Geometry and symmetry* (Dover, repr. 1988). A further useful work is by M. B. Boisen and G. V. Gibbs *Mathematical crystallography* as volume 15 in *Reviews of Mineralogy* (Mineralogical Society of America, 1985).

Crystallography is an area of culture somewhat neglected by most non-scientific commentators, although the great popularity of polyhedra and geometrical problems shows that ideas of mathematical symmetry are fundamental to our civilization. However, there are now a number of books which examine the importance of mathematical symmetry in general culture: these are Hermann Weyl's *Symmetry* (Princeton U. Pr., 1952) and *Symmetry in science and art* by A. V. Shubnikov and V. A. Koptsik (1974), now available in English (Plenum, 1974). The prints by M. C. Escher have achieved great popularity in several collections and an edition by Caroline MacGillavry, *Symmetry aspects of M. C. Escher's drawings* (IUC, 1965), is designed for the instruction and entertainment of crystallographers. M. F. C. Ladd has produced an excellent work for crystallographers *Symmetry in molecules and crystals* (Ellis Horwood, 1989).

Tables of extensions of space-group theory have been published by V. A. Koptsik *Shubnikov groups* (Moscow U. Pr., 1966) representing the work of a group of Russian crystallographers, but the material will also be found in the new edition of the *International Tables*. Complete tables: *Crystallographic groups of four-dimensional space* by H. Brown *et al.* were published by Wiley in 1978 and are slowly finding some application.

Computing

Computing is one of the major factors revolutionizing crystallography and underlies most experimental work, the Fast Fourier Transform being an important theoretical advance and the Cray computer (with the beginnings of parallel processing) as the current state of the art. The literature is much behind, but following F. R. Ahmed's book *Crystallographic computing* (Munksgaard, 1970) we have D. Sayre (ed.) *Computational crystallography* (Clarendon, 1982) presenting up-to-date practice. Computer graphics in crystallography has not yet led to special books, but there is a *Journal of Molecular Graphics*. Computer program abstracts appear regularly in *Journal of Applied Crystallography* and provide a rapid means of communicating up-to-date information on new programs which have been reviewed by one or two members of the IUCr Commission of Crystallographic Computing.

Diffraction theory

The main features of diffraction theory were worked out remarkably fully by the pioneers of X-ray crystallography, such as Ewald, von Laue and C. G. Darwin, but their works are now little used, most readers preferring more modern representation. With the advent of the laser, X-ray interferometry, radio astronomy and electron microscopy there has been a renaissance in physical optics, and all wavelengths are seen as showing similar phenomena in their interactions with periodic structures.

The internal logic of the diffraction by crystals enables crystal structures to be solved automatically by 'direct methods'. M. F. C. Ladd and R. A. Palmer (eds.) in *Theory and practice of direct methods in crystallography* (Plenum, 1980) document this, as does H. Hauptman, one of the pioneers of this technique, in *Direct methods in crystallography* (edited proceedings, 1978). C. A. Taylor has produced a text promoting the unified outlook with *Images: a unified view of diffraction and image formation with all kinds of radiation* (Wykeham, 1978).

An interesting use of new technology is found in the book by S. C. Wallwork *X-ray crystallography* (Royal Society of Chemistry, 1990) which provides a slim volume supplemented by an audio cassette.

Structure analysis and general crystallography

A major general textbook (in four volumes) has been produced from the Institute of Crystallography in Moscow, the largest such centre in the world. It is *Contemporary crystallography* (*Sovrememnaya kristallografiya*) (Nauka, Moscow, 1979–1981) and is edited by B. K. Vainshtein *et al.*; it is already appearing in English translation under the Springer imprint. The volumes are: 1. *Symmetry of crystals and methods of structural crystallography*, 2. *The structure of crystals*, 3. *The formation of crystals* and 4. *Physical properties of crystals*.

In general, there are now plenty of good textbooks, and mention may be made of a selection beginning with the classic by M. M.

Woolfson (1970) *Introduction to X-ray crystallography* (CUP) whilst the earlier text by H. Lipson and W. Cochran (1966) *The determination of crystal structures* (3rd edn., Cornell U. Pr.) is still valid. Other excellent texts include J. P. Glusker and K. N. Trueblood (1985) *Crystal structure analysis* (2nd edn., OUP); M. F. C. Ladd and R. A. Palmer (1985) *Structure determination by X-ray crystallography* (2nd edn., Plenum); G. H. Stout and L. H. Jensen (1989) *X-ray structure determination* (2nd edn., Wiley-Interscience) and a work in the IUCr book series by A. Domenicano and I. Hargittai (1992) *Accurate molecular structure: determination and importance* (OUP).

J. P. Glusker (1985) has also produced an elementary text on the subject *Crystal structure: a primer* (2nd edn., OUP) and a further introductory work is by D. M. McKie and C. McKie (1986) *Essentials of crystallography* (Blackwells). A more specialized area is by B. G. Hyde and S. Andersson (1988) *Inorganic crystal structures* (Wiley). An interesting work on the identification of crystals is by P. Darling (1991) *Crystal identifier* (Apple) whilst A. W. Vere has produced *Crystal growth: principles and progress* (Plenum, 1987).

Crystal chemistry

The pervading influence of Linus Pauling's book *The nature of the chemical bond and the structure of molecules and crystals* (Cornell U. Pr., 3rd edn., 1960) was marked by the appearance of *Structure and bonding in crystals*, edited by M. O'Keeffe and A. Navrotsky (Academic, 1981), which commented that there has been, until recently, little progress in dealing with complex silicates since the formulation of *Pauling's rules* 50 years ago. A further historical, but still useful text is by C. W. Stillwell (1983) *Crystal chemistry* (McGraw-Hill). A new handbook, *Mineraly*, edited by F. V. Chukhrov (Nauka, Moscow) is slowly appearing. Two volumes (III.2 and III.3) on certain classes of silicates came out in 1981.

As a compendium, the massive *Structural inorganic chemistry*, by A. F. Wells, now in its 5th edition (OUP, 1984) is highly recommended and stands as a monument to the author's life work. His monographs on polyhedra, *Three-dimensional nets and polyhedra* (Wiley, 1977) and *Further studies of three-dimensional nets* (ACA Monograph No. 8, 1979), are also useful references as is his *The third dimension in chemistry* (Clarendon, 1956). H. Jaffe has produced two useful textbooks both published by CUP, 1988; *Introduction to crystal chemistry* is an elementary text formed from the first 11 chapters of *Principles of crystal chemistry and refractivity*.

Crystal physics

Crystal physics was a field which languished for many years, becoming almost confined to the physics of quartz, but with the establishment of the theory of dislocations and defects and with the production of a large variety of materials as large single crystals, the subject is now of much importance.

A useful survey is R. Newnham (1975) *Structure-property relations* (Springer). R. M. Hazen and L. W. Finger's (1982) *Comparative crystal chemistry – temperature, pressure, composition and the variation of crystal structure* (Wiley) deals also with experimental work.

The formal physics of anisotropic media is covered by J. F. Nye's *Physical properties of crystals* of which a paperback edition has been produced with corrections and new material (Oxford Science Publications, 1985) and by S. Bhagavantam and T. Venkatarayudu's *Theory of groups and its application to physical problems* (Academic, 1969).

Imperfections, morphology and growth

With the rise of solid state electronics the production of single crystals has become an industry of great importance with a corresponding literature. F. R. N. Nabarro, one of the pioneers in dislocation work, has edited eight volumes covering *Dislocations in solids* (North-Holland, 1979–1989). Nabarro has also produced an excellent text on the same subject *Theory of crystal dislocations* (Clarendon, 1967).

B. R. Pamplin is responsible for a number of important books on crystal growth: (as editor) *Inorganic biological crystal growth* (Pergamon, 1982) and six volumes, *Progress in Crystal Growth and Characterization* (Pergamon, 1977–1983). A. M. Keesee and her collaborators have compiled a large *Crystal growth bibliography* (IEI/Plenum, 1979). There are very many other works, including W. A. Tiller *The science of crystallization: macroscopic phenomena and defect generation* (CUP, 1991).

Liquid crystals

Liquid crystals were first discovered in 1888 by Friedrich Reinitzer and cover the state of matter occurring on the phase diagram between the crystal and the liquid phases. As such it is a fascinating state because it combines properties of both phases. After early work in the 1920s and 1930s the subject became somewhat unfashionable until there was a great resurgence of interest in the 1970s. The study of liquid crystals is the acknowledged speciality of G. W. Gray who, with G. W. Goodby, has produced *Liquid crystals: identification, classification and structure* (Pergamon, 1984) (as well as earlier volumes). There is a large *Handbook of liquid crystals* by H. Kelker and R. Hatz (Verlag Chemie, 1980). Other texts include S. Chandrasekhar (1977) *Liquid Crystals* (CUP) and the recent book by P. J. Collings (1990) *Liquid crystals: nature's delicate phase of matter* (Hilger). The Royal Society of Chemistry is responsible for a work as volume 435 in the American Chemical Society series (1991) *Liquid crystalline polymers*.

Electron microscopy and diffraction

In 1980 a book marking the discovery of electron diffraction by Davisson and Germer and by G. P. Thompson in 1926 appeared: *50 years of electron diffraction*, edited by P. Goodman (IUC, 1981). We might note also *Nobel Symposium 47 – direct imaging of atoms in crystals and molecules*, edited by L. Kihlborg (Royal Swedish Academy, 1979), since then there has been very great progress in electron microscopy which, through Abbe's theory of the microscope and image processing, has become assimilated thoroughly to the mainstream of crystallography.

Neutron diffraction and synchrotron radiation

A major development in crystallography has been the use of electron synchrotrons for the production of extremely intense beams of X-rays of accurately controllable wavelength. The use of nuclear reactors for the production of neutron beams has now been extended by neutron spallation sources such as those at the Rutherford Appleton Laboratory in the UK and the Los Alamos Scientific Laboratory in the USA. The principal European facility for neutron studies is at the Institute Laue-Langevin in Grenoble, but in the UK, work started at AERE, Harwell. U. W. Arndt and B. T. M. Willis early produced an authoritative account of the necessary apparatus as *Single crystal diffractometry* (CUP, 1966). Technical descriptions of the various synchrotron facilities are available from the establishments – Rutherford and Daresbury Laboratories in the UK, DESY in Hamburg and Brookhaven and Los Alamos in the USA (also Orsay, Italy, Tsukuba in Japan and Novosibirsk in the former USSR). H. Winick and S. Doniach have edited a general volume on

Synchrotron radiation research (1980). The classic text by G. E. Bacon (1975) *Neutron diffraction* (3rd edn., Clarendon) has been supplemented by S. W. Lovesey (1984) *Theory of neutron scattering from condensed matter* (2 vols, Clarendon) and many other works.

Biological structures

The application of X-ray diffraction, the electron microscope and now, neutron methods to biological structures is proceeding at a tremendous rate, as the sheer weight of the principal periodical *The Journal of Molecular Biology* (Academic) attests. Serials such as the *Cold Spring Harbour Symposium on Quantitative Biology* (now about 40 volumes) and *The Nobel Symposium 11, Symmetry and function of biological systems at the macromolecular level*, edited by A. Engstrom and B. Strandberg (Wiley, 1969) are authoritative but quickly overtaken.

As regards textbooks, R. E. Dickerson and I. Geis (1969) *The structure and action of proteins* (Harper and Row), and more recently, *Haemoglobin – structure, function, evolution and pathology* (Benjamin/Cummings, 1982), by the same authors, are strongly recommended as model productions. T. L. Blundell and L. N. Johnson (1976) *Protein crystallography* (Academic) was one of the first manuals directed towards protein crystal structure analysis but there are now others. S. Neidle has begun to edit an extensive series of reports on nucleic acids with *Topics in nucleic acid structure* (Macmillan, 1981). J. P. Glusker has added to her extensive output in a work which aims to allow chemists and biologists to acquaint themselves with the principles of crystal structure analysis, J. P. Glusker *et al. (1992) Crystal structure analysis for chemists and biologists* (VCH).

References

Allen, F. H., Kennard, O. and Taylor, R. (1983) Systematic analysis of structural data as a research technique in organic chemistry. *Accounts of Chemical Research*, **16**, 146–53.

Bassi, Gerard C. (ed.) (1973) *Journal of Applied crystallography*, **6**(4), 309–46.

Bernstein, F. C. *et al.* (1977) The protein data bank: a computer-based archival file for macromolecular structures. *Journal of Molecular Biology*, **112**, 535–42.

Brown, I. D. (1983) The standard crystallographic file structure. *Acta Crystallographica*, **A39**, 216–24.

Garfield, E. (1974) *Current Contents*, **37**, (11 September).

Hawkins, D. P. (1980) Crystallographic literature: a bibliometric and citation analysis. *Acta Crystallographica*, **A36**, 775–482.

Von Laue, M., Friedrich, W. and Knipping, P. (1912) *Sitzungsberichte der Bayerische Akad. der Wissenshaften*, pp. 303–322 and 363–373.

CHAPTER NINE

Electricity, magnetism and electromagnetism*

TORE T. FJÄLLBRANT AND NANCY FJÄLLBRANT

Describing and understanding the behaviour of materials in terms of their electrical and magnetic qualities, and formulating underlying theories, is the main business of current research in the physics of electrical and magnetic phenomena. Since this activity is largely directed towards the possibility of applications, and since certain branches of electrical engineering are becoming increasingly theoretical, the difference between the two disciplines is gradually diminishing. The sources described below reflect this trend.

Electricity and magnetism are also intimately connected with other branches of physics, and separating them from solid state physics, quantum theory or electronics is well-nigh impossible. Therefore, the reader in search of information sources would do well to scan the other relevant chapters as well, particularly Chapters 10, 16, 18 and 19.

General reference

ABSTRACTS JOURNALS

The major abstracting services in English covering the whole of electricity and magnetism are: *Physics Abstracts* (PA), *Electrical and Electronics Abstracts* (EEA) and *Computer and Control Abstracts* (CCA). For further details see Chapter 2.

A corresponding publication which covers the whole field of engineering is the *Engineering Index Monthly* produced by

---

* based on the second edition chapter written by Judit Brody

Engineering Information, USA, with annual cumulations and four-year cumulations. The corresponding database COMPENDEX is available on a number of online hosts. There is also a CD-ROM version COMPENDEX PLUS from 1989 (quarterly updates), which uses the excellent DIALOG software. Other abstracting services are *Physics Briefs* (FIZ Karlsruhe and AIP), *Pascal Folio. Part F21. Électrotechnique* and *Pascal Explore. Part F20. Électronique et Télécommunications* (INIST, CNRS). The latter were previously published as *Bulletin Signalétique* part 140: *Électrotechnique*, mostly about electric power apparatus, and part 145: *Électronique*, about electromagnetic waves and telecommunications in addition to electronics. It covers all fields of physics and related topics.

The *PASCAL* folios can be searched on the PASCAL database, which is available on the QUESTEL, ESA-IRS and DIALOG online systems. There is a CD-ROM version PASCAL CD-ROM which is from 1987 onwards. *Physics Briefs* is available for online searching on STN International. There is considerable overlapping between *PA* and *Physics Briefs*.

The main sections of *Referativnyi Zhurnal*, corresponding to VINITI (Database of the All-Union Institute for Scientific and Technical Information) which should be mentioned here are *Fizika* and *Elektrotekhnika*. Abstracts of former Soviet and East European work can be found in English in *USSR Report* under the section for electronics and electrical Engineering (Joint Publications Research Service (JPRS)). The annual two-volume *Index to IEEE Publications* has a useful conference index and with its use the reader may retrieve publications otherwise difficult to find.

CURRENT AWARENESS PUBLICATIONS

Keeping up to date with the ever increasing amount of published literature is of great importance to most scientists and engineers, There are a number of tools specifically designed to help in this. Many of these are or will be available in electronic form. For current awareness, the bimonthly *Current Papers in Electrical and Electronics Engineering* and *Current Papers in Physics* (produced by INSPEC) can be consulted. These have a classified arrangement but carry no abstracts. Other important current awareness publications are the ISI (Institute of Scientific Information) Current Contents series *Current Contents: Engineering and Applied Sciences* and *Current Contents: Physical, Chemical and Earth Sciences* available both in weekly paper form or on diskettes. The latter facilitates searching on specific subject topics. Another current awareness product which covers the whole field of engineering, including electronics and electrical engineering is the CD-ROM PAGE 11 produced by Engineering Information, USA. This uses the DIALOG software and is updated monthly.

The *Digest of INTERMAG* contains photographically reproduced abstracts from the authors' copy and is published as a programme for the annual IEEE International Magnetics Conference.

HANDBOOKS, DICTIONARIES, DATA, ABBREVIATIONS

General reference works for the entire field of physics normally contain separate chapters or volumes on electricity and magnetism. *Encyclopedia of physics* edited by R. G. Lerner and G. L. Trigg (VCH, 2nd edn., 1991) is a recent reference work containing short articles on all aspects of physics, including those relating to electricity and magnetism. The *Encyclopedia of modern physics* edited by R. A. Meyers (Academic, 1990) is intended to provide a survey of the most rapidly advancing fields of theoretical and applied physics. It includes, for example, review articles on electrodynamics, ferromagnetism, plasma science, superconductivity mechanisms etc. The *Encyclopedia of applied physics* (VCH, 1991) edited by G. L. Trigg *et al.* is published in five volumes, vol. 3 has the title *Calibration and maintenance test and measuring equipment to collective phenomena of solids*. Flügge's *Handbuch der physik* – a systematic treatise of the whole field of physics – includes in the second edition published between 1956 and 1968 the following:

Volume 16: *Electric fields and waves* (1958)
Volume 17: *Dielectrics* (1956)
Volume 18: *Magnetism. Ferromagnetism* (1966–68)
Volumes 19–20: *Electrical conductivity* (1956–57)
Volumes 21–22: *Electron emission. Gas discharges.* (1956)
Volume 23: *Electrical instruments* (1967)

Additional material will be found in vol. 2 (parts 6,7,9 and 10) of Landolt, H. H. and Börnstein, R. *Zahlenwerte und Funktionen aus Naturwissenschaften und Technik – Numerical Data and Functional Relationships in Science and Technology* (6th edn., 1961–, Springer, Neue Serie) This series is bilingual and contains electrical and magnetic data of materials both in macroscopic and microscopic description.

The annual *CRC Handbook of Chemistry and Physics* (CRC, 1914–) the 'Rubber Bible' is an outstanding collection of data and tables and an essential volume for any physics laboratory. Kaye, G. W. C. and Laby, T. H. (1986) *Tables of physical and chemical constants, and some mathematical functions* (Longmans, 15th edn.) covers the whole range of physical and chemical data and includes a section 1.8 on electricity and magnetism. NBS (National Bureau of Standards) special publication 396–4: *Critical surveys of data sources: electrical and magnetic properties of metals* (NBS, 1976) is a detailed review of data sources available, covering mostly the 1960s.

Handbooks containing basic physical data meant primarily for the use of electrical and electronics engineers are: *Standard handbook for electrical engineers*, edited by D. G. Fink and H.W. Beaty (McGraw-Hill, 12th edn., 1987) and the companion *Electronics engineers' handbook*, edited by D. G. Fink and D. Christiansen (McGraw-Hill, 3rd edn., 1989). Other useful handbooks are the *Electronic engineer's reference book* (Butterworths, 6th edn., 1989), edited by F. Mazda, and the *Electrical engineer's reference book* (Butterworths, 14th edn., 1985), edited by M. A. Laughton and M. G. Say. The 15th edition, edited by G. R. Jones, M. A. Laughton and M. G. Say, is announced to be published in 1993. A data sourcebook is Graf, R. F. (1988) *Electronic databook* (TAB, 4th edn.).

Simple lists of terms in several languages, and those supplying definitions as well, are usually aimed at electrical and electronics engineers but also contain information for the physicist. *Elsevier's electrotechnical dictionary in six languages*, compiled by W. E. Clason (Elsevier, 1965), is still a standard work. More recent multi-lingual dictionaries are: *Dictionnaire CEI multilingue de l'électricité* (*IEC multilingual dictionary of electricity*) published in two volumes (CEI, 1983). Vol.1 contains entries and definitions of all terms in French followed by entries and definitions in English with equi-valents in Russian, German, Spanish, Italian, Dutch, Polish, and Swedish, and vol. 2 contains alphabetical indexes in each language. Luginsky, Y. N. *et al.* (1985) *Dictionary of electrical engineering* contains some 8000 terms in English followed by German, French, Dutch and Russian equivalents. P. Macura's *Russian–English dictionary of electrotechnology and allied sciences (with supplement)* (Krieger, 1986) is a reprint of an earlier 1971 edition with additional material, and P.-K. Budig's *Dictionary of electrical engineering and electronics* (Elsevier, 1985) are for those seeking information from Russian and German literature, respectively.

The *IEEE standard dictionary of electrical and electronics terms* (4th edn., 1988) is an authoritative collection of terms and defini-tions derived from the IEEE (and other) standards. In addition, it contains a list of abbreviations, symbols and acronyms. Wennrich, P. (1992) *International dictionary of abbreviations and acronyms of electronics, electrical engineering, computer technology and informa-tion processing* (Saur) provides an up to date reference tool. For those engaged in indexing or information work the *INSPEC thesaurus* (IEE, 1991) is invaluable.

There is an abundance of journals devoted to electricity and electronics; however, most of these are addressed to practical and theoretical engineering problems. Those listed below contain a significant amount of information on electricity, magnetism and electromagnetism which is relevant to physics.

IEE Proceedings (INSPEC) are currently divided into nine separate sections. Part *A Science, Measurement and Technology* covers physical science, measurement and instrumentation, management, education, and reviews; parts *B* to *I* deal with applications. Special issues appear irregularly, sometimes in addition to the bi-monthly issue.

The *Proceedings of the IEEE* (IEEE) publishes review and research papers on subjects of broad interest to the members of the Institute. The various professional groups and societies of the IEEE publish a whole series of journals and transactions; most have the occasional special issue devoted to a single topic only.

Archiv für Elektrotechnik – Archive of Electrical Engineering (Springer), bi-monthly, includes both electrical and electronics engineering. Papers are published in both German and English, all have both German and English titles and abstracts.

Electronics & Communications in Japan (Scripta) in three parts: I – Communications; II – Electronics; III – Fundamental Electronic Science. This journal contains translations of original papers from the *Transactions of the Institute of Electronics, Information and Communications Engineers of Japan*. Part 2 covers opto- and wave electronics, quantum and ultrasonics electronics, electronic materials and devices. Each part has 12 issues per year.

Soviet Journal of Communications Technology & Electronics (Scripta), formerly *Radio Engineering and Electronic Physics*, is a translation of the Russian journal *Radiotekhnika i Elektronika*. There are 16 issues per year.

Journals covering more specialized fields will be found listed under their appropriate subsection below.

There are a number of journals which deal with the teaching of physics and electrical engineering, for example *Physics Education* (IOP), *The Physics Teacher* and *The American Journal of Physics* (American Association of Physics Teachers), and the *International Journal of Electrical Engineering Education* (Manchester U. Pr.). These contain a fair amount of material on the teaching of electricity and magnetism at various levels. It is interesting to note that they contain a number of descriptions of software programs for teaching.

Another source of software for use in higher education is *The*

CHEST Software Directory (Bath U. Pr., 5th edn., 1992). (CHEST is the acronym for the Combined Higher Education Software Team). A general journal is *Computers & Education* (Pergamon). In the United Kingdom, the UK Computers in Teaching Initiative (CTI) has led to the production of a number of projects, such as the use of computers in the teaching of physics at the CTI Physics Centre at the University of Surrey.

Electromagnetism

GENERAL INTRODUCTORY TEXTBOOKS

There is a very wide choice indeed of textbooks on electricity and magnetism. This is not surprising in view of the fact that, generally, both science and engineering students are required to be familiar at least with Maxwell's equations. Most books are inductive in their approach, using empirical laws to lead up to Maxwell's equations, and some show how the reverse can be carried out.

A presentation of Maxwell's work is to be found in J. Hendry's monograph *James Clerk Maxwell and the theory of the electromagnetic field* (Hilger, 1986). This book focuses on Maxwell's theories and relates them to the current intellectual movements of the time. Another monograph which examines the historical development of electromagnetic theory is Buchwald, J. Z. (1985) *From Maxwell to microphysics. Aspects of electromagnetic theory in the last quarter of the nineteenth century* (Chicago U.Pr.). In this work, there is a comparison of the prevailing theories – the 'macroscopic' British model with the micromodels popular on the European continent. For sources on Maxwell the reader is referred to the *Catalogue of books and papers in electricity and magnetism belonging to the Institute for the History of Electricity (at Chalmers University of Technology)* by S. Ekelöf (Chalmers University of Technology, 1991).

The following short list is a selection of undergraduate textbooks, many containing exercises and problems with solutions. The list is arranged in alphabetical order by author:

Bleaney, B. I. and Bleaney, B. (1989) *Electricity and magnetism* (OUP, 3rd edn.).
Choudhury, M. H. (1989) *Electromagnetism* (Ellis Horwood). A coherent presentation suitable for final year undergraduates.
Christopoulos, C. (1990) *An introduction to applied electromagnetism* (Wiley). Emphasis is placed on the grasp of physical processes, the development of models for these processes, and practical applications.

Cottingham, W. N. and Greenwood, D. A. (1991) *Electricity and magnetism* (CUP). This provides a concise text which focuses on the theory and coherence of the Maxwell equations.

Di Bartolo, B. (1991) *Classical theory of electromagnetism* (Prentice-Hall). This book gives a basic background in classical electricity and magnetism to enable students to deal with experimental problems in electromagnetism and with the quantum mechanical treatments of radiation and matter.

Duffin, W. J. (1990) *Electricity and magnetism* (McGraw-Hill, 4th edn.). A concise work, one of the standard textbooks.

Grant, I. S. and Phillips, W. R. (1990) *Electromagnetism* (Wiley, 2nd edn.). This book is intended as a general course on electromagnetism for physics students. It includes examples illustrating how the laws of electromagnetism are applied to practical problems.

Kovetz, A. (1990) *The principles of electromagnetic theory* (CUP) in which electromagnetism is presented as a classical theory. This book shows the manner in which electromagnetism is linked to mechanics and thermodynamics.

Kraus, J. D. (1992) *Electromagnetics* (McGraw-Hill, 4th edn.). This presents the basic elements of electromagnetics in a series of steps from the simplest to more general cases.

Nasar, S. A. (1992) *2000 solved problems in electromagnetics* (McGraw-Hill). This is a volume in Schaum's 'Solved Problems' series.

Nayfeh, M. H. and Brussel, M. K. (1985) *Electricity and magnetism* (Wiley). This a basic textbook with some 300 detailed problem-examples.

Reitz, J. R., Milford, F. J. and Cristy, R. W. (1992) *Foundations of electromagnetic theory* (Addison-Wesley, 4th edn.).

The following books are specially written for the engineering student:

Balanis, C. A. (1989) *Advanced engineering electromagnetics* (Wiley). This emphasises the use of numerical methods.

Hammond, P. (1986) *Electromagnetism for engineers* (Pergamon, 3rd edn.) which provides the physical basis of electromagnetism for engineering students.

Hayt, W. H. (1989) *Engineering electromagnetics* (McGraw-Hill, 5th edn.). Introductory course based on the central theme of Maxwell's equations.

Solymar, L. (1984, rev. edn.) *Lectures on electromagnetic theory: a short course for engineers* (OUP) adopts a deductive approach, i.e. postulates Maxwell's equations. Written in a lively style and adorned with quotations by famous scientists.

Having mastered the basic theory, the student can progress either to a theoretical treatment of electromagnetic fields and waves or to applications:

Bladel, J. van (1985) *Electromagnetic fields* (Hemisphere). This book deals with the calculation of electric and magnetic fields in the presence of ponderable bodies at rest.

Bladel, J. van (1991) *Singular electromagnetic fields and sources* (Clarendon). The main emphasis is on the discussion of the various 'infinities' which occur in electromagnetic fields and sources.

Cheng, D. K. (1989) *Field and wave electromagnetics* (Addison-Wesley, 2nd edn.). This a basic textbook for undergraduate students.

Davis, J. L. (1990) *Wave propagation in electromagnetic media* (Springer). This is the second of two volumes on the phenomena of wave propagation in non-reacting and reacting media. The book is concerned with wave-propagation in reacting media and in particular electromagnetic materials.

Hirose, A. and Lonngren, K. E. (1985) *Introduction to wave phenomena* (Wiley). This is a textbook which presents a unified approach to wave phenomena and includes both sound and electromagnetic waves.

Lax, P. D. and Phillips R. S. (1989) *Scattering theory* (Academic, rev. edn.). This monograph treats the Lax-Phillips scattering theory based on the wave equation and examines results from the last 20 years. Bibliographical references are included.

Lorrain, P., Corson, D. R. and Lorrain, F. (1988) *Electromagnetic fields and waves* (Freeman, 3rd edn.). A comprehensive textbook.

Ogilvy, J. A. (1991) *Theory of wave scattering from random rough surfaces* (Hilger). This is appropriate for graduates familiar with techniques used in wave scattering theory.

Read, F. H. (1980) *Electromagnetic radiation* (Wiley). Includes both classical and quantum electrodynamics.

Sander, K. F. and Reed, G. A. L. (1986) *Transmission and propagation of electromagnetic waves* (CUP, 2nd edn.). This work provides the necessary knowledge of electromagnetic waves to facilitate application to line, waveguide and radio systems and optical transmission.

Tiersten, H. F. (1990) *A development of the equations of electromagnetism in material continua* (Springer). This book provides an introduction to the interaction of the electric and magnetic fields with deformable solid criteria.

Yamashita, E. (ed.) (1990) *Analysis methods for electromagnetic wave problems* (Artech). The purpose of this book is to cover basic analysis methods now in use for analysing electromagnetic wave problems.

PERIODICALS

The following journals are of interest in this field:

Electromagnetics (Hemisphere, 1981–) published quarterly, accepts papers on all aspects of the subject.

IEEE Transactions on Electromagnetic Compatibility deals with all aspects of electromagnetic compatibility, including standards, measurement and undesired sources.

IEEE Translation Journal on Magnetics in Japan, published monthly under the auspices of the IEEE Magnetics Society, covers Japanese research and development activities within many fields such as soft magnetic materials, bubbles and other memory technologies, microwave magnetics, permanent magnet materials and technologies, magnetic materials – properties and processing and theory of the magnetization process.

Journal of Electromagnetic Waves and Applications (VSP, 1987–) aims to cover all aspects of electromagnetic wave theory and its various applications. It includes both original papers and review articles.

Radio Science (AGU)

COMPUMAG, the Conference on the Computation of Magnetic Fields, is published in the *IEEE Transactions on Magnetics*. The most recent conference in this series was held in St Louis in April 1992. Other conference papers are published in the same journal, for example the Twelfth International Conference on Magnet Technology, Leningrad, June 1991; the Fourth Biennial Conference on Electromagnetic Field Computation, Toronto, October 1990.

The Electromagnetic Wave Propagation Panel of the NATO Advisory Group for Aerospace Research and Development (AGARD) holds regular meetings, and these appear in the *AGARD Conference Proceedings*.

A series of monographs is the IEE Electromagnetic Wave Series (Peregrinus), each dealing with a specific topic.

In many areas of solid state physics the use of synchrotron radiation as a powerful source for polarized uv radiation between 5 eV and about 1500 eV has produced many interesting results. Synchrotron radiation has been used in more than 20 laboratories throughout the world, including CERN and the Stanford Linear

Accelerator Center (SLAC). A book which describes the development of the Stanford Synchrotron Radiation Laboratory (SSRL) and applications of synchrotron radiation is *Synchrotron radiation research*, edited by H. Winick and S. Doniach (Plenum, 1980). The proceedings of a workshop held at the Centro Stefano Francscini, Ascona, Switzerland, in 1990 *Synchrotron radiation: selected experiments in condensed matter physics*, edited by W. Czaja (Birkhäuser, 1991), describes synchrotron radiation in relation to magnetic properties, electronic structure, interfaces and crystal structure determination. *Application of synchrotron radiation*, edited by C. R. A. Catlow and G. N. Greaves (Blackie, 1990), describes recent applications . Two recent volumes on *Synchrotron radiation research*: vol. 1. *Techniques* and vol. 2. *Issues and technology* have been published by Plenum in 1992.

The proceedings of the series of National Conferences on Synchrotron Radiation Instrumentation are published in the journal *Nuclear Instruments and Methods in Physics Research. Section A: Accelerators, Spectrometers, Detectors and Associated Equipment.* A recent conference dealing with *Light, lasers and synchrotron radiation: a health risk assessment* (eds. M. Grandolfo, A. Rindi and D. H. Sliney) was published in the NATO Applied Science Institute Series (ASI) in 1991.

Electrostatics and magnetostatics

Electrostatics has been described as the study of phenomena recognized by the presence of stationary or moving electrical charges and the interactions due to the existence of these charges but not to their motion. The difference between electrostatics and other branches of electromagnetism is the absence of magnetic effects. Electrostatics is of considerable importance in the design and analysis of a wide range of applications ranging from the copying machine to nuclear fusion reactors.

The quarterly *Journal of Electrostatics* (Elsevier) was brought out in 1975. The series *International Conference on Electrostatic Phenomena* is published by the Institute of Physics in its conference series. The most recent conference in this series is *Electrostatics 1991: invited and contributed papers from the eighth international conference held at the University of Oxford, 10–12 April, 1991* (edited by B. C. O'Neill).

A classic work on electrostatics is the three-volume *Electrostatique*, by E. Durand (Masson, 1964–1966).

Crowley, J. M. (1986) *Fundamentals of applied electrostatics* (Wiley) gives a comprehensive introduction to electrostatics. The book consists of four parts: electrostatic fields, electrostatics of particles, electrostatics of materials, and electrostatics of circuit elements. There is an epilogue on areas of application. A bibliography is included in each chapter. The book is suitable for both undergraduate and postgraduate students, as well as practising engineers. J. Cross (1987) *Electrostatics: principles, problems and applications* (Hilger) provides an alternative text. Another textbook at the postgraduate level, is Smythe, W. R. (1989) *Static and dynamic electricity* (Hemisphere, 3rd rev. edn.).

Since electronic equipment was first developed, static electricity has presented problems for users and designers. In the last few years electrostatic discharge (ESD) has become one of the most insidious and dangerous problems in electronic systems. This is because newer devices, such as semiconductor integrated circuits, are more susceptible to ESD than older equipment such as vacuum tubes. This is a particularly important area of knowledge for the design of reliable systems. There are a number of recent books on this topic:

Boxleitner, W. (1989) *Electronic discharge and electronic equipment: a practical guide for designing to prevent ESD problems* (IEEE). This book explains how ESD is generated, how it affects electronic equipment, how to design equipment to prevent ESD problems and methods of testing for ESD problems.

McAteer, O. J. (1990) *Electrostatic discharge control* (McGraw-Hill) describes the principles of static generation, material variables, induction, resistivity, fields, potential, energy, capacitance and capacitive coupling, and the various aspects of ESD control and management. Electrostatic phenomena exert a major influence on the design of spacecraft, and regular conferences are held on how to avoid the hazard of spacecraft charging.

In the series 'Progress in Aeronautics and Astronautics' there is the following volume (viii): Fredrickson, A. R. *Spacecraft dielectric material properties and spacecraft charging* (AIAA). For magnetostatics, see E. Ducand (1968) *Magnétostatique* (Masson).

Electric properties of materials in the condensed state

The literature on physics of semiconductors is presented by R. J. Nicholas in Chapter 18.

Dielectrics

The physics of dielectrics has developed as a branch of solid state physics and deals with the properties of dielectric materials, that is, substances which have exceptionally low conductivities. The imposition of an external electric field on a dielectric (as named by Faraday in 1839) is 'the production of a polarized condition of the material arising from the perturbation of the constituent charges. This usually occurs without significant transport of charge through the substance' (Scaife, 1989). Dielectrics can exist in solid, liquid or gaseous states. There is, at present, considerable interest in the applications of dielectrics in devices and components in the electrical and optical industries.

PERIODICALS

Papers encompassing all aspects of current research are published in the bi-monthly *IEEE Transactions on Electrical Insulation* (IEEE Electrical Insulation Society).

There are a number of conferences which deal with various aspects of dielectrics, for example:

IEEE Conference on Electrical Insulation and Dielectric Phenomena
IEEE International Conference on Conduction and Breakdown in Solid Dielectrics

Another series of conference proceedings (triennial), published by the Institution of Electrical Engineers (IEE) is: *International Conference on Dielectric Materials, Measurements and Applications*. These started in 1970. *The Digest of Literature on Dielectrics* has been published annually since 1947 by the US National Academy of Sciences.

BOOKS

A recent book by B. K. P. Scaife (1989) *Principles of dielectrics* (Clarendon) provides a detailed introduction to the basic principles of the theory of dielectrics. This work is at the level of final year undergraduates in the physical sciences. There is a general bibliography as well as a comprehensive list of references.

The monographs that follow discuss various aspects of dielectric physics:

Blythe, A. R. (1979) *Electrical properties of polymers* (CUP).
Bunget, I. and Popescu, M. (1984) *Physics of solid dielectrics* (Elsevier) which deals with the fundamental problems of solid dielectrics.
Goodman, C. H. L. (ed.) (1980) *Physics of dielectric solids* (IOP).
Hasted, J. B. (1973) *Aqueous dielectrics* (Chapman and Hall).
Herbert, J. M. (1985) *Ceramic dielectrics and capacitors* (Gordon and Breach).

Keldysh, L. V., Kirzhnitz, D. A. and Maradudin, A. A. (eds.) (1989) *The dielectric function of condensed systems* (North-Holland). This is volume 24 in the series 'Modern Problems in Condensed Matter Sciences'.

Nelson, D. F. (1979) *Electric, optic and acoustic interactions in dielectrics* (Wiley) treats a broad segment of research on macroscopic physics of dielectrics from a unified and deductive point of view and thus combines the attributes of a research monograph with those of a textbook.

O'Dwyer, J. J. (1973) *The theory of electrical conduction and breakdown in solid dielectrics* (Clarendon) presents the theory and experimental work to illustrate some applications of the theory.

Pohl, H. A. (1978) *Dielectrophoresis* (CUP) describes the behaviour of neutral matter in non-homogeneous electric fields.

Properties of dielectrics

Piezoelectricity, the linear coupling between electrical, mechanical and thermal states, is produced by applying mechanical pressure to dielectric crystals. Of the 32 crystal classes, 20 demonstrate piezoelectricity, and of these, 10 are characterized by permanent polarization in the absence of an applied electric field. Called spontaneous polarization, this frequently cannot be detected by charges on the surface. However, in these pyroelectric (or polar) materials, polarization is temperature dependent and that is observable.

A subclass of the pyroelectric materials is ferroelectrics, that is, their polarization can be changed by an electric field. Since the 1950s the field of pyroelectric and ferroelectric studies has grown rapidly, stimulated by interest in applications. Ferroelectric liquid crystals possess properties that give them potential for commercial applications in flat panel displays, and in optical processing and computing. Data on ferroelectrics is published in the Landolt-Börnstein *Numerical Data and Functional Relationships in Science and Technology,* new series, group III: *Crystal and solid state physics.*

PERIODICALS

There are a number of conference series that deal with the properties of dielectrics:

International Conference on Dielectric Materials, Measurements and Applications, which is published as an IEE conference publication. The sixth conference in this series was held at Manchester in 1992.

International Conference on Properties and Applications of Dielectric Materials, sponsored by the IEEE Dielectrics and Electrical Insulation Society. The third conference in this series was held in Tokyo, Japan, July 8–12, 1991.

IEEE International Symposium on Electrets sponsored by the IEEE Dielectrics and Electrical Insulation Society. The seventh symposium was held in Berlin in 1991.

Ferroelectrics (Gordon and Breach) is an international journal devoted to the theoretical, experimental, and applied aspects of ferroelectrics and related materials. It regularly contains bibliographies and literature guides. Since 1982 it has had a separate *Ferroelectrics Letters Section. Ferroelectrics* now carries the proceedings of several sequences of conferences such as the International Meeting on Ferroelectricity, while the Japanese International Meetings on Ferroelectric Materials are reported in the *Japanese Journal of Applied Physics.*

Another conference series is the *IEEE International Symposium on Applications of Ferroelectrics.* Another journal which includes articles on ferroelectrics is *IEEE Transactions on Ultrasonics, Ferroelectrics and Frequency Control.* This is published bimonthly.

Dielectric properties are well covered bibliographically. Bibliographies are to be found in the journal *Ferroelectrics.* There is also the multi-volume *Solid State Physics Literature Guides*, edited by T. F. Connolly, with its volume 1: *Ferroelectric materials and ferroelectricity* (1970) and a continuation as volume 6: *Ferroelectrics literature index* (1974). *Sourcebook of pyroelectricity* by S. B. Lang (Gordon and Breach, 1974) describes the fundamentals of dielectric properties and includes a substantial bibliography.

BOOKS

An important series is 'Ferroelectricity and Related Phenomena' (Gordon and Breach) with:

Vol. 1. *An introduction to the physics of ferroelectricity* (1970)
Vol. 2. *Sourcebook of pyroelectricity* (1974)
Vol. 3. *Ferroelectrics and related materials* (1984)
Vol. 4. *Piezoelectricity* (1985)
Vol. 5. *Medical applications of piezoelectric polymers* (1988)
Vol. 6. *Defects and structural phase transitions* (1988)
Vol. 7. *Ferroelectric liquid crystals: principles, properties and applications* (1990)
Vol. 8. *The photovoltaic and photorefractive effects in noncentrosymmetric materials* (1992)

The following is a selected list of monographs:

Burfoot, J. C. and Taylor, G. W. (1979) *Polar dielectrics and their applications* (Macmillan).

Hilczer, B. and Malcki, J. (1986) *Electrets* (Elsevier). This book provides a wide-ranging review of the fundamental properties of dielectrics in general and electrets in particular.

Rosen, C. Z., Hiremath, B. V. and Newnham, R. (eds.) (1992) *Piezoelectricity* (AIP). This is a collection of 33 review papers which examine the physics of piezoelectricity. It provides a guide to defining the criteria for preparing and choosing piezoelectric materials and in designing signal processing devices.

Sessler, G. M. (ed.) (1987) *Electrets* (Springer, 2nd edn.) deals with the phenomenon named by Heaviside. The term electret refers to a dielectric exhibiting a quasi-permanent electric charge which can be either a real charge, a polarization charge or a combination of both. Strictly speaking, all piezoelectric materials are electrets; this book excludes non-polymeric piezoelectrics.

Xu, Y. (1991) *Ferroelectric materials and their applications* (North-Holland). This book provides a very useful introduction to ferroelectric materials. It presents the basic physical properties and structures of ferroelectric materials, and describes methods of synthesis and how the ferroelectrics could be of use in numerous practical devices. The major emphasis is placed on oxide ferroelectrics.

Superconductivity

Superconductors are materials possessing strange electric and magnetic properties which allow them to become perfect conductors of electricity under certain conditions such as chilling to extremely low temperatures. Superconductivity was discovered by Kamerlingh Onnes of Leiden University, Holland, in 1911, in connection with his work on the liquefaction of helium. This was followed by much research and attempts to explain superconductivity within the current theories of physics. Superconductivity at temperatures above the liquid helium range were first shown by Ascherman, Friedrich, Justi and Kramer of Germany, in 1941, for niobium nitride, which was superconducting at about 15K. The field of superconductivity started to grow extremely rapidly after the remarkable discoveries of Johannes Georg Bednorz and Karl Alex Müller, at the IBM Zürich Laboratory, in April 1986, of 'high-temperature' superconductors where the maximum critical temperatures increased from 23K to 90K and higher. Bednorz and Müller were awarded the Nobel Prize in October 1987. Their discoveries led to an enormous amount of research and the collection of experimental data on materials that exhibit superconductivity. This resulted in a corresponding rapid growth in publication. It has been estimated that some 4000 papers were published on high-temperature superconductivity during 1987–1988 and that some 17 000 papers were published up to the end of 1990. The field of high-temperature superconductivity is one of the most rapidly growing research areas in physics, but there is as yet no generally accepted theoretical explanation for this behaviour. Literature on the purely cryogenic and quantum mechanical aspects of superconductors will be found in Chapters 11 and 17.

BOOKS

An excellent review of the historical roots and development of superconductivity is given by P. F. Dahl *Superconductivity. Its historical roots and development from mercury to the ceramic oxides* (AIP, 1992). This describes the early years of superconductivity from 1911 to the Meissner effect in the mid-war years and the materials development in the 1950s. An eminently readable account of research on high-temperature superconductors in the late 1980s is that of R. M. Hazen *The breakthrough.The race for the superconductor* (Summit, 1988). Bob Hazen, who worked as a physicist at the Geophysical Laboratory, Washington, D.C. gives an insider's view of the scientific process and the intense research efforts involved in the search for high-temperature superconductor materials. A clear general introduction can be found in the *Engineer's guide to high-temperature superconductivity* by J. D. Doss (Wiley, 1989). This provides a brief history of superconductivity, together with a presentation of fundamental considerations on the phenomena of superconductivity and developments in high-temperature superconductivity. Applications are discussed in some detail. There is an extensive selected bibliography covering books, monographs and conference proceedings.

A considerable number of textbooks on superconductivity have been, understandably, published during the last few years. These include *Fundamentals of superconductivity* by V. Z. Kresin and S. A. Wolf (Plenum, 1990). This focuses on qualitative aspects of superconductivity and is aimed at a wide group of readers. A short, yet comprehensive, introduction to the experimental aspects of superconductivity is provided in *Superconductivity, fundamentals and applications* by W. Buckel (VCH, 1991) based on an earlier German edition. An interesting series of 10 superconductor applications is described in *Superconducting technology. 10 case studies*, edited by K. Fossheim (World Scientific, 1991). A description of similarities in superfluidity and superconductivity is given by D. R.

Tilley and J. Tilley in *Superfluidity and superconductivity* (Hilger, 3rd edn., 1990). This is aimed at the graduate student level. A recent introduction is *High-temperature superconductivity: an introduction* by G. Burns (Academic, 1992). This provides a concise overview of the field of high-temperature superconductivity focussing both on experimental results and theoretical issues.

An overview and comparison of organic superconductors and oxide superconductors is to be found in *The physics and chemistry of organic superconductors*, edited by G. Saito and S. Kagoshima (Springer, 1990). The superconducting properties of organic super-conductors (synthetic organic metals) are surveyed in *Organic superconductors (including fullerenes) – synthesis, structure, properties and theory* by J. M.Williams *et al.* (Prentice Hall, 1992).

PERIODICALS

There are now many periodicals which contain articles on super-conductivity. The following is a selection:

Applied Physics Letters (AIP)
Applied Superconductivity (Pergamon, 1993–)
IEEE Transactions on Applied Superconductivity (IEEE, 1991–)
Journal of Low Temperature Physics (Plenum, 1969–)
Journal of Superconductivity (Plenum, 1988–)
Physica. C: Superconductivity (North-Holland, vol. 152-, 1988–)
Physical Review B (AIP)
Superconductivity Review (Gordon and Breach,1992–)
Superconductor Science & Technology (Institute of Physics, 1988–)

The multi-volume series *Progress in Low Temperature Physics* (North-Holland) contains articles on superconductivity. So do the proceedings of the conference series: *International Conference on Low Temperature Physics*. Proceedings of the Applied Super-conductivity Conference appear in the *IEEE Transactions on Magnetics*.

The scientific review provides a valuable starting point for research workers and graduate students. Two in the NATO Advanced Study Institute series, which, taken together, represent a detailed review of the physical foundations of superconducting tech-nology are S. Foner and B. B. Schwartz (eds.) *Superconducting machines and devices – large systems applications* (Plenum, 1974) and B. B. Schwartz and S. Foner (eds.) *Superconductor applications: SQUIDS and machines* (Plenum, 1977). A third NATO Advanced Study Institute by the same editors is *Superconductor materials science* (Plenum, 1981).

An important series is the Proceedings of the NATO Advanced

Study Institute on High Temperature Superconductors – Physics and Materials Science – *Physics and Materials Science of High Temperature Superconductors*. I – Bad Winsheim, FRG, August, 1989 and II – Porto Carras, Greece, August, 1991.

Annual conferences on *Superconductivity and Applications* have been organized by the New York State Institute on Superconductivity, from 1987 onwards.

A very useful series devoted to the *Physical properties of high temperature superconductors* has been edited by D. M. Ginsberg and published by (World Scientific: vol. I 1989, vol. II 1990, vol. III, 1992). These volumes contain comprehensive authoritative reviews, and include many references.

ELECTRONIC NEWSLETTERS AND DATABASES

High-Tc Update is the high-Tc superconductivity information electronic exchange newsletter. This is published for the Division of Materials Sciences, Office of Basic Energy Sciences, USDOE. This electronic newsletter is available over BITNET and the Internet. Enquiries and contributions should be addressed to the editor Dr Ellen Feinberg at: feinberg@alisuvax.bitnet or feinberge@vaxld.ameslab.gov.

The Westinghouse STC database of high-Tc superconductivity references contains source and keyword references for some 7000 papers published from 1987 to 1990. This database exists in both Macintosh and IBM-compatible versions for downloading, and comes complete with a search program. For details consult the *High-Tc Update* editor. (Note the database has been published in two parts in the *Journal of Superconductivity* vol. 2 (1), pp. 1–210 (1989) and vol. 3 (1), pp. 1–155 (1990).

Thermoelectricity

A potential difference can be created by subjecting an electrical conductor to a temperature gradient. Applications for electric power generation will be found in the section below entitled 'Production and storage'.

A special publication of the American Society for Testing and Materials (ASTM) by D. D. Pollock *Thermoelectricity: theory, thermometry, tool* (ASTM, 1985) treats the subject of thermoelectricity from both theoretical and practical aspects. The book explains the theory and bases for thermoelectric materials in general and those materials commonly used for thermoelectric purposes at elevated temperatures in particular. References are given after each chapter.

A recent conference on thermoelectrics was held in Annaheim, California in May 1991. This resulted in the publication *Modern perspectives on thermoelectrics and related materials*, edited by D. D. Allred, C. B. Vining and G. A. Slack (Materials Research Society, 1991).

Magnetohydrodynamics

Magnetohydrodynamics (MHD) is concerned with the behaviour of electrically conducting fluids in a magnetic field. Dynamo theory, which is a branch of MHD, explains the generation and maintenance of magnetic fields by these fluids. MHD ranges from phenomena concerned with liquid metals or ionized gases to the behaviour of the mass of conducting matter which make up planets or stars. In the last few years diverse applications of MHD have become increasingly important in geophysics, space physics and astrophysics.

A recent introductory text is *Fundamentals of magnetohydrodynamics* by R. V. Polovin and V. P. Demutskii (Consultants/Plenum, 1990). This introduces the reader to the basic principles of magnetohydrodynamics and the methods used in MHD. The book is mainly theoretical and contains an extensive subject index. Another textbook on MHD is *Ideal magnetohydrodynamics* by J. P. Freidberg (Plenum, 1987). This provides a detailed introduction to ideal magnetohydrodynamics. The treatment has a theoretical emphasis.

Other monographs on magnetohydrodynamics include:

Ghil, M. (1987) *Topics in geophysical fluid dynamics: atmospheric dynamics, dynamo theory, and climate dynamics* (Springer).
Krause, F. and Rädler, K.-H. (1980) *Meanfield magnetohydrodynamics and dynamo theory* (Pergamon) which gives a systematic introduction and general survey of results.
Lifschnitz, A. E. (1988) *Magnetohydrodynamics and spectral theory* (Kluwer). (Included in the series 'Developments in Electromagnetic Theory and Applications').
Moreau, R. (1990) *Magnetohydrodynamics* (Kluwer). This book is a cross between a course and a monograph. It starts by introducing the essential principles and equations of MHD followed by a brief analysis on the influence of movement in a magnetic field and analyses of various flow regimes subjected to the influence of a magnetic field.

The use of MHD in electrical power generation is beyond the scope of this book (see below in the section on 'Production and storage').

The journal *Magnetohydrodynamics: An International Journal* was published by Hemisphere starting in 1989. The Beer-Sheva Seminar on *MHD Flows and Turbulence* is held every three years in Israel. Other recent conferences include:

Lielpeteris, J. and Moreau, R. (eds.) (1989) *Liquid metal magneto-hydrodynamics* (Kluwer). This is a collection of papers from an IUTAM symposium held in Riga, Latvia.

Priest, E. R. and Hood, A. W. (eds.) (1991) *Advances in solar system magnetohydrodynamics* (CUP). This is a collection of the papers presented at the Symposium on Geophysical and Astrophysical MHD – held in St Andrews in 1990.

Ionization and discharge

Discharges

Journals that treat this subject are:

Journal of Vacuum Science and Technology. A. Vacuum, surfaces, and films

Journal of Vacuum Science and Technology. B. Microelectronics processing and phenomena

Vacuum, which is the international journal and abstracting service for vacuum science and technology.

There are a number of conferences dealing with discharges, for example: International Symposium on Discharges and Electrical Insulation in Vacuum which is held biennially (14th, Santa Fe, New Mexico, Sept. 1990; 15th, Dormstadt, Sept. 1992) and International Conference on Gas Discharges and their Applications.

Monographs on arcs and sparks in vacuum and in gases and applications are:

Chapman, B. (1980) *Glow discharge processes: sputtering and plasma etching* (Wiley).

Hirsch, M. N. and Oskam, H. J. (eds.) (1978–) *Gaseous electronics, vol. 1 Electrical discharges* (Academic). This series is planned to encompass a dozen volumes.

Lafferty, J. M. (ed.) (1980) *Vacuum arcs* (Wiley) describes research and development on the high-power vacuum circuit interrupter.

Latham, R. V. (1981) *High-voltage vacuum insulation: the physical basis* (Academic) gives a concise account of the basic physical concepts and is directed towards the researcher and the development engineer.

Meek, J. M. and Craggs, J. D. (eds.) (1978) *Electrical breakdown of gases* (Wiley, 2nd edn.) includes laser-induced breakdown but excludes lightning discharges.

Plasmas

The name 'plasma', first used by I. Langmuir in 1928, describes a collection of electrically energized matter in a gaseous state where the positive and negative charges are nearly equally balanced. In a plasma, the energy of the particles is so great that the electric forces which bind the atomic nucleus to its electrons are overcome. The properties of plasma are so distinctive that it is sometimes called the 'fourth state' of matter. Most of visible matter in the universe exists as plasma. Lightning and the aurora are the only natural manifestations of the plasma state on Earth. The collection of positively and negatively charged particles can be thought of as a gas, but the waves that occur in it have much in common with waves in electrically conducting solids and its bulk behaviour can be described using terminology and equations developed for fluids. Plasma can be produced by electric discharges in a gas but also by intense laser radiation, chemical processes or heating of matter to high temperatures. Completely ionized plasmas play a central role in fusion research.

PERIODICALS

The main journals are:

IEEE Transactions on Plasma Science (IEEE Nuclear and Plasma Sciences Society), quarterly.
Journal of Plasma Physics (CUP), bi-monthly.
Contributions to Plasma Physics (Akademie), bi-monthly, articles are in English, French, German, and Russian.
Nuclear Fusion (IAEA).
Physics of Fluids B: Plasma Physics (AIP).
Plasma Physics and Controlled Fusion (Pergamon).

Several series of conferences are held and published in diverse formats: the European Physical Society Plasma Physics Division Conference; International Conference on Phenomena in Ionized Gases; International Conference on Plasma-Surface Interactions in Controlled Fusion Devices; the International School of Plasma Physics; and the International Conference on Plasma Physics and Controlled Nuclear Fusion: this last one appears as a special supplement volume to the journal *Nuclear Fusion* (IAEA). The latter also contains other special supplements *World Survey of Activities in Controlled Fusion Research* of which the most recent was published in 1991.

The *Handbook of plasma physics* (general editors M. N. Rosenbluth and R. Z. Sagdeev) published by North-Holland consists so far of three volumes: vol. 1 *Basic plasma physics I* (1983), vol. 2 *Basic plasma physics II* (1984) and vol. 3 *Physics of laser plasma* (1991). The first two volumes are devoted to fundamental plasma physics. Subsequent volumes are to be devoted to more specialized topics.

BOOKS

Plasma physics is an important area of current research, with a prolific literature in several of its aspects. The following is a selected list:

Birdsall, C. K. and Langdon, A. B. (1991) *Plasma physics via computer simulation* (Hilger). This book is on particle simulation of plasmas and is aimed at developing insight into the essence of plasma behaviour.

Bittencourt, J. A. (1986) *Fundamentals of plasma physics* (Pergamon). This is a general introduction presenting a comprehensive logical and unified treatment of the fundamentals of plasma physics based on statistical kinetic theory.

Cairns, R. A. (1985) *Plasma physics* (Blackie). This book describes some of the most important and characteristic properties of a plasma, at a level suitable for advanced undergraduates or postgraduate students. The range of problems discussed illustrates how different fluid and kinetic levels of description are appropriate in various circumstances. Application of the results to the problem of controlled nuclear fusion is emphasized throughout the text.

Chandrasekhar, S. (1989) *Plasma physics, hydrodynamic and hydromagnetic stability, and applications of the tensor-virial theorem. Selected papers. Vol. 4.* (Univ. of Chicago). Part 1 of this collection of papers is on plasma physics and belongs to the period of Chandra's work on hydrodynamic and hydromagnetic stability.

Dendy, R. O. (1990) *Plasma dynamics* (Clarendon). This book explains the fundamental concepts of plasma physics and provides an introduction to plasma particle dynamics, plasma waves, magnetohydrodynamics, plasma kinetic theory, two-fluid theory and non-linear plasma physics. Emphasis is placed on the underlying physical principles.

Ichimara, S. (1986) *Plasma physics: an introduction to statistical physics of charged particles* (Benjamin). This is an investigation into the fundamental principles of plasma phenomena making use of basic statistical physics.

Ichimaru, S. (1992) *Statistical plasma physics* (Addison-Wesley). This is no. 87 in the series Frontiers in Physics.

Swanson, D. G. (1989) *Plasma waves* (Academic). This is an advanced text for graduate students.

Aspects of plasma interactions are described in:

Baumgärtel, K. and Sauer, K. (1987) *Topics on non-linear wave-plasma interaction* (Birkhäuser).

Bobin, J. L. (1985) *High intensity laser plasma interaction* (North-Holland).

Hora, H. (1991) *Plasmas at high temperature and density: applications and implications of laser-plasma interaction* (Springer).

For plasma instabilities, see the following:

Cap, F. F. (1976–1982) *Handbook on plasma instabilities*, 3 vols (Academic). This gives a comprehensive picture with an extensive bibliography.

Mikhailovskii, A. B. (1992) *Electromagnetic instabilities in an inhomogeneous plasma* (IOP). The aim of this book is a comprehensive presentation of the theory of electromagnetic instabilities in a magnetized inhomogeneous plasma.

Sitenko, A. G. (1982) *Fluctuations and non-linear wave interactions in plasmas* (Pergamon) is a translation of the 1977 Russian edition.

Plasmas for fusion are discussed in the following:

Chen, F. F. (1984) *Introduction to plasma physics and controlled fusion*, 2 volumes. (Plenum, 2nd edn.).

Miyamoto, K. (1989) *Plasma physics for nuclear fusion* (MIT, rev. edn.).

Nishikawa, K. and Wakatani, M. (1990) *Plasma physics: basic theory with fusion applications* (Springer).

Shaw, E. N. (1990) *Europe's experiment in fusion: the JET joint undertaking* (North-Holland).

Tajima, T. (1989) *Computational plasma-physics: with applications to fusion and astrophysics* (Addison-Wesley) This is no. 72 in the series Frontiers in Physics. Each chapter contains problems and references.

Wesson, J. (1987) *Tokamaks* (Clarendon).

Magnetic properties

Magnetism

Current research on magnetism has the double aspect of finding out about the magnetic properties of materials and of formulating a theoretical framework. Titles of books and conferences often reflect this duality. Research in magnetism has evolved from the more practical applications into a testing ground for advanced statistical physics. Magnetic fields have, in one way or another, been involved in many of the important discoveries in modern physics.

JOURNALS AND CONFERENCE PROCEEDINGS

The important general journals on magnetism are:

IEEE Transactions on Magnetics (IEEE Magnetics Society), bi-monthly
Journal of Magnetism and Magnetic Materials (North-Holland)
IEEE Translation Journal on Magnetics in Japan.

It is easy to get lost in the maze of the various different congresses, conferences and symposia on magnetism. Occasionally, parts of the same conference are published in several different journals, some appear individually as books, some join forces in certain years and many have regular name changes. The International Conference on Magnetism (ICM) is sometimes referred to as the International Congress on Magnetism. The proceedings of the 1991 Conference (ICM '91), held in Edinburgh in September 1991, was published in 1992 in the *Journal of Magnetism and Magnetic Materials* as four volumes (104–107). Then there is the International Magnetics Conference (INTERMAG) which is published in the journal *IEEE Transactions on Magnetics*. The proceedings of the Thirty-Fifth Annual Conference on Magnetism and Magnetic Materials (MMM), held in San Diego in 1990, was published in the *Journal of Applied Physics* (1991) 69 (8). There is also the Joint Magnetism and Magnetic Materials – International Magnetics Conference, the fifth of which was held in Pittsburgh in 1991 and published partly in the *IEEE Transactions on Magnetics* and partly in the *Journal of Applied Physics*, both journals containing a complete table of contents. The joint IMs are supposed to take place when the ICM is held outside the American continent. The *IEEE Transactions on Magnetics* includes proceedings of other conferences, for example the Twelfth International Conference on Magnet Technology (MT), held in Leningrad in June 1991, which was published in the January 1992 issue. The *Journal of Magnetism and Magnetic Materials* also publishes other conference proceedings, for example, the proceedings of the Tenth International Conference on Soft Magnetic Materials and selected papers from the International Colloquium on Magnetic Fields and Surfaces, were published in vols. 112 and 113, in 1992. Another journal that publishes proceedings of conferences on magnetism is *Physica B+C* (Elsevier).

BOOKS

A comprehensive introduction to magnetism and magnetic materials is given in *Introduction to magnetism and magnetic materials* by D. Jiles (Chapman & Hall, 1991). This book covers the following areas: magnetic fields, magnetization, magnetic properties, magnetic measurements, magnetic domains and domain walls, domain processes, magnetic order within the domains, electronic magnetic moments, quantum theory of magnetism, soft magnetic materials, hard magnetic materials, magnetic recording technology, superconductivity and materials evaluation. The chapters contain a number of problems (and their solutions). Extensive references to the principal publications in magnetism are given at the end of each chapter. The quantum theoretical model is developed in *Quantum theory of magnetism* by R. M. White (Springer, 2nd edn., 1983).

The subject of surface magnetism is now rapidly evolving as a field of solid state physics. An introduction to this field is provided by the book *Introduction to surface magnetism* by T. Kaneyoshi (CRC, 1991). A number of the unsolved problems in this area are posed and a good selection of references is included.

Magnetic properties of layered transition metal compounds, edited by L. J. de Jongh (Kluwer, 1990), provides a survey of the main trends in two-dimensional magnetism research.

A multi-volume series which is intended to supply a comprehensive description of the various methods currently in use for the investigation of magnetic materials is *Experimental magnetism*, edited by G. M. Kalvius and R. S. Tebble (Wiley, 1979–).

The theory of magnetism, volume I, *Statics and dynamics,* by D. C. Mattis (Springer, 2nd edn., 1981) treats the physics of magnetic materials and the criteria for macroscopic co-operative behaviour and includes a good historical chapter. Volume II, *The theory of magnetism II. Thermodynamics and statistical mechanics* by the same author, published in 1985, deals with the statistical mechanics of magnetism. The theory of spin glass, the Kosterlitz-Thouless phase transition and the 2D Ising model are discussed and extensive information on the 3D Ising model is given.

Strong and ultrastrong magnetic fields and their applications, edited by F. Herlach (Springer, 1985), provides a review of experiments with strong magnetic fields.

Magnetic materials

Magnetic properties can be exhibited both by crystalline and by amorphous materials. During the last few years magnetism has expanded into a variety of different areas of research, comprising the magnetism of several classes of novel materials. Examples are quadrupolar interactions, magnetic superconductors, quasicrystals and magnetic semiconductors.

JOURNALS AND CONFERENCE PROCEEDINGS

The two main journals in this field are *Journal of Magnetism and Magnetic Materials* and *IEEE Transactions on Magnetics*. Series of specialized conferences are regularly published in these journals. For example, selected papers from the [thirteenth] International Colloquium on Magnetic Films and Surfaces, held in Glasgow in 1991, and the proceedings of the 5th International Conference on Magnetic Fluids, held in Riga in 1989, were published in the *Journal of Magnetism and Magnetic Materials*. In addition, there are many specialized conferences and workshops, for example those in the NATO Advanced Study Institute series: *Magnetic molecular materials*, edited by D. Gatteschi *et al.* (Kluwer, 1991), *Supermagnets, hard magnetic materials*, edited by G. J. Long and F. Grandjean (Kluwer, 1991), and *Science and technology of nanostructured magnetic materials*, edited by G. C. Hadjipanayis and G. A. Prinz (Plenum, 1991). Another conference series is the biennial *International Conference on Physics of Magnetic Materials*, the fifth of which was held in Mjadralin, Poland, in 1990.

BOOKS

Bibliographies covering the period up to the mid 1970s are *Solid state physics literature guides*, volume 5: *Bibliography of magnetic materials and tabulation of magnetic transition temperatures* published in 1972, and a bibliography covering the years 1950–1976 published in the *Journal of Materials Science* (vol. 12, 1977), under the title: 'Amorphous magnetism and magnetic materials'.

An important series intended as a comprehensive reference work and which can also be used as a textbook was published under the title 'Ferromagnetic Materials', edited by E. P. Wohlfarth (North-Holland, 1980–). This multi-volume work now (1993) includes six volumes, the last of which was published in 1991 with a change of title to *Handbook of magnetic materials* edited by K. H. J. Buschow. The new series title reflects a change in emphasis to cover the newer specialized areas of magnetism. A textbook covering magnetic materials is *Introduction to magnetism and magnetic materials* by D. Jiles (Chapman & Hall, 1991) also mentioned in the previous section. During the last decades the magnetic properties of the d-transition elements have increased rapidly. A useful evaluated data collection *Magnetic properties of metals d-elements, alloys and compounds*, edited by H. P. J. Wijn, has been published in 1991 (Springer). *Introduction to the magnetic properties of solids* by A. S. Chakravarty (Wiley, 1980) analyses in chronological order the development of different theoretical models, their failures and successes. *Magnetism and metallurgy of soft magnetic materials* by C.-W. Chen (North-Holland, 1977) divides neatly into two parts, namely magnetism and metallurgy. However, metallurgy has to be understood in a broad sense, since magnetic non-metallic compounds and amorphous thin films are also included.

Magnetic domains

By applying a suitable, temporary, pulsed magnetic field to a thin single-crystal layer it is possible to obtain stable, isolated, symmetrical magnetic domains. These magnetic bubbles have found applications in electronic devices, but they are also of considerable interest as manifestations of intriguing physical principles.

BOOKS

A selection of monographs is listed below:

Eschenfelder, A. H. (1981) *Magnetic bubble technology* (Springer, 2nd edn.).
Jouve, H. (ed.) (1986) *Magnetic bubbles* (Academic).
O'Dell, T. H. (1981) *Ferromagneto-dynamics: the dynamics of magnetic bubbles, domains and domain walls* (Macmillan). This is about the spatial distribution of magnetization and its time dependence, and includes a history of the subject.

Permanent magnets

The proceedings of the regularly held International Conference on Magnet Technology are published in various formats, some as books and some as parts of journals.

Recent books dealing with permanent magnets include:

Andriessen, F. and Terpstra, M. (eds.) (1989) *Rare earth metals based permanent magnets* (Elsevier). This is a literature study.
Koper, G. H. M. and Terpstra, M. (eds.) (1991) *Improving the properties of permanent magnets. A study of patents, patent applications and other literature* (Elsevier).
McCaig, M. and Clegg, A. G. (1987) *Permanent magnets in theory and practice* (Wiley, 2nd edn.) includes a useful historical introduction and fundamental concepts, followed by chapters on the

fundamental causes of permanent magnetism, permanent magnet materials, design methods and applications.

Parker, R. J. (1990) *Advances in permanent magnetism* (Wiley) provides a unified and comprehensive treatment of permanent magnetism from the origin of permanent magnet behaviour to the use of magnet components in energy conversion devices.

Production and storage of electrical energy

This subject is fully treated in a companion volume in this series: *Information sources in engineering*, 2nd edn., 1985 (editor L. J. Anthony), Chapter 19 'Electric power systems and machines' by W. T. Norris.

Measurements and instrumentation

The literature covered in this section includes both measurement of electrical and magnetic quantities and the theory of electrical appliances that are used in *all* areas of physical and engineering experimentation. Publications range from undergraduate laboratory manuals to treatises on general principles for research workers.

PERIODICALS AND CONFERENCES

Abstracting journals in this field are *Key Abstracts: Measurements in Physics* and *Electronic Instrumentation* (INSPEC). The quarterly *IEEE Transactions on Instrumentation and Measurement* (IEEE Instrumentation and Measurement Society) caters for the specialist in the subject and regularly publishes proceedings of conferences on electrical and electronic measurements. *Measurement Science and Technology* (formerly the *Journal of Physics E: Scientific Instruments*) published monthly by the Institute of Physics, London, is devoted to the theory and practice of measurement and to the apparatus involved in the pursuit of physics, chemistry, engineering and the life sciences. Another journal in this field is *Measurement Journal of IMECO*, the International Measurement Confederation (UK Institute of Measurement and Control). The French journal *Mesures* deals with both instrumentation and industrial automation.

A conference series is the IEEE Instrumentation and Measurement Technology Conference (IMTC) held annually. These conferences cover instrumentation and measurement in a number of different fields and are also important as a meeting place with an exhibition showing instruments, measurement systems and engineering services.

Two journals which are translations of Russian journals are: *Measurement Techniques* (Consultants), corresponding to *Izmeritel'naya Tekhnika*; and *Instruments and Experimental Techniques* corresponding to *Pribory i Tekhnika Éksperimenta* (Consultants).

There are a number of useful trade directories concerned with electrical instruments and apparatus:

Electrical and electronics trades directory (Peregrinus) – the 'Blue Book' which was first published in 1883, is an annual comprehensive guide to companies in the electrical and electronics industries, showing their main sources of supply and industry. Emphasis is British.

Dial 87: electrical, electronics (Dial, 5th edn., 1985) is a buyers' guide to over 8000 electrical and electronic products and services.

Where to buy electrical equipment and services, edited by L. Gale (Where to Buy), is a buyers' guide arranged alphabetically under product, component or service. For further sources the reader is referred to the books by Noltinck and Moore in the following section.

BOOKS

Books and manuals which deal with measurement and instrumentation, in relation to physics, include:

F. Kohlrausch *Praktische physik* (edited by D. Hahn and S. Wagner, Teubner, 23rd edn., 1985–86) is a three-volume comprehensive laboratory manual which includes electricity and magnetism.

B. P. Petley (1985) *The fundamental physical constants and the frontier of measurement* (Hilger). The author, from the National Physical Laboratory (UK), examines the measurement of fundamental physical constants mainly from an experimental point of view and discusses what it is possible to measure meaningfully.

S. Flügge (ed.) (1967) *Handbuch der physik*, vol. 23, is entitled *Electrical instruments*.

The series *Handbook of measurement science*, edited by P. H. Sydenham in three volumes (Wiley, 1982–83), is a collection of fundamental concepts applicable in various branches of science and engineering.

J. P. Bentley in *Principles of measurement systems* (Longman, 2nd edn., 1988). This book is divided into three main sections: an introduction to general measurement systems, followed by the principles and characteristics of typical sensing, signal conditioning, signal processing and data presentation elements, and finally examples of specialized measurement systems.

Another text is Doebelin, E. O. (1983) *Measurement systems. Application and design* (McGraw-Hill, 3rd edn.).

Instrumentation is not a clearly defined subject but forms an important part of all scientific and engineering disciplines. The *Instrumentation reference book*, edited by B. E. Noltingk (Butterworths, 1988), provides a good technical introduction to many aspects of instrumentation as well as a section on 'Directories and commercial information' with addresses of manufacturers, academic institutions and learned societies and engineering institutions. The book has references at the end of the chapters as well as short reviews of some general books on the subject. A useful practical handbook by J. Moore *et al.* is *Building scientific apparatus. A practical guide to design and construction* (Addison-Wesley, 2nd edn., 1989). Chapter 6 has the title 'Electronics'. The book includes a list of manufacturers and suppliers (in USA).

The following is a selection of books dealing with electrical and electronic measurements:

Cooper, W. D. and Helfrick, A. D. (1985) *Electronic instrumentation and measurement techniques* (Prentice-Hall, 3rd edn.).

Gregory, B. A. (1981) *Introduction to electrical instrumentation and measurement systems* (Macmillan, 2nd edn.).

Jones, L. D. and Chin, A. F. (1991) *Electronic instruments and measurements* (Prentice-Hall, 2nd edn.).

Mazda, F. F. (1987) *Electronic instruments and measurement techniques* (CUP).

Reissland, M. U. (1989) *Electrical measurements. Fundamentals, concepts, applications* (Wiley).

Scroggie, M. G. (1990) *Radio and electronic laboratory handbook* (Newnes-Butterworths, 9th edn.).

The electrical means for measurement of temperature is of importance in many aspects of physics. There are a number of recent books on this topic, all of which include descriptions of the use of electrical and electronic methods of measurement. The following are useful exemplars:

McGee, T. D. (1988) *Principles and methods of temperature measurement* (Wiley).

Michalski, L., Eckersdorf, K. and McGhee, J. (1991) *Temperature measurement* (Wiley).

Quinn, T. J. (1990) *Temperature* (Academic, 2nd edn.). This is one of a series of Monographs in Physical Measurement.

Finally, a compilation of previously published papers by staff of the National Bureau of Standards (NBS) is NBS Special Publication 300 entitled *Precision measurement and calibration*, volumes 3 and 4 being *Electricity – low frequency* (1968) and *Electricity – radio frequency* (1970), respectively.

Reference

Scaife, B. K. P. (1989) *Principles of dielectrics*. Oxford: Oxford University Press.

CHAPTER TEN

Computer hardware and electronics for physics*

P. J. PONTING

It is rare, in this final decade of the twentieth century, to find a physicist or engineer working completely alone. The desktop personal computer or workstation, linked through a communications network to a central server has become commonplace. The combination provides access to massive computing power for the analysis of physics data and has also created a revolution in the methods of designing experiments. Computer tools have become essential factors in the specification and implementation of even small configurations.

Tools, inevitably, reflect the skills of the operator and to be used effectively require a sound fundamental knowledge coupled with regular reviews as technology advances. The example of the development of Very Large Scale Integration (VLSI) assemblies for use as elements in modern experimental configurations forcibly illustrates the need of expertise in both domains. In these designs the derivation of a complex algorithm is followed by its implementation through the application of a series of sophisticated engineering tools on a powerful workstation. Much of this type of software can run on disparate computer platforms due to the unifying nature of modern operating systems. Consequently, the application is given more emphasis than pure computer hardware in the following text.

The sources mentioned in this chapter are certainly not exhaustive. They do, however, provide a reasonable balance between tutorial literature and the identification of modern design methods.

---

*based on the second edition chapter written by I. Pizer

Books

Identifying the topic

The range of subjects embraced by the specialist literature of electronics and computing hardware for physics is wide and it is worthwhile to take a little time to obtain a broad view before attempting to focus on a specific topic. Three books which between them offer a general introduction, especially in electronics, are:

Ferbel, T. (1991) *Experimental techniques in high energy physics* (World Scientific, 2nd edn.), which provides good coverage of detectors and noise.
Dunlap, R. A. (1988) *Experimental physics: modern methods* (OUP) which introduces instrumentation for experiments.
Chorafas, D. N. (1990) *The new technology: a survival guide to new materials, supercomputers and global communications for the 1990s* (Sigma) which offers a comprehensive guide to computing hardware essentials: photonics, databases, networks, computers, etc.

Electronics

The analogue circuitry required to receive and adapt signals from physics detectors is as varied as the detectors themselves and a sound knowledge of component primitives is essential. The adapted data is usually passed to digital logic, often implemented as pre-packaged functionality, for further processing. The following selection of literature describing the basic elements of electronic components and design, acts as an excellent starting point:

Hayes, T. C. and Horowitz, P. (1989) *Student manual for the art of electronics* (CUP). Basic concepts of analogue and digital electronics.
Horowitz, P. and Hill, W. (1989) *Art of electronics* (CUP, 2nd edn.). Wide, detailed coverage of transistors, amplifiers, filters, microprocessors, high frequencies, signal processing, digital circuits.
Vassos, B. H. and Ewing, G. W. (1985) *Analog and digital electronics for scientists* (Wiley, 3rd edn.). Signals, components, operational amplifiers, transistors, transducers, microprocessors, logic.
Havill, R. L. and Walton, A. K. (1985) *Elements of electronics for physical scientists* (Macmillan, 2nd edn.). Semiconductors, diodes, junctions, components, feedback, oscillators, switching, digital logic, integrated circuits, Darlingtons, FETs.
Stanier, B. J. (1985) *Modern electronics and integrated circuits*

(Hilger). Semiconductors, gates, amplifiers, fabrication of analogue and digital LSI.

Before proceeding further it must be stated that the problem of noise is inseparable from the design process. Many types and sources of noise exist and each has an influence in the experimental environment. A study of the topic is important. Van der Zial, A. (1986) *Noise in solid state devices and circuits* (Wiley) focuses on the understanding and reduction of systematic noise in the component chain.

Moving on from the basics a set of books introducing the concepts of electronic algorithms and design aids is of great importance. This class of book offers a fine entry point to the practical design environment of today:

Pelloso, P. (1986) *Practical digital electronics* (Wiley). TTL and CMOS integrated circuits, binary and BCD arithmetic logic.

Cham, Kit Man *et al.* (1986) *Computer-aided design and VLSI device development* (Kluwer). MOS integrated circuits, simulation and production.

Williams, T. W. (ed.) (1986) *Advances in CAD for VLSI*, vol 5. *VLSI testing.* (North-Holland). Semiconductors, memories, fault modelling.

McCanny, J. V. and White, J. C. (eds.) (1987) *VLSI technology and design* (Academic). Integrated circuits, NMOS, CMOS, CAD tools, arrays, parallel processing.

Mead, C. (1989) *Analog VLSI and neural systems* (Addison-Wesley). Transistors, amplifiers, signals.

A term which has achieved a deserved popularity in recent years is that of signal processing. It is usually taken to mean real-time processing of digital signals but does, in fact, have a much wider scope as most signals from a detector are processed several times during their progress through both the analogue and digital domains. Priemer, R. (1991) *Introductory signal processing* (World Scientific) describes time-continuous, time-discrete, linear and correlation principles; Laplace, Fourier and Z transforms; digital filters and microprocessors. Roberts, A. R and Mullis, C. T. (1987) *Digital signal processing* (Addison-Wesley) offers basic introductory material on linear discrete time systems and methods for solving approximation problems for digital filters. Also presents the design and application of numerical algorithms that new technologies have invoked.

A new field which has been brought about by the complexity of modern circuits is that of evaluation, testing and fault diagnosis. Miller, D. M. (ed.) (1987) *Developments in integrated circuit testing*

(Academic) discusses CMOS technology, the concepts of matrices and cells, and error detection.

Developments involving the direct mounting of analogue signal processing circuits on detectors are creating a surge of interest in this field. Additionally, these circuits may be required to work in radiation harsh environments. Two books to read are: Gregorian, R. and Temes, G. (1986) *Analog MOS integrated circuits for signal processing* (Wiley) and Ma, T. P. and Dressendorfer, P. V. (1989) *Ionising radiation effects in MOS devices and circuits* (Wiley).

Computers

The books mentioned in this section have a direct link to electronics and hardware. The first set, on languages and operating systems, invoke two distinct groups. The generic group are general programming languages which contain commands and constructs which allow easy access to hardware, bits in data registers for example:

Kernighan, B. W. and Ritchie, D. M. (1989) *The C programming language* (Prentice-Hall, 2nd edn.).

Kelley, A. and Pohl, I. (1990) *A book on C* (Benjamin/Cummings, 2nd edn.) (ANSI C).

Stroustrup, B. (1987) *The C++ programming language* (Addison-Wesley).

Sobell, M. G. (1989) *A practical guide to UNIX system: system V and BSD 4.3.* (Benjamin/Cummings, 2nd edn.).

A further group comprises languages which are destined to assist in the development and precise specification of designs from components to complete systems:

Lipsett, R., Schaefer, C. F. and Ussery, C. (1991) *VHDL: hardware description and design* (Kluwer). Emphasizes the use of the language to develop VLSI circuitry.

Thomas, D. E. and Moorby P. *The Verilog hardware description language* (Kluwer). Less specific and has its roots in a top-down approach to system design and simulation.

Harman, T. L. (1989) *The Motorola MC68020 and MC68030 microprocessors: assembly language, interfacing, and design* (Prentice-Hall). Important in that these devices are extensively used as embedded processors in acquisition systems. In addition, the family is native to VMEbus, also extremely popular amongst physicists.

The next set of books has less immediate connection with hardware but does provide valuable insights into techniques which are often applied in an *ad hoc* manner in data acquisition systems. Partitioning, parallelism, buffers, interfaces, flow control; all are essential to the construction of an effective data handling environment. Some overlap is inevitable but, in general, these volumes are complementary to each other.

Meyer, R. (1988) *Object oriented software construction* (Prentice-Hall). Covers modularity, reusability, program construction, inheritance, interfaces, memory management, languages.

Tello, E. R. (1991) *Object-oriented programming for Windows* (Wiley). More specialist, covers tools, interfaces, graphics, windows design.

Lewis, E. G., El-Rewini, H. and Kim, I. (1992) *Introduction to parallel computing* (Prentice-Hall). Caters for shared and distributed memory, data flow, scheduling.

Treleaven, P. C. (ed.) (1990) *Parallel computers:object-oriented, functional, logic* (Wiley). Includes details of Eüropean industry and object oriented languages.

Potter, J. L. (1992) *Associative computing:a programming paradigm for massively parallel computers* (Plenum). Gets down to the basics of input/output, ASCII, data, Prolog design.

Axford, T. (ed.) (1989) *Concurrent programming:fundamental techniques for real-time and parallel software design* (Wiley). Covers yet another range of topics with switches, sets, kernels, buffers, and reliability.

Raynal, M. and Helary, J-M. (1990) *Synchronization and control of distributed systems and programs* (Wiley). Deals with phases, networks, waves, wave sequences, and logic pulsing.

An aspect of computing which is often critical to the achievement of good physics results is that of performance. Again, several viewpoints can be taken including operating efficiencies, monitoring and diagnostics, and user interfaces:

Simmons, M., Koskela, R. and Bucher, I. (eds.) (1989) *Instrumentation for future parallel computing systems* (Addison-Wesley). Concerned with performance and monitoring.

Gelenbe, E. (1989) *Multiprocessor performance* (Wiley). Deals with architecture, structures, and networks.

Gillies, A. C. (1991) *The integration of expert systems into mainstream software* (Chapman and Hall). Covers CASE, software engineering design, and interfaces.

Handbooks

In contrast to the books mentioned in previous sections the handbook caters to the need to extract specific information on an occasional basis. The following selection will satisfy the majority of searchers for instant information:

Nagy, P. and Tarjan, G. (1988) *Elsevier's dictionary of microelectronics in five languages: English, German, French, Spanish and Japanese* (Elsevier).

(Annually) *The ARRL handbook* (The American Radio Relay League). Practical guide to many topics such as filter design, fundamentals, transmission, Standards, OSI, etc.

Walsh, B. C. *et al.* (1986) *Computer users' data book* (Blackwell). Covers numbers, representation, chips, tapes, discs: hard and floppy, networks, communications, microprocessors, and operating systems.

The software catalog: systems software; a directory of software for the computer professional (Elsevier, 1986). Gives many details of suppliers, micros, minis, etc.

Einspruch, N. G. (ed.) (1985) *VLSI handbook* (Academic) is essential to the VLSI designer. It covers design, automation, MOS and bipolar technologies, microprocessors, RAM, ROM, silicon, lithography, and radiation effects.

Sinnadurai, F. N. (ed.) (1985) *Handbook of microelectronics packaging and interconnection technologies* (Electrochemical Publ.) is complementary to the previous document; this one has a manufacturing bias and covers semiconductors, chips, thin and thick films, hybrids, reliability, printed circuit boards, and temperature.

Periodicals

The market is bursting with periodicals covering every aspect of electronics and computing related to physics. The selection given below is chosen to demonstrate the range available, from the frequency of publication to the style and profundity of the contents. Some journals dedicate particular issues to specific topics and these are mentioned when appropriate. The titles of the majority of journals published weekly or monthly are not particularly meaningful, while the scope of the contents is wide and often random. Readers must, therefore, make their own selections from the list.

Physics and related topics

Physics Today. August issue, part 2, is an excellent buyers' guide.
Nuclear Instruments and Methods in Physics Research (North-Holland). *Section A: Accelerators, spectrometers, detectors and associated equipment. Section B: Beam interactions with materials and atoms.*
Lightwave. Journal of fibre optics, general and commercial in nature, but good for news.

Electronics

Analogue Dialogue 23–3 (1989). Grounding for high-frequency operation.
Electronics.
Electronics Design. The issues of August 6 and 20, 1992, contain excellent articles on the testing of analogue to digital converters.
International Journal of High Speed Electronics.
Electronics World and Wireless World. Rather engineering oriented.
Electronics Weekly. Early reviews of products and trends.
Microelectronics Journal (Elsevier).

Computers

BYTE. (McGraw-Hill) Occasional articles on topics such as neuron studies.
Computing.
Computer Design.
Computer.
Computer Weekly. Companion of *Electronics Weekly.*

Magazines devoted to specific families of machines are popular and contain articles of surprising relevance to the physicist. The more important are: *PC Magazine* and *PC Computing* for the IBM (clone) world, *Macworld* for the Macintosh, and *SunWorld* for Sun. *UNIX/world* fits nicely into this pattern with its emphasis on this universal product.

Institutional publications

Two institutions involved in the extensive publishing of learned papers in the fields of electronics and computers are the Institution of Electrical and Electronics Engineers (IEEE) in USA and the Institution of Electrical Engineers (IEE) in UK. The more relevant titles are given below with the full range obtainable from the appropriate institution:

Publications Catalog and Standards Bearer (IEEE). Regular issues giving details of all IEEE documents.

IEEE Transactions on Circuits and Systems. Part I: Fundamental Theory and Applications. Part II: Analog and Digital Signal Processing.

IEEE Transactions on Signal Processing.

IEEE Micro covers a wide range of topics, as examples: an excellent summary of IEEE P1596, SCI, was published in February 1992, the April issue contained an in-depth description of a second generation RISC processor and in August that same year there was an article on standards in the field of electronic design.

Journal of Lightwave Technology (IEEE).

IEE Proceedings. Part E: Computers and Digital Techniques. Part J: Optoelectronics.

In addition to the foregoing publications, the European Organization for Nuclear Research (CERN) in Geneva produces three newsletters at regular intervals for the information of a large community of physicists, engineers, computer scientists and programmers. These are:

CSENL. Computing Support for Engineering Newsletter.

Online. The newsletter of data acquisition and computing for experiments.

CERN Computer Newsletter. Covering central computing and networks.

Conferences

Conference records are, undoubtedly, a rich source of information. In particular, as a means of recruiting colleagues with common interests. The conferences listed hereafter most satisfactorily fulfill this requirement for electronics and computing for physics:

Nuclear Science Symposium (Annual) (IEEE). Proceedings published as conference record and special issue of *IEEE Transactions on Nuclear Science*. Detectors, circuits, data acquisition, triggers, calorimeters.

Real Time '91 on Computer Applications in Nuclear, Plasma and Particle Physics, Aachen (IEEE). Applications, data acquisition, parallel processing, high speed processing, bus architectures.

Open Bus Systems '92 in Research and Industry, Zurich (CERN, ESONE, *et al.*). Bus-based architectures, high speed serial buses, real-time systems, bus models, links, components.

CHEP92, Computing for High Energy Physics, Annecy (LAPP,

CERN). Largely software oriented but good overview on high speed interconnects, SCI, HIPPI, ATM.

CAMAC '91, Dubna (Innovation, Moscow) and *CAMAC '92*, Warsaw (ESONE). Data acquisition systems, buses, modules.

International Symposium on Nuclear Electronics 14 (1990) Warsaw (JINR, Dubna). Data acquisition systems.

European Symposium on Semiconductor Detectors 5 (1989) Munich (North-Holland). New developments in radiation detectors.

International Conference on the Impact of Digital Microelectronics and Microprocessors on Particle Physics (1988) Trieste (World Scientific).

International Conference on VMEbus in Research (1988) Zurich (Esone, North-Holland).

International Symposium on Measurement of Electrical Quantities 1 (1986) Como (Imeko, Budapest). Noise in electrical measurements.

Aegean Workshop on Computing (1986) Loutraki (Springer). VLSI algorithms and architectures.

Databases and abstract services

WORLDWIDEWEB (W^3), a networked hypertext information service covering many topics: computing, astronomy, etc.

COMPUTER AIDED PRODUCT SELECTION (CAPS), a component database (Cahner).

SCIENTIFIC ABSTRACTS ON DISKETTE (ISI).

Buyer's guides

VME.VXI.Futurebus+ Compatible products directory (VITA, 1991).

Wedgwood, C. G. (ed.) (1986) *International electronics directory: the guide to European manufacturers and agents* (North-Holland).

Manufacturer's literature and tools

Mathcad applications packs (Mathsoft). Advanced mathematics and electrical engineering problem solving.

The programmable gate array design handbook (1992) (Xilinx). Just one example of programmable logic. Similar documents are available from Altera, Actel, Lattice, among others. Each family has unique properties but all are well suited to physics applications.

SPICAD (Auris). This item is quoted rather than the usual *PSpice* as it is expanded to include mixed analogue and digital simulation and, as such, has found favour by designers of front-end logic.

Instrumentation Newsletter (National Instruments), vol. 4–3, *LABVIEW*, an emulator enabling the creation of virtual instrument sets, facilitating the later incorporation of a real set in an experimental configuration.

DEC User and *DEC Professional* are house journals of the Digital Equipment Corporation.

Standards

The term 'Standard' has many interpretations: to an engineer it could mean consistency and support while a physicist may think of it as a compromise to save time and money. However, selecting the correct standards and ensuring an appropriate implementation has proved to be the definitive method of guaranteeing compatibility of hardware and software from different sources as found on the immense experimental installations of today. Contemporary hardware standards commonly used or being assessed by the physics community include:

Standard backplane bus specification for multiprocessor architectures:Futurebus+ (IEEE, 1990). ANSI/IEEE 896.1

Standard FASTBUS modular high speed data acquisition and control system (1989) ANSI/IEEE Std 960. Also available as IEC 935. Large form factor boards and ECL technology used extensively for front-end channel electronics.

CAMAC. A modular instrumentation system for data handling. A family of documents available as EUR 8500 from the EEC, Luxembourg, SHO8482 from the IEEE, and Publication 516, etc. from the IEC. Very popular system for both data acquisition and control of smaller experiments.

VMEbus (1987) IEEE 1014. TTL-based, has achieved much commercial support and is the preferred bus system where considerable processing power is needed.

Scalable Coherent Interface. IEEE P1596. A new standard, still at the project stage, whereby data packets can be moved at Gigahertz rates between the functional elements of experiments proposed for the next decade.

A standard which targets an entirely different field to those above is: *Standard Test Access Port and Boundary-Scan Architecture* (IEEE, 1149–1990). Specification of procedures for the introduction of auxiliary logic into electronic circuits for diagnostic purposes. Already incorporated into many commercial devices.

More general reading regarding the evolution of standards can be found in:

Hitchcock, S. (ed.) *Bus standards update* (Butterworth). A special issue, 12 (3), 1988, reviews microprocessors and microsystems.

CHAPTER ELEVEN

Experimental heat and low temperature physics

R. BERMAN

The properties of matter which are discussed in this chapter have many important applications and often have to be measured over a wide temperature range. They all have in common the necessity of temperature (or temperature difference) measurement and if the results are to be useful to others, the temperature must be determined according to some agreed scale and certainly for comparison with the theory of the property this must be the temperature based on its thermodynamic definition. Such a temperature is defined in terms of a single fixed point, the triple point of pure water which is given the exact value of 273.16 K (K denotes degrees Kelvin). In general, temperature is measured by a relatively simple instrument appropriate to the range to be covered and this must be calibrated in some way which will yield the Kelvin temperature. If the temperature or temperature difference is wrongly determined the value derived for the property will be incorrect and will correspond to the wrong temperature. An example of this problem arose in the mid 1950s in the case of the specific heats of the noble metals copper, silver and gold. It was found that the values at liquid helium temperatures did not conform to the theoretically predicted simple form $C_v = aT + bT^3$. At these temperatures the thermometric parameter is usually the vapour pressure of liquid ^{4}He and one possibility for the disagreement was that the relation between vapour pressure and temperature (the liquid helium temperature scale) then in use was incorrect. Corrections to the scale were proposed (W. S. Corak *et al.*,1955) which would give the expected temperature dependence of the specific heats, and these were similar to corrections proposed for other reasons.

It is thus appropriate to preface discussion of thermal properties with a brief account of the present internationally agreed temperature scale. Quite apart from correlation with theory, if everyone uses the same temperature scale, intercomparisons between different experimenters will have some validity.

Temperature

The ratio of two temperatures on the Kelvin scale is identical to the ratio of the ideal gas scale temperatures, given by the ratios of pV for a fixed mass of an ideal gas at the two temperatures. If the same single assigned reference temperature is used for both the gas and the thermodynamic scales, then the numerical values of all other temperatures will be identical on the two scales. Unfortunately, gas thermometry of the required accuracy is difficult in practice and many corrections have to be applied for the non-idealities of the gas and of the apparatus, so that such work is confined to only a few laboratories. By using one of the available ways of deriving Kelvin temperature (gas thermometry, velocity of sound, magnetic susceptibility, pyrometry), the temperatures of a number of phase transitions – the fixed points – are determined and relatively convenient instruments for measuring temperature are calibrated. The 'everyday' user is provided with the means for calibrating his own thermometers in terms of their indications at suitable fixed points.

The organization which has the responsibility of recommending temperature scales is the Conférence Général des Poids et Mesures (CGPM) which meets at the Bureau International des Poids et Mesures (BIPM) in Sèvres, near Paris. The previous International Practical Temperature Scale, IPTS-68 was adopted in 1968. It was defined in terms of three instruments, a sufficiently pure platinum resistance thermometer, a Pt v. Pt-10% Rh thermocouple and an optical pyrometer. The lowest fixed point was the triple point of equilibrium hydrogen given the value 13.81 K, the highest was the freezing point of gold at 1337.58 K. It has been usual with temperature scales that by the time they are recommended and are in use, current work has shown that they could be improved upon. This was also the case with IPTS-68; for example, the ratio of the hydrogen and helium boiling points as given in IPTS-68 was incorrect and even the boiling point of water was found to be less than 100 K above the freezing point (i.e. the ratio $(pV)_{bp}/(pV)_{fp}$ for an ideal gas is less than the value assumed in determining the value to be given to the freezing point and hence to the triple point) and a Provisional Temperature Scale (Echelle Provisoire – EPT-76) was introduced in 1976 which extended down to 0.5 K. Work up to that time was outlined in the previous edition of this book and was thoroughly covered in the first edition of T. J. Quinn's book *Temperature*

(Academic, 1983). Other defects in the scale have also shown up and there is now a new International Temperature Scale ITS-90 which is, this time, thought to be accurate enough to last at least 20 years. This scale is described in the second edition of Quinn's book (Academic, 1990) and the experimental basis for it is also given in R. L. Rusby *et al.* (1991). The main improvements result from the choices of the fixed points (triple points rather than boiling points at 'low' temperatures; even the boiling point of water has been replaced by freezing points on either side of it, those of Ga and In) and the extension of the platinum resistance thermometer as the standard instrument into the range previously defined by the thermocouple (up to the freezing point of silver at 1234.93 K; the thermocouple extended up to the freezing point of gold at 1337 K). ITS-90 extends down to 0.65 K, which is a temperature conveniently measured by the vapour pressure of ^{3}He. Much of the work connected with the temperature scale is published by BIPM itself and also appears in the journal *Metrologia* (Springer).

Thermophysical properties

There are many compilations of data on the thermophysical properties of solids, liquids and gases and a large number of journals which publish articles in the field. However, if values are required for a material which differs slightly from that found in tabulations of properties one must have some appreciation of the theory of the property, to be able to estimate how it will be affected by small changes in composition, as there are some properties which are very sensitive to this and some which are not. For example, an 'average' diamond which may contain tenths of 1% of nitrogen has a thermal conductivity at room temperature 'only' about twice that of copper, while the type (IIa) which is almost free of nitrogen has a conductivity about five times that of copper (the basis for its use as a heat 'sink'). However, a chemically pure diamond made from carbon with practically no ^{13}C (natural diamond contains about 1% of the heavier isotope) has a conductivity about eight times that of copper. Although these are enormous differences in conductivity, it requires extremely high accuracy to detect any differences between the specific heats.

An introduction (at first-year undergraduate level, according to the author) to the properties of matter can be found in A. J. Walton *Three phases of matter* (Clarendon, 1983), while some books dealing with the properties of the separate phases are: C. Kittel *Introduction to solid state physics* (Wiley, 6th edn., 1986); N. W. Ashcroft and N. D. Mermin *Solid state physics* (Holt-Rinehart, 1976); J. S. Rowlinson and F. L. Swinton *Liquids and liquid mixtures* (Butterworths, 3rd edn., 1982); R. C. Reid, J. M. Prausnitz and B. E. Poling, *The properties of gases and liquids* (McGraw-Hill, 4th edn., 1987), which has many tables of properties, E. H. Kennard *Kinetic theory of gases* (McGraw-Hill, 1938); R. D. Present *Kinetic theory of gases* (McGraw-Hill, 1958) and J. S. Rowlinson *The perfect gas* (Pergamon, 1963). The second edition of L. B. Loeb's *Kinetic theory of gases* has been reprinted with a summary of later progress (with references) by Dover (1961).

Methods of measuring thermophysical properties of solids are described in the two volumes edited by K. D. Maglic *et al. Compendium of thermophysical property measurement methods* (Plenum); volume 1 (1984) is a survey of measurement techniques, while volume 2 (1992) gives recommended practices for determining thermophysical properties.

Tabulations of values appear in *Equilibrium properties of fluid mixtures* (Plenum): volume 1 (1975) is *A bibliography of data on fluids of cryogenic interest* and volume 2 (1982) is *A bibliography of experimental data on selected fluids*. IUPAC's *International thermodynamic tables of the fluid state* (Pergamon) and the NBS (now NIST) series of monographs on Thermophysical Properties (US Government Printing Office, Washington) also contain tables of values. *The Journal of Physical and Chemical Reference Data* (ACS and AIP for NIST) contains articles on the establishment of standards and occasionally lists recent data compilations. The journal *Pure and Applied Chemistry* (Pergamon) carries a section in each issue on recommendations and reports. It has included a *Catalogue of reference materials from national laboratories* with addresses of the relevant institutions (Cali, 1976). It has also given the conversion of thermodynamic properties measured on the IPTS-68 temperature scale to the basis of ITS-90 (Goldber and Weir, 1992). *The Journal of Physical and Chemical Reference Data* has supplements on the JANAF (Joint Army, Navy and Air Force) *Thermochemical tables* which are also published separately by ACS; those appearing up to 1985 are available in two separate volumes.

Publications which are entirely devoted to tabulations of thermophysical data are the TPRC *Thermophysical properties of matter* in thirteen volumes (Plenum, 1970–1975), *Thermophysical properties of high temperature solid materials* in six volumes (Macmillan, 1967) and the TPRC *Thermophysical properties research literature retrieval guide 1900–1980,* published in seven volumes (Plenum, 1982). *Physical property data for the design engineer* (C. F. Barton and G. F. Hewitt (eds.), Hemisphere, 1989) has tables of density, specific heat and thermal conductivity for solids and, in addition viscosity for liquids. Probably the best known compilation, Landolt-

Börnstein *Zahlenwerte und Funktionen aus Naturwissenschaft und Technik* is now appearing in a new series (neue Serie). As for the TPRC compilations, the volumes in which data relevant to this chapter can be found will be given later when considering the individual properties.

The bi-monthly journal *High Temperatures-High Pressures* (Pion) publishes the papers presented at the European Thermophysical Conferences. For example, the 12th ETPC in Vienna completely filled the 1991 issues, but when not reporting these conferences the articles correspond more closely to the title of the journal.

Among the Task Groups set up by CODATA there was one on Thermophysical Properties of Solids, the report of which, by M. L. Minges, was issued as *CODATA Bulletin* no. 60 (Pergamon, 1986) with the title *Evaluation of thermophysical property measurement methods and standard reference materials.* This gave a critical evaluation of the properties of a small number of materials, so that anyone setting up apparatus for such measurements could check its working by comparing results on these standard reference materials.

The thermal properties of polymers are discussed in vol. 16 of *The encyclopedia of polymer science and technology*, edited by H. F. Mark *et al.* (Interscience, 1970), and data can also be found in *Polymer handbook*, edited by J. Brandrup and E. H. Immergut (Wiley, 3rd edn., 1989).

There are also journals with titles defining conditions under which properties are measured, such as:

Cryogenics (Butterworths)
Journal of Low Temperature Physics (Plenum)
The Russian *Journal of Low Temperature Physics* (translated as *Low Temperature Physics*) (AIP)
The Russian *Thermophysics of High Temperatures* (translated as *High Temperature*) (Consultants)

We now give more detail on the individual properties

Specific heat

The specific heat of solids always has a component from the coupled vibrations of the atoms. On classical theory this would be $3R$ at all temperatures, but because of quantum effects there is always a fall-off at 'low' temperatures and $C \to 0$ as $T \to 0$. The Debye theory contains several simplifying assumptions on the basis of which the specific heat would be a universal function of the ratio T/Θ_D, where Θ_D is a temperature characteristic of each substance, (proportional to the maximum vibrational frequency) ranging from 25 K for helium solidified at the lowest possible density to 2000 K for diamond. This is why the specific heat of diamond does not obey the Dulong and Petit rule at room temperature; the specific heat at T/Θ_D = 1/7 (room temperature) is about 1/5 of the classical value and has only reached 90% of $3R$ at 1400 K. In metals there is also an electronic component which is linear in temperature (γT). The theory of specific heats and tables of Θ and γ for elements and some compounds are given in E. S. R. Gopal *Specific heats at low temperatures* (Heywood, 1966). Graphs of specific heat of metals and values of γ can be found in Landolt-Börnstein (H. R. Schober and P. H. Dederichs, vol. III/13a, 1981). Specific heats of liquid systems are given by J. D'Ans, H. Surawski and C. Synowietz (Landolt-Börnstein, vol. IV/1b, 1977). Specific heats are also given in *Handbook of high temperature compounds: properties, production, applications*, edited by T. Ya. Kosolapova (Hemisphere, 1990) which has over 1100 references. Values for polymers are given in *Polymer handbook*, edited by J. Brandrep and E. H. Immergut (see above), and *Properties of polymers* by D. W. Van Krevelen and P. J. Hoftyzer (Elsevier, 1976). Much detail of experimental methods for measuring specific heat over a wide temperature range for solids, liquid and gases can be found in two volumes published for IUPAC by Butterworths: *Experimental thermodynamics*, vol. I, *Calorimetry of non-reacting systems*, edited by J. P. McCullough and D. W. Scott (1968) and vol. II *Experimental thermodynamics of non-reacting fluids*, edited by B. Le Neindre and B. Vodar (1975). Together, these two volumes contain several thousand references. The volumes of the TPRC series which contain data on specific heat are vol. 4 (*Metallic elements and alloys*), vol. 5 (*Nonmetallic solids*), and the two parts of vol. 6 (*Nonmetallic liquids and gases*).

Thermal expansion

In a real solid the frequency of a vibrational mode changes with volume. If the frequency for all modes varied in the same simple way, $\nu_i \propto V^{-\gamma}$, then the volume coefficient of expansion would be given by $\beta = (\gamma C_v)/(VB_T)$, where γ is a constant and B_T is the isothermal bulk modulus. In fact, γ_i is a function of ν_i and its equivalent (the Grüneisen parameter) is $(\Sigma\gamma_i C_i)/C$, where C_i is the contribution of mode i to the specific heat. There is also an electronic contribution in metals related to the electronic specific heat. The temperature variation of β thus follows that of C to some extent and results of measuring β are often expressed in terms of the variation of the effective γ with temperature. For the rare gas solids, γ does not vary by more than 10 per cent from the melting point down, while for the rubidium halides, for example, it is ~1.5 at $T = \Theta_D$ and ~0 below $\Theta_D/10$. The theory, experimental methods and results on thermal expansion in solids are given by R. S. Krishnan *et al. Thermal expansion of crystals* (Pergamon, 1979) which contains over 2000 references and very extensive tables. The review article *Thermal expansion of solids at low temperatures* by Barron *et al.* (1980) contains about 500 references and also discusses standards for expansion coefficient. Thermal expansion is also covered in the handbook by Kosolapova (1990) mentioned previously (p. 222).

The volumes in the TPRC Data Series concerned with thermal expansion are vol. 12 (*Metallic elements and alloys*) and vol. 13 (*Nonmetallic solids*), and a new volume, 1–4, is in preparation, edited by R. E. Taylor.

The CODATA Task Group has evaluated existing data on the reference materials Cu, W, Al_2O_3 and Si *CODATA Bulletin* no. 95 (1985).

Heat transfer

The literature appropriate to individual means for heat transfer will be discussed separately but there are some journals and conferences which cover all the mechanisms:

International Journal of Heat and Mass Transfer (Pergamon). This occasionally gives a bibliography of Soviet work and a review of the previous year's literature (over 2000 references to 1991 work in 1992)

Heat Transfer – Soviet Research (Scripta, translated)

Heat Transfer – Japanese Research (Scripta, translated)

Wärme und Stoffübertragung (Springer)

Transactions of the ASME. J. Heat Transfer (ASME). Some issues print the papers given at conferences on heat transfer.

Journal of Thermal Insulation (Technomic)

Journal of Thermophysics and Heat Transfer (AIAA inc.)

Previews of Heat and Mass transfer (Rumford and Pergamon) is a bi-monthly journal giving a survey of literature taken from over 100 journals. In addition, there is a two-volume *Handbook of heat transfer* (McGraw Hill, 2nd edn., 1985); vol. 1, *Fundamentals* (W. M. Rohsenow and J. P. Hartnett); vol. 2. *Applications* (W. M. Rohsenow, J. P. Hartnett and E. N. Ganic). Review articles can be found in *Advances in Heat Transfer* (Academic) and in *Progress in Heat and Mass Transfer* (Pergamon).

Thermal conductivity and diffusivity.

All materials conduct heat through the transfer of energy by some sorts of interaction; for solids, through the coupling between the atomic vibrations and, in metals, by means of the electrons, in gases the interactions are usually termed collisions between molecules, and both sorts can be considered to act in liquids. When conditions are not steady, the rate of change of temperature is determined by the ratio of thermal conductivity to heat capacity per unit volume λ/C, the thermal diffusivity. The majority of measurements made above room temperature are now determinations of thermal diffusivity.

The theory of conduction in solids is discussed in J. E. Parrott and A. D. Stuckes *Thermal conductivity of solids* (Pion, 1975) and R. Berman *Thermal conduction in solids* (OUP, 1976), while a long article by G. A. Slack, *The thermal conductivity of nonmetallic crystals* (1979), discusses several subtle points which are not considered in any detail in these books. Details of experimental methods can be found in R. P. Tye *Thermal conductivity*, 2 vols (Academic, 1969), while the classic work on the solution of mathematical problems which arise in practice is H. S. Carslaw and J. C. Jaeger *Conduction of heat in solids* (OUP, 2nd edn., 1959). Roder (1981 and 1982) has given a description of a transient hot-wire apparatus for measurements on liquids and gases and an example of its use in the case of oxygen. The theoretical background for gases and liquids is given in N. V. Tsederberg *Thermal conductivity of gases and liquids* (Arnold, 1975). There is a biennial conference on thermal conductivity in North America, the proceedings of which are now published by Plenum with the title *Thermal conductivity* (and the conference number – in the low twenties in 1992). The publication of papers given at the biennial European Thermophysics Conferences has already been mentioned.

In the TPRC Data Series the relevant volumes are:vol. 1 (*Metallic elements and alloys*), vol. 2 (*Nonmetallic solids*), vol. 3 (*Nonmetallic liquids and gases*) and vol. 10 (*Thermal diffusivity*). The articles in Landolt-Börnstein are in Gruppe III/15c (1991); *Thermal conductivity of pure metals* (G. K. White), *Thermal conductivity of alloys* (P. G. Klemens and G. Neuer), *Effect of pressure on thermal transport properties of metals and alloys* (B. Sundqvist) and *Thermal conductivity of pure semimetals and their dilute alloys* (C. Uher). *The thermal conductivity of minerals and rocks* by V. Cermak and L. Rybach is in V/1a (1981).

The CODATA Task Group has evaluated data on electrolytic iron, tungsten, a stainless steel and polycrystalline graphite and the conclusions are contained in *CODATA Bulletin* no. 60 (1986). Standard reference materials (SRMs) for thermal conductivity held by NIST are: tungsten, electrolytic iron and stainless steel.

Convection

In convective heat transfer, bulk movement of a fluid controls the rate of heat flow. In natural convection the movement is driven by gravity, while in forced convection the movement is provided mechanically. Mixed convection involves a combination of the two. J. H. Lienhard *A heat transfer textbook* (Prentice-Hall, 1981) provides a good introduction to the subject, while L. C. Burmeister, *Convection heat transfer* (Interscience, 1983) is a treatment at postgraduate level. Applications are described in J. M. Coulson and J. F. Richardson *Chemical engineering*, vol. 1 (Pergamon Press, 1982).

The annual reviews of the heat transfer literature published in the December issues of *The International Journal of Heat and Mass Transfer* provide a view of the development of the subject; the 1991 review (35(12), 1992) contains over 1200 references to work on convective heat transfer and an additional 100 references to developments in experimental technique as well as brief reviews of all relevant conferences held during 1991. One of the databases of the literature *The Heat Transfer and Fluid Flow Service Digest* is edited by P. Hinklin at Harwell. *Previews of Heat and Mass Transfer*, mentioned earlier is also useful for keeping abreast of the literature.

Radiation

The application of radiation to temperature measurement is described in Quinn's book, and heat transfer by radiation is discussed in R. Siegel and J. R. Rowell *Thermal radiation heat transfer* (McGraw-Hill, 1972).

The rate of heat transfer by radiation depends not only on temperature and geometry, but also on the nature of the radiating surfaces, which can range between almost 'black' to almost completely reflecting. The emissivity (or absorptivity) is also a function of wavelength so that the overall emissivity is an average over the emissivity for a particular wavelength, weighted by the contribution of the wavelength to the energy radiated. For any particular surface the emissivity must be known if the radiation is to be estimated. There are three volumes of the TPRC Data Series on Thermal Radiative Properties (emittance, reflectance, absorptance, transmittance); vol. 7 (*Metallic elements and alloys*), vol. 8 (*Nonmetallic solids*) and vol. 9 (*Coatings*).

Thermopower

When there is a temperature gradient along a normal (non-superconducting) metal there is an emf produced across its ends. This cannot just be measured by connecting the ends to a voltmeter since the emf would then be annulled over the whole circuit. For thermometry, two junctions are made between different metals and the overall emf, which is the difference between the emfs of the two materials, is a measure of the temperature difference between the two junctions. The principles of thermoelectricity are explained in D. K. C. MacDonald *Thermoelectricity: an introduction to the principles* (Wiley, 1962), J. M. Ziman *Electrons in metals* (Taylor and Francis, 1963), R. D. Barnard *Thermoelectricity in metals* (Taylor and Francis, 1972) and F. J. Blatt *et al. Thermoelectric power of metals* (Plenum, 1976).

Thermopower is very sensitive to impurities and although, according to the third law of thermodynamics, thermopower must tend to zero as $T \to 0$, and the sensitivity of many thermocouples decreases monotonically with decreasing temperature, others with magnetic impurities (e.g. Fe in Au) have thermopowers which increase again below 20 K and can be used for thermometry down to below 1 K.

There are tables of thermoelectric voltages for seven thermo-couples in the first edition of Quinn's book but, as has been mentioned, the Pt v Pt + Rh thermocouple is no longer the standard instrument for the high-temperature range below the gold point. *NBS Reports* 9712 and 9719 (1968) give tables for various 'low'-temperature thermocouples, including some with Au + 0.06% Fe and Au + 0.02% Fe as the active arm. If the second arm is made of a material with thermopower opposite in sign to that of these alloys, such as chromel, the thermopowers add and a thermocouple can be used from below 1 K to above room temperature with a sensitivity never less than 10 μV/K. In Britain, Johnson-Matthey produce gold wire with 0.03% Fe and the thermopower of thermocouples which include this material, as well as the effect of a magnetic field, can be found in Berman and Kopp (1971).

Although for thermometry the thermopower of a combination of materials is used with one junction at a known temperature, it is necessary to know the absolute thermopower of a material in order to interpret the phenomenon. This requires knowledge of the absolute thermopower of one material, and lead (Pb) has become the international standard for this. If the *reversible* heat (μ) evolved or absorbed when unit current flows through a conductor in which there is unit temperature gradient is measured, then the absolute thermopower is given by

$$S = \int_0^T \mu / T \, dT.$$

Results of such measurements for lead are given by Roberts (1977, 1982) for the ranges 7 to 350 K and 300 to 550 K, respectively. The latter article also gives the absolute thermopower of Cu, Au and Pt to 900 K. Below 7 K, lead is a superconductor and has zero thermopower, so a simple electrical measurement on thermo-couples in which one arm is a superconductor will give the thermo-power of other conductors.

The relevant articles in Landolt-Börnstein are: III/15b (1984), C. L. Foiles *et al. Thermopower of pure metals and dilute alloys*, and III/15a, J. Bass *Size dependence of thermoelectric power.*

High-temperature superconductors

Before 1986, the subject of superconductivity would have been con-fined to the very low temperatures of liquid hydrogen and liquid helium. According to the Bardeen-Cooper-Schrieffer (BCS) theory the expression for the transition temperature to the superconducting state for phonon-assisted formation of Cooper pairs of electrons is

$$T_c \sim (\hbar\omega_g/k) \exp (-2/V' g(E_F))$$

where ω_g is a lattice frequency, $g(E_F)$ is the density of electron states at the Fermi surface and V' is the electron-phonon coupling con-stant. It appeared that for real metals these three parameters vary with one another in such a way that the maximum value that could be expected to be possible for the transition temperature would be about 30 K. However, in 1986, Bednorz and Müller discovered superconductivity in the oxygen-deficient Ba-La-Cu-O system with transition temperature up to about 40 K, and transition tempera-tures well above 100 K are now attainable. Because of the scientific interest and possible practical applications of this discovery, there is now a vast amount of work on the subject, which is mentioned in Chapter 17 on Physics of materials. *Physica C* (North-Holland) is devoted to superconductivity, mainly high temperature super-conductors. Although the interest is not primarily in the thermal properties, it is worth mentioning that several thermal properties of these materials are being investigated as part of the effort to under-stand the phenomenon. Here we mention just two reviews of thermal properties: N. E. Phillips *et al.* (1992) on the specific heat and G. K. White (1992) on thermal expansion, and an article describing work on specific heat and thermal conductivity (Hussey *et al.*, 1992).

Low-temperature physics

Much of what has been written above covers thermometry and properties of matter at low temperatures, but various additional techniques are required for working below temperatures (~0.5 K) which can be reached by pumping on liquid ^{3}He, with temperature measured by using vapour pressure scales and other standard types of measurement (e.g. resistance thermometry). The method of cooling to below 1 K by adiabatic demagnetization of a paramagnetic salt, magnetized isothermally at the lowest temperature which could be reached with liquid ^{4}He, was proposed independently by Debye and by Giauque in 1927 and the first experiments were reported in 1933. Soon afterwards, Kurti and Simon started on similar experiments and demonstrated the use of such salts to determine the temperatures reached, by direct application of the second law of thermodynamics (Kurti and Simon, 1935). Since then, cooling to still lower temperatures has been achieved by the analogous method of demagnetizing nuclear paramagnets; in this case the temperature can be determined by the angular distribution of γ-rays emitted from nuclei which are oriented at these temperatures. The ^{3}He-^{4}He dilution refrigerator has become a fairly standard piece of equipment for reaching these sorts of temperatures.

There is a growing number of books which describe the techniques used at low temperatures, both above and below 1 K, among which are the following: G. K. White *Experimental techniques in low-temperature physics* (Clarendon, 3rd edn., 1987); R. C. Richardson and E. N. Smith *Experimental techniques in condensed matter physics at low temperatures* (Addison-Wesley, 1988); O. V. Lounasmaa *Experimental principles and methods below 1 K* (Academic, 1974); D. S. Betts *Refrigeration and thermometry below one Kelvin* (Sussex U. Press, 1976) and D. S. Betts *An introduction to Millikelvin technology* (CUP, 1989). All the techniques are mentioned in a less technical way in R. G. Scurlock (ed.) *History and origins of cryogenics* (Clarendon, 1992) which covers developments in many parts of the world.

Superconducting magnets are used in many applications at low temperatures and their use has spread to other domains, the best known of which is probably in NMR body scanners. The problems involved in their design and application are given in M. N. Wilson *Superconducting magnets* (Clarendon, 1983). Superconductors have also found applications in the measurement of very small magnetic fields, using superconducting quantum interference devices, SQUIDS, and in other ways (J. H. Hinken *Superconductor electronics*, (Springer, 1988) and V. Kose, (ed.) *Superconducting Quantum Electronics* (Springer, 1989)).

Soon after Kamerlingh Onnes first liquefied helium in 1908, he discovered superconductivity in mercury, and research on superconductivity has formed a large part of work at low temperatures. The focus of attention on superconductivity has now shifted to higher temperatures, as discussed in the previous section. It was only in about 1935 that the peculiar properties of liquid ^{4}He below 2.2 K began to be uncovered and there is now great interest in the super-fluid properties of the light isotope which occur at much lower temperatures. These developments are described in: P. V. E. McClintock, D. J. Meredith and J. K. Wigmore *Matter at low temperatures* (Blackie, 1984); D. S. Betts and J. Wilks *An introduction to liquid helium* (Clarendon, 1987); D. R. Tilley and J. Tilley *Superfluidity and superconductivity* (Hilger, 1990); D. Vollhardt and P. Wölfle *The superfluid phases of helium-3* (Taylor and Francis, 1990) and W. P. Halperin and L. P. Pitaevskii *Helium three* (North-Holland, 1990). Nuclear cooling and nuclear orientation are discussed by K. Andres and O. V. Lounasmaa (1982) and N. J. Stone and H. Postma *Low temperature nuclear orientation* (North-Holland, 1986).

There are International Conferences on Low Temperature Physics every three years and the proceedings are either published by the organizing institution or appear in *Physica* (North-Holland) or published separately by the same publishers. The 1993 conference (LT 20) was in Eugene, Oregon. There is an International Cryogenic Engineering Conference every two years. A number of publishers have been involved in printing the proceedings, the last (Kiev, 1992) was published as a supplement to vol. 32 of *Cryogenics*. The 1994 conference (ICEC15) will be in Genoa, Italy.

Journals which are devoted to low-temperature physics are *Journal of Low Temperature Physics* (Plenum), the Russian *Journal of Low Temperature Physics* (translated as *Low Temperature Physics (Soviet Journal of . . .* dropped from January 1993) (AIP) and *Cryogenics* (Butterworths). As most of the work at very low temperatures is concerned with condensed matter, articles also appear in the main condensed matter journals, such as *Physical Review B* (AIP), *Journal of Physics: Condensed Matter* (IOP) which, from 1989, has replaced *Journal of Physics C* and *F*, *Physica B* and *Zeitschrift fur Physik B* and *Physica Status Solidi A and B*, applied and basic research respectively.

The biennial USA Cryogenic Engineering Conferences are now combined with the International Cryogenic Materials Conferences, the proceedings being published by Plenum. The UK Institute of Physics publishes a *Directory of low temperature research and development in Europe* (P. C. McDonald, IOP, 7th edn., 1992).

Acknowledgements

I would like to thank Dr. T. W. Davies, Professor O. V. Lounasmaa, Professor J. S. Rowlinson, Professor C. A. Swenson and Dr. G. K.

White for their advice on what to include in the topics on which they are experts.

References

Andres, K. and Lounasmaa, O. V. (1982) *Progress in Low Temperature Physics*, ed. D. F. Brewer, **8**, 221. London: North-Holland.

Barron, T. H. K., Collins, J. G. and White, G. K. (1980) *Advances in Physics*, **29**, 609.

Bednorz, J. G. and Müller, K. A. (1986) *Zeitschrift für Physik*, B**64**, 189.

Berman, R. and Kopp, J. (1971) *Journal of Physics F*, **1**, 457.

Cali, J. P. (1976) *Pure and Applied Chemistry*, **48**, 503.

Corak, W. S. *et al.* (1955) *Physical Review*, **98**, 1699.

Debye, P. (1926) *Ann. d. Physik*, **81**, 1154.

Giauque, W. F. (1927) *Journal of the American Chemical Society*, **49**, 1864, 1870.

Goldber, R. N. and Weir, R. D. (1992) *Pure and Applied Chemistry*, **64**, 1545

Hussey, N. E., *et al.* (1992) *Proc. Int. Workshop on Electronic Properties and Mechanisms of High T_c Superconductors*, Tsukuba, 1991, p. 329. North-Holland.

Krishnan, R. S., Srinivasan, R. and Devanarayanan, X. (1979) *Thermal expansion of crystals*. Pergamon.

Kurti, N. and Simon, F. E. (1935) *Proceedings of the Royal Society*, A **149**, 152.

Maglic, K. D., Cezairliyan, A. and Peletsky, V. E. (1948, 1992) *Compendium of thermophysical property measurement methods*, vols. 1 and 2. Plenum.

Mark, H. F. *et al.* (eds) (1989) *Encyclopedia of polymer science and technology*. 2nd edn. Wiley.

Phillips, N. E., Fisher, R. A. and Gordon, J. E. (1992) *Progress in Low Temperature Physics*, ed. D. F. Brewer, **13**, 267. London: North-Holland.

Roberts, R. B. (1977) *Phil. Mag.*, **36**, 91.

Roberts, R. B. (1982) *Phil. Mag.*, **43**B, 1125.

Roder, H. M. (1981) *J. Res. Nat. Bur. Stand.*, **86**, 457.

Roder, H. M. (1982) *J. Res. Nat. Bur. Stand.*, **87**, 279.

Rusby, R. L., *et al.* (1991) *Metrologia*, **28**, 9.

Slack, G. A. (1979) The thermal conductivity of nonmetallic crystals. In: *Solid State Physics*, eds. H. Ehrenreich, F. Seitz and D. Turnbull, **34**, 1. New York: Academic.

White, G. K. (1992) *Studies of High Temperature Superconductivity*, ed. A. V. Narlikar, **9**, 121. Nova.

CHAPTER TWELVE

Geophysics, astrophysics and meteorology

E. MARSH

The boundaries around the areas of physics were never so complete that each discipline was a separate entity. Plasma could be considered as a laboratory phenomenon in laser and plasma physics or as an astrophysical phenomenon in the larger laboratory of the Universe. Astrophysics and particle physics have long shared an interest in cosmic rays but this is now extended to include such areas as solar neutrinos and big-bang phenomena. The subjects within this chapter give us a two-way view, in one we are looking out from Earth to an astrophysical view of the Universe and in the second looking in from beyond the atmosphere to see our place in that Universe.

In the 35 years since the first artificial satellites we can see the results of this two-way view. Telescopes, such as the Hubble Telescope, taken beyond our atmosphere or satellites such as the Infra Red Astronomical Satellite (IRAS) give a clearer view of the stellar medium and other satellites, using remote-sensing techniques give us a clearer view of the Earth and its atmosphere. In this chapter the definition of astrophysics as the application of modern physics to the problems of astronomy and of geophysics as the similar application to the study of the Earth will be followed. In order to remain consistent, this definition will also be applied in meteorology and there will be little reference to observational techniques in any of the three disciplines.

The previous edition noted two particular trends in the previous decade; the growth in the use of online bibliographic databases and the growth of databanks, both accessed over international networks. It was noted that possession of a computer terminal and telephone modem provided access for even the smallest group to resources which were previously only enjoyed by the largest libraries. This trend caused publishers of abstracting journals to worry about declining sales and librarians, who had been accustomed to providing a free information service, were faced with extra costs at a time of falling budgets. Although there has been no reversal in the use of databases and databanks these are increasingly provided in CD-ROM format either as stand-alone systems or available over local area networks. The cost of equipment and the annual cost of subscription can be justified in the wider availability of the information. The continuing cost of online access and the complexity of the command languages make end user searching impractical in most institutions. With CD-ROM systems there is little further cost after the initial charges for hardware and the regular subscription for material. There is also a range of more user-friendly software. Access to the systems by end users is limited only by availability of terminals and queuing is becoming a problem in some libraries.

The second point is the continuing trend towards publication in English which is discussed by Dennis Shaw in Chapter 2.

Abstracting journals

In addition to the relevant abstracting journals, reference has been made in this chapter to their availability as databases online or in CD-ROM format. The four main physics abstracting journals all have substantial coverage for these subjects. *Referativnyi Zhurnal* is the major Russian abstracting service published by VINITI and has two appropriate sections, Astronomya and Geofizika. The service has wide coverage but only limited use to non-Russian speakers, as many of the papers are abstracted in English-language publications even before the cover-to-cover translations are available. However, it can be a useful source of information on such obscure publications as the observatory reports from Eastern bloc countries. A similar comment could be made for the French INIST database *PASCAL*. There are two main contenders within the field of English physics abstracting journals: *Physics Abstracts*, and *Physics Briefs* (for fuller details, see Chapter 3). It would appear singularly unfortunate that these two publications have not been able to combine their resources to the production of a single service. There is a considerable overlap in the coverage, and many librarians must object to the cost of two subscriptions. Both publications cover a similar number of abstracts each year and have a similar delay in abstracting papers. *Physics Briefs* covers a slightly wider range of publications by including more report and thesis material. *Physics Abstracts* is available on all the major online hosts as INSPEC and is also available as a CD-ROM; *Physics Briefs* is available on the INKA system.

The major specialized abstracting service for astrophysics is *Astronomy and Astrophysics Abstracts*, which has appeared since 1968 and is a continuation of *Astronomischer Jahresbericht*. The older publication was in German and its replacement is in English. The journal is very comprehensive in its coverage, being prepared under the auspices of the International Astronomical Union (IAU) by the Astronomisches Rechen Institut at Heidelberg. It is published by Springer as two bound volumes a year, each covering the literature published during one half of the year. This can lead to delays of up to a year in the publication of a particular abstract. Each volume has an author index and a detailed subject index for the retrieval of individual articles. The abstracts are arranged into more than one hundred general subject categories and the six-monthly publication makes this an ideal way of getting a wide overview of the literature of a particular subject but is not ideal for immediate scanning.

Geophysics and Tectonic Abstracts (formerly *Geophysical Abstracts*) is published commercially by Geo Abstracts. About 3000 abstracts are noted in six issues produced each year, covering solid-earth geophysics and tectonics. There is an annual author and subject index and the material is also available online as part of the much wider coverage of the database GEOBASE. *Meteorological and Geoastrophysical Abstracts* contains material from 250 journals and is published monthly by the American Meteorological Society and lists over 7000 abstracts a year; the considerable delay in publication of the abstracts which was mentioned in the previous edition of this article has not, unfortunately, changed. Each issue contains an author, subject and geographical area index. The online database GEOREF, produced by the American Geological Institute, covers worldwide literature of geology and geophysics and corresponds to several publications, including *Geophysical Abstracts*.

The above abstracting journals generally cover only the formally published literature but much information is now published in report form. The space programme has generated a massive output of reports in each of the three subject areas and the National Aeronautics and Space Administration (NASA) has co-ordinated the publication of two guides both issued fortnightly.

International Aerospace Abstracts is issued by the American Institute of Aeronautics and Astronautics (AIAA) and contains some astrophysics papers and reports. The papers are mostly included in the abstracting journals mentioned above and the report material is generally included in *Scientific and Technical Aerospace Reports (STAR)*. This abstracting journal provides a very comprehensive survey of the report literature in space sciences and so includes much that is of interest. NASA funds much research in the remote-sensing fields which is applicable in geophysics and meteorology as well as the more obvious astrophysics. With both publications the delay time for publishing abstracts is about six to 12 months, and both journals include extensive author and subject indexes in each issue. These two abstracting journals are available as an online database through the Dialog system. However, the database is only available in the USA and Canada unless approval is given by the American Institute of Aeronautics and Astronautics who do not appear to answer letters of request.

Some publications are now only produced electronically and not on paper, including QUAKELINE, a bibliographic database covering eathquake engineering literature, produced by the National Center for Earthquake Engineering Research which has a very strong American bias with less than 10% of material in foreign languages, mostly Japanese. There is a wide coverage of types of material with some 400 items a month, including videocassettes, slides and reports with the more conventional books and journals.

Other secondary sources

The Institute of Scientific Information produces a weekly indexing periodical *Current Contents – Physical, Chemical and Earth Sciences* which reproduces the contents pages of a wide range of periodicals, including many appropriate to these fields. Each issue contains author and subject indexes and a reprint supply service is offered (OATS). The Meteorological Office, Bracknell, provides a library accession list which reflects their wide holdings. The list is issued monthly and contains approximately 800 items; although no abstracts are given, each paper is assigned a UDC number. The journal *Solar Physics* includes a list by subject groupings of papers of interest in other journals; the list is included at nine-monthly intervals. The National Oceanographic and Atmospheric Administration (NOAA) issues a mammoth *Catalog of the Atmospheric Sciences Collection* in the Library of NOAA and lists acquisitions from 1890 onwards in the form of photographed catalogue cards.

The Royal Meteorological Society, London, has produced a useful publication in 1982 entitled *Union catalogue of rare books on meteorology* available through the National Meteorological Library. Lang has collaborated in the preparation of a sourcebook (K. R. Lang and O. Gingerich (eds.) (1979) *A sourcebook in astronomy and astrophysics* (Harvard U.P.)) on the classic papers in astronomy and astrophysics published 1900–1975. A source of specialized materials is *Research on meteorology and climatology: a catalogue of doctoral dissertations*, published by University Microfilms.

As many journals are now produced electronically, resulting in the text held in machine-readable form, there is potential for immediate access to author and title indexes and to author-produced abstracts. An EC-funded project in mathematics has been noted but further work is required.

Primary journals

Although much new material is published in report form, the refereed paper in a scientific journal is still the preferred means of communication. Some articles may appear in rapid-publication journals, especially *Nature*, and a small number in general science journals such as *Scientific American, Science* and *New Scientist*. There is an overlap of interest in journals dealing with instrumentation, so that *Applied Optics, Journal of Physics E (Scientific Instruments)* and *Review of Scientific Instruments* will all contain papers in the topics for this chapter as well as for other subjects.

Within each subject area there will probably be a national journal devoted to each subject, and these are frequently in English: *Acta Astronomica* is the astrophysical journal for Poland and is normally in English; *Contributions to Atmospheric Physics (Beitrage zur Physik der Atmosphäre)* is published by Vieweg for the Deutsche Meteorologische Gesellschaft. This journal previously only accepted contributions in English but now makes very occasional exceptions for papers in German. *Revista Mexicana de Astronomia y Astrofisica* is a free publication issued by Universidad National Autónoma de México with the text in English and summaries in Spanish. The main Japanese journal dealing with astrophysics, *Publications of the Astronomical Society of Japan*, is published almost entirely in English. An important development in recent years has been the creation of European journals formed from the amalgamation of leading national journals. *Astronomy and Astrophysics*, published by Springer, is formed from an amalgamation of *Annales d'Astrophysique* (France), *Bulletin Astronomique* (France), *Arkiv for Astronomi* (Sweden), *Zeitschrift für Astrophysik* (Germany), *Bulletin of the Astronomical Institutes of the Netherlands* to which was added the *Bulletin of the Astronomical Institutes of Czeckoslovakia* in 1991. In addition to the 24 parts issued each year, there is also a four-volume supplement series usually giving accumulations of data, thus forming a substantial publishing programme. *Annalae Géophysicae* covering atmospheres, hydrospheres and space science was formed in 1983 as the official organ of the European Geophysical Society from an amalgamation of *Annales de Geophysique* and *Annali di Geofisica*. For both of these Europhysics journals, which reverse the normal trend of publishing more titles, the commonly used language is English. A leading journal which occupies a similar position to the above for meteorology is *Tellus* (*A: Dynamic Meterology and Oceanography* and *B: Chemical and Physical Meteorology*), published by Munksgaard for the Swedish Geophysical Society; *Pure and Applied Geophysics* (formerly *Geofisica*) published by Birkhauser is another journal in this category. India has an active space programme and this is reported in the *Indian Journal of Radio and Space Physics*, which is published in English by the Council of Scientific and Industrial Research.

The American Geophysical Union produces three main cover-to-cover translations of Russian journals: *Geomagnetism and Aeronomy* (translated from *Geomagnetizm i Aeronomiya), Physics of the Solid Earth (Fizika Zemli)* and *Atmospheric and Oceanic Physics (Fizika Atmosfery i Okeana*). A further Russian publication, *Geophysical Journal*, is translated from *Geofizicheskii Zhurnal* on a commercial basis by Gordon and Breach, while Plenum produce *Astrophysics* from the Armenian journal *Astrofizika*. The American Institute of Physics translating programme includes two of the major Soviet journals, *Soviet Astronomy* (*Astronomicheskii Zhurnal*) and *Soviet Astronomy Letters (Pisma v Astronomicheskii Zhurnal).* Somewhat more specialized is *Solar System Research (Astronomicheskii Vestnik*), published quarterly as part of the translation series from Consultants Bureau, whilst Allerton produce *Soviet Metereology and Hydrology (Meteorologiya i Gidrologiya).* There is, as yet, no clear indication of publication changes following the dissolution of the Soviet Union. The learned societies can reasonably be expected to print a significant proportion of the output of scientific research, sometimes reaching back more than a century. *Astrophysical Journal* (U. of Chicago Pr.) occupies a position in astrophysics comparable to *Physical Review* in some other areas of physics, and reflects the leading position which the United States occupies in astrophysical research. In addition to the main journal, there is a supplement series, mostly for data compilations and *Astrophysical Journal Letters,* which should not be confused with *Astrophysical Letters and Communications,* published commercially by Gordon and Breach. The American Geophysical Union (AGU) has a similar publishing schedule with *Journal of Geophysical Research (A: Space Physics; B: Solid Earth; C: Oceans and D: Atmospheres)*, whose standing again emphasizes the importance of American research, although scientists of other nationalities do publish in American journals. The AGU also publishes a letters journal, *Geophysical Research Letters*, and a newspaper, *Eos*, which contains a useful calendar of forthcoming events, book reviews and abstracts of papers presented at conferences.

The American Meteorological Society (AMS) produces the *Journal of Climate and Applied Meteorology* and *Journal of the Atmospheric Sciences*. The *Bulletin of the AMS* is useful for its book reviews and for its calendar of forthcoming meetings and conferences.

Comparable journals from British learned societies include the *Monthly Notices of the Royal Astronomical Society*, which has a preponderance of astrophysical papers. The *Geophysical Journal* of the Royal Astronomical Society merged in 1988 with the *Journal of Geophysics* (previously *Zeitschrift für Physik*) and *Annales Geophysicae* to produce *Geophysical Journal International* which is published on a monthly basis by Blackwells Scientific for the RAS, the Deutsche Geophysikalische Gesellschaft and the European Geophysical Society. The Royal Meteorological Society publishes a *Quarterly Journal* and is also responsible for the popular magazine *Weather*, which is aimed at the interested layman in addition to the professional meteorologist. Another British journal, *The Observatory*, edited from the Royal Greenwich Observatory, contains short notes which are often on astrophysical topics and a wide range of book reviews. Reports from research groups and observatories which are often difficult to obtain from other sources can be found in the *Quarterly Journal of the RAS*.

As in other areas of physics, there is a large number of commercially produced journals. These are sometimes general in their contents, as are the society journals; for example, *Astrophysics and Space Science* (Kluwer) covers virtually every branch of astrophysics. But the commercial journals concentrate usually on a specific subject area. Thus *Icarus* (Academic) and *Planetary and Space Science* (Pergamon) deal with solar system studies, as does *Solar Physics* (Kluwer), which covers solar research and solar-terrestrial relations. The journal *Geochimica and Cosmochimica Acta* (Pergamon), although covering a narrow subject area, crosses the boundaries between astrophysics and geophysics. Similarly, *Journal of Atmospheric and Terrestrial Physics* (Pergamon) is interdisciplinary to geophysics and meteorological physics. Three specialized journals which have excellent reputations are *Dynamics of Atmosphere and Oceans* (Elsevier), *Boundary-Layer Meteorology* (Kluwer) and *Physics of the Earth and Planetary Interiors*, also published by North-Holland. Kluwer (which now incorporates Reidel) publishes extensively in astronomy and astrophysics and a further specialized journal on comparative planetology is *Moon and the Planets*.

Review journals

The primary purpose of the journals described above is the publication of original research, although some of them may include occasional review articles. However, the growth in scientific publishing has been such that many scientists find great difficulty in keeping up with the literature of their subjects. This has given rise to a number of journals whose admirable purpose is to present an overview of a subject, hopefully by someone respected in the field and containing an extensive bibliography.

Some reviews of the subjects of this chapter may appear in general review journals such as *Reports on Progress in Physics* (IOP) and *Reviews of Modern Physics* (AIP). Various review journals cover specific subject areas, including *Reviews of Geophysics and Space Physics* (AGU), *Space Science Reviews* (Kluwer) and *Vistas in Astronomy* (Pergamon), which has an irregular publication schedule. A more recent journal, *Earth Oriented Applications of Space Technology* (Pergamon), has now dropped the initial *'Advances in'* from its title and falls between the division of primary and review journals. The *Astronomy and Astrophysics Review* (Springer) provides a primary source for reviews of the world literature of astronomy and astrophysics reviewing all fields, with frequency of review determined by the level of activity. In a slightly different category is *Comments on Astrophysics*, part C of *Comments on Modern Physics* (Gordon and Breach), which aims to provide critical comment on current controversial matters in astrophysics rather than reviewing them in depth. Springer publish a continuing series of reviews, and the volumes on geophysics (S. Flügge (ed.) (1956–) *Handbuch der Physik*, Bd. 47–49. *Geophysik I-III*, edited by J. Bartels and K. Rawer) are well recommended (but seem to have been delayed in publication).

Annual volumes of reviews often have a wider scope covering the whole of the discipline and the Annual Reviews series has a number of excellent titles. Two appropriate titles are *Annual Review of Astronomy and Astrophysics* and *Annual Review of Earth and Planetary Sciences*. Some papers relating to atmospheres and oceans as fluids will be found in *Annual Review of Fluid Mechanics*, and material on cosmic rays may be found in *Annual Review of Nuclear and Particle Science*. *Advances in Geophysics* (Academic) is a somewhat less regular publication, with some supplementary volumes and some issues devoted to the proceedings of major conferences.

Conference proceedings

Some conferences are one-off affairs, but many of the interesting volumes form part of a continuing series. Some sequences of conferences are only linked by the fact that they may have the same sponsorship. Thus the National Aeronautics and Space Administration (NASA) and the European Space Agency (ESA) sponsor a wide range of conferences, including remote-sensing and astrophysical topics which they publish as reports in special series (NASA-CP and ESA-CP). NATO is also a major conference organizer but the proceedings are usually published through a variety of independent publishers.

Within astrophysics, the International Astronomical Union (IAU) sponsors symposia and colloquia at frequent, but not fixed, intervals. Each conference is devoted to a specific topic and the resulting volumes contain mixed review and research material often representing the most comprehensive discussion of the topic available; the *IAU Symposia* are published by Kluwer. Other conferences which cover a wide subject area are issued as parts of *Advances in Space Research* (Pergamon), which is the official journal of the Committee on Space Research (COSPAR). Each issue contains the proceedings of a COSPAR-sponsored conference. The International Union of Geodesy and Geophysics holds a regular general assembly with an extensive programme of meetings, although the proceedings take some time to publish.

In common with most American professional associations, the AGU and the AMS have an extensive programme of conferences with their resultant proceedings. The International Association of Meteorology and Atmospheric Physics (IAMAP) has a smaller conference list, many being held in conjunction with other international bodies such as COSPAR and the World Meteorological Organization (WMO).

General series of conference, not specifically within these subject areas should not be ignored; examples include the conferences from the Society of Photo-Optical and Instrumentation Engineers in which *SPIE* vol. 996 (1990) covers high data rate atmospheric and space communications and *SPIE* vol. 1492 is the proceedings of a conference on earth and atmospheric sensing, whilst those from the American Institute of Physics include *AIP Conference Proceedings* no. 232 (1991), the International Symposium on Gamma ray Line Astrophysics.

Owing to the increased cost of book publication and to the extra delay which frequently results, it is becoming increasingly popular to publish the proceedings of a conference as a special issue of one of the journals.

Reports

A very high proportion of the world's report literature results from US government-funded research which is listed in *US Government Reports Announcements and Index (GRA &I)* and in *Scientific and Technical Aerospace Reports (STAR)*. Within our subject area, NASA produces a considerable volume, including conference proceedings, data compilations and visual material in addition to technical reports. Some report material is in preprint form, later appearing as a formal publication, whilst a considerable number of American reports are of a form required for quarterly or annual accounting.

In addition, there is the ESA report series which contains multidisciplinary material. Astrophysics generates a large proportion of the report literature and major centres include the European Southern Observatory, Munich, the Center for Astrophysics, Cambridge, USA, the Owens Valley Radio Observatory, Pasadena, USA, and the Institute of Astronomy, Cambridge, UK. Many of these reports are not listed in the abstracting journals and difficulty can be experienced in tracing them. This is discussed fully by John Chillag in Chapter 19.

Reference works and handbooks

In a subject with a high proportion of jargon a useful reference work is *Acronyms and abbreviations in astronomy, space sciences and related fields* (1991), edited by A. Heck and published by the Observatoire Astronomique, Strasbourg, containing 45 000 terms. The same organization and editor have also produced *Astronomy, space sciences and related organizations of the world* as a two-volume directory, ASpScROW 1991.

The *Astronomical Almanac* has replaced both the *American Ephemeric and Nautical Almanac* and the *Astronomical Ephemeris of the Royal Greenwich Observatory*, UK. The publication is printed by the US Government Printing Office with some material from the UK. The *Handbook of the British Astronomical Association* is an annual publication giving ephemerides and predictions for the coming year.

Dictionaries can be a useful source of information and the *Concise dictionary of astronomy*, edited by J. Mitton (OUP, 1991), covers terms and names peculiar to astronomy but with terms from physics and space sciences. A work within geophysics is the *Collins dictionary of geology* (1990), edited by D. R. Coates and D. F. Lapidus, which has grown from the *Facts of file dictionary of geology and geophysics* (1987) although less authoritative than the *Glossary of geology* from the American Geological Institute (3rd edn, 1987). Within meteorology, the 6th edition of *The meteorological glossary* was published by HMSO in 1991 and there is the *Meteorology source book* (1988), edited by S. P. Parker as part of the McGraw-Hill Science Reference Series, with a mini-encyclopaedic approach. For space sciences, there is the *Dictionary of space technology*, edited by M. Williamson (Hilger, 1990), which provides an excellent illustrated introduction to 1600 terms on launch vehicles, orbits and materials.

Specialized publications

As part of the International Geophysical Year in 1957, world data-centres were established to collate solar and ionospheric data. The National Geophysical and Solar-Terrestrial Data Centre, Boulder, USA, produces a monthly report in two sections, tabling data on such topics as solar flares and cosmic rays. The World Data Centre, Chilton, UK, specializes in ionospheric data and there are other centres in Russia and Japan. Some geophysical and meteorological material will be found in reports from NOAA which are readily available and are indexed in several abstracting journals.

Springer have long been noted for the quality of their data compilations, and a new series contains astronomy and astrophysics: H. H. Landolt and R. Börnstein (eds.) *Landolt-Börnstein numerical data and functional relationships in science and technology*, New Series, Group VI, vol. 2. *Astronomy and astrophysics*, edited by K. Schaifers and H. H. Voigt, 3 vols, (1981–1982, Springer). Within astrophysics the most widely used source of information is C. W. Allen *et al.* (1973) *Astrophysical quantities* (Athlone), of which there has regrettably not been a later edition. A newer work is *Handbook of space astronomy and astrophysics* by M. V. Zombeck of which CUP published the 2nd edition in 1990. It is intended as a working tool for researchers, technicians and graduate students with the advantage that diskettes for IBM-compatible personal computers are available on request and contain much of the material in the book. A complementary work to that by C. W. Allen is K. R. Lang (1974) *Astrophysical formulae* (Springer), stressing the physics rather than the astronomy of the subject. The material on stars is now formally published in R. Burnham (1978) *Burnham's celestial handbook*, 3 vols, (Dover), parts of which circulated privately for some years. WMO sponsors the *Monthly Climatic Data for the World* which is published by NOAA.

The main databank for information concerning the properties of stars is held at the Stellar Data centre at the University of Strasbourg Observatory. The magnetic-tape versions of the catalogues are now made available on the Starlink computer network. The IAU produces a small folded news sheet, *Information Bulletin on Variable Stars*, and small cards which are issued from the Central Bureau of Astronomical Telegrams and results from recent discoveries.

Positional material may be needed for astrophysical research, and from the multitude of star catalogues the *Yale bright star catalogue* (D. Hoffleit and C. Jaschek), *The bright star catalogue* (Yale University Observatory, 4th edn., 1982) and the *Smithsonian star catalog* are recommended. Many galaxies and star clusters are known by their NGC number given in J. L. E. Dreyer (1953) *New general catalogue of nebulae and clusters of stars* (RAS). Modern technology is to be found in the presentation of imaging data from the Einstein Observatory in CD-ROM format, produced by the Smithsonian Astrophysical Laboratory as the third part of *Imaging and X-ray astronomy – a decade of the Einstein Observatory achievements*, edited by M. Elvis (CUP, 1990). Remote sensing data, including SPOT data is available in CD-ROM format from a number of American and French centres. Geophysical and meteorological observational data can be obtained in report format from the US Geological Survey, and from a useful catalogue of geophysical anomalies in encyclopedic format: W. R. Corliss (1982) *Lightning, auroras, nocturnal lights and related luminous phenomena (Sourcebook)*.

Monographs

Rather than attempting to produce an exhaustive bibliography of these three subjects the author's aim has been to include a selection of standard texts and recent books which have been well reviewed. The introductory texts are given first, with more specialized material later. An inter-disciplinary book giving an overall view of the connections between the environment of Man and Earth and of Earth in the cosmos is R. S. Kandel (1980) *Earth and cosmos* (Pergamon).

Although Stephen Hawking has written one of the most popular texts on the universe for the general reader with *A brief history of time* (Isis, reprinted 1989), this should be preceded by C. A. Ronan *The natural history of the universe* (Doubleday, 1991) which is an intelligent layman's guide to the origin, structure and possible future of the universe.

Of general interest to scientist and layman is the excellent book by R. W. Smith (1989) *The space telescope* (CUP) which presents a carefully documented history of one of the largest science projects ever undertaken. E. Böhm-Vitense (1989–1992) *Introduction to stellar astrophysics* (CUP) provides a comprehensive text at undergraduate level and L.H. Aller (3rd edn., 1990) *Atoms, stars and nebulae* is a completely new version of a standard undergraduate level textbook. A standard astronomy text which is encyclopedic in scope, G. O. Abell *Exploration of the universe* (Saunders, 6th edn., 1991), is intended for students, and uses the technique of a different typeface to isolate more complex ideas from the rest of the text. A. Unsold and B. Baschek *Der neue Kosmos* (Springer, 4th English edn., 1992) is considered an unrivalled guide to present-day astrophysics and is a translation of the 5th German edition (1991) which is an update of the completely revised edition of 1988. M. Harwit (1973, reprinted 1982) *Astrophysical concepts* (Wiley) has aimed for the physical concepts which are important in astrophysics in the hope that the book will still be useful when further discoveries have invalidated more fashionable theories. A similar wish is expressed by R. Kippenham and A. Weigert (1990) for *Stellar structure and evolution* (Springer) which is a basic physics and mathematical treatment of the theory by two pioneers in the field.

Works which go into many editions can usually be considered to fill a need, and B. J. Bok and P. F. Bok *The Milky Way* (Harvard U.P., 5th edn., 1981) has obviously done this with its treatment of the galaxy as an organic entity rather than as constituent parts. W. Hamblin has also met a continuing need with *The earth's dynamic systems* (Macmillan, 5th edn., 1989). M. Rowan-Robinson (1990) *Universe* (Longman) with S. A. Kaplan (1982) *The physics of stars* (Wiley) have been well recommended as introductory texts. C. R. Hitchin has provided good coverage of instruments and practical techniques for undergraduates and researchers in his *Astrophysical techniques* (Hilger, 2nd edn., 1991). F. H. Shu (1991) considers physical principles in *Physics of astrophysics* (University Science Books), whilst P. Lena (1988) is concerned with techniques in a book translation from the French, *Observational astrophysics* (Springer). A rather specialized area of astrophysics which still remains just within the general discussion is high-energy astrophysics. At a slightly lower level of pre-course reading is J. V. Narlikar (1982) *Violent phenomena in the universe* (OUP), with M. S. Longair (1981) *High energy astrophysics* (CUP) providing a well-written and stimulating book which is somewhat more advanced. It would not be possible to ignore the monumental series edited by

G. P. Kuiper (1960–1975) *Stars and stellar systems*, 9 vols (U. of Chicago Pr.), which, although dated in parts, is a valuable general collection.

A text which provides an interesting if difficult link is C. W. Gordon *et al.* (eds.) (1978) *Handbook of astronomy, astrophysics and geophysics,* vol. 1 (Gordon and Breach), of which only the first volume appears to have been issued. A fascinating account of geophysics for both the interested general reader and the professional geophysicist is C. C. Bates *et al.* (1982) *Geophysics in the affairs of man* (Pergamon), with a detailed history of twentieth-century geophysics and its effects on mankind. A more conventional text for both undergraduates and researchers, which provides a comprehensive survey of both theoretical and practical aspects of geophysical methods, is D. S. Parasnis *Principles of applied geophysics* (Chapman and Hall, 4th edn., 1986). An aspect of geophysics which has been covered in a well-reviewed book is W. D. Parkinson (1983) *Introduction to geomagnetism* (Scottish Acad. Pr.), whilst A. M. Jessop (1990) has written *Thermal geophysics* (Elsevier) as an introduction to the subject for students and researchers. Geodesy is a specialized area of geophysics involving the study of the shape and gravity field of the earth. A work aimed at graduate students is K. Lambeck (1988) *Geophysical geodesy: the slow deformations of the Earth* (Clarendon), whilst a more advanced text is H. Moritz (1990) *The figure of the Earth* (Wichmann).

K.-N. Liou (1980) *An introduction to atmospheric radiation* (International Geophysics Series, 26) (Academic) is useful both as a textbook and as a reference book. It requires a substantial background in mathematics and physics and gives problems at the end of each chapter. R. G. Fleagle and J. A. Businger *An introduction to atmospheric physics* (Academic, 2nd rev. edn., 1981) is the second edition and a substantial review of a familiar and much-used undergraduate textbook. An approach which requires only elementary mathematics for an undergraduate work on atmospheric physics is J. V. Iribane and H.-R. Cho (1980) *Atmospheric physics* (Reidel). An advanced textbook on atmospheric physics is the monograph by T. Tohmatsu (1990) *Compendium of aeronomy* (Kluwer) which was translated from the original Japanese. The study of the upper atmosphere and solar-terrestrial relations has grown in importance, and J. K. Hargreaves (1979) *The upper atmosphere and solar-terrestrial relations* (Van Nostrand) presents an integrated account of the wide range of phenomena in the Earth's outer environment. A specific area within atmospheric studies is covered by T. Godish with *Air quality* (Lewis, 2nd edn., 1991) which covers atmospheric pollution, stratospheric ozone and global warming at an advanced undergraduate level. R. P. Wayne was moved to produce a further edition of *Chemistry of atmospheres; an introduction to the chemistry of the atmospheres of Earth, the planets and satellites* (Clarendon, 2nd edn., 1991) by the discovery of ozone depletion, although the title is somewhat misleading as only one chapter is devoted to other planets. An excellent work which covers the main areas involved in the physics of neutral atmospheres is J. T. Houghton *The physics of atmospheres* (CUP, 2nd edn., 1986). This same author, (a former Director-General of the UK Meteorological Office) wrote two excellent books in 1984: *Remote sounding of atmospheres* (Cambridge Planetary Sciences 5, CUP) and *The global climate* (CUP).

As meteorology is so much in evidence in our daily lives, it is of interest to both amateur and professional. B. W. Atkinson (ed.) (1981) *Dynamical meteorology* (Methuen) is a collection of essays in which well-qualified authors describe some basic questions of dynamical meteorology. The book is an outgrowth of a series of articles in the magazine *Weather*, and done at that level. E. K. Lutgens and E. J. Tarbuck *The atmosphere* (Prentice-Hall, 5th edn., 1992), R. G. Barry *et al. Atmosphere, weather and climate* (Methuen, 5th edn., 1987) and H. J. Critchfield *General climatology* (Prentice-Hall, 4th edn., 1983) are standard undergraduate works on climate and meteorology which have run into several editions. R. A. Anthes *et al. The atmosphere* (Merrill, 3rd edn., 1981) provides more varied subject matter and L. J. Battan *Fundamentals of meteorology* (Prentice-Hall, 2nd rev. edn., 1984) also considers the meteorology of the Southern Hemisphere. The work of H. H. Lamb in modern meteorology is difficult to ignore and his *Climate, history and the modern world* (Methuen, 1982) gives a history of the transformation in thinking on climatology. K. I. Kondratiev (1988) looks at natural and man-made effects in *Climate shocks: natural and anthropogenic* (Wiley).

S. Weinberg (1977) *The first three minutes* (Deutsch) is now reissued in paperback form and was received with great enthusiasm and has provided the basis of continuing discussion between astrophysicists and particle physicists. The interstellar medium has been well covered by a number of excellent books: J. E. Dyson and D. A. Williams (1980) *The physics of interstellar matter* (Halsted) and L. Spitzer (1982) *Searching between the stars* (Mrs Hepsa Ely Silliman Memorial Lectures, 46) (Yale U. Pr.) are of an intermediate level, with L. Spitzer (1978) *Physical processes in the interstellar medium* (Wiley) for the more advanced student. Interplanetary dust is covered in P. W. Hodge (1981) *Interplanetary dust* (Gordon and Breach) or, better still, because it is more complete, I. Halliday and B. A. McIntosh (eds.) (1980) *Solid particles in the solar system* (IAU Symposium 90) (Reidel). V. C. Reddish (1974) *The physics of stellar interiors* (Edinburgh U. Pr.), although somewhat older, is a classic work on stellar interiors. Our own particular star, the Sun, has an extensive publications list, including E. R. Priest (1982) *Solar magnetohydrodynamics* (Reidel), and Z. Svetstka (1976) *Solar flares* (Geophysics and Astrophysics Monographs, 8; Kluwer). S. I. Akasofu and Y. Kamide (1987) have produced an excellent work for graduate students, *The solar wind and the Earth* (Reidel) which avoids advanced mathematics. The subject of cosmic magnetic fields is exhaustively treated in E. N. Parker (1979) *Cosmical magnetic fields* (OUP). An advanced text edited by A. G. Petschek (1990) *Supernovae* (Springer) is a compilation of 10 articles with disparate styles giving the most up to date information on the subject.

The visitation from Halley's comet in 1986 resulted in several useful works, including an IAU colloquium (1991) *Comets in the post-Halley era*, (2 vols, Kluwer) and J. Mason (ed.) (1990) *Comet Halley: investigations, results, interpretations* (2 vols, Ellis Horwood). A comprehensive review of theories on the origins of comets for professionals and students will be found in M. E. Bradley *et al.* (1990) *The origin of comets* (Pergamon). The listing of comets has been a popular activity and D. K. Yeomans (1991) *Comets, a chronological history of observation, science, myth and folklore* (Wiley) is an entertaining work for both amateurs and professionals. A fascinating though specialized work is J. Williams (1987 reprint of the original 1871 edition) *Observations of comets: from 611BC to AD1640 extracted from the Chinese annals* (Science and Technology).

Two books which cross the divide between astrophysics and geophysics are P. S. Wesson (1978) *Cosmology and geophysics* (Hilger) and K. Y. Kondratyev and G. E. Hunt (1982) *Weather and climate on planets* (Pergamon). The first links cosmology and geophysics with the theory of the expansion of the Earth, a predicted consequence of several cosmological theories. The second discusses the atmosphere of other planets in the light of recent space missions and provides detailed comparisons of related processes in the Earth's atmosphere. R. A. Wells (1979) *Geophysics of Mars* (Elsevier) is still a standard work of reference for the geophysics of Mars. An advanced postgraduate text which is heavily mathematical is provided by B. Bertotti and P. Farinella (1990) *Physics of the Earth and solar system* (Kluwer) which covers dynamics and evolution, space navigation and space-time structures.

The increase in observational precision which results from the use of satellites requires further theoretical developments, and these are well presented in H. Moritz (1981) *Theories of nutation and polar motion: reports of geodetic science, 309 and 318* (Ohio State U. Pr.). Another work which discusses the nature of the Earth's rotation and the geophysical mechanisms responsible is K. Lambeck (1980) *The Earth's variable rotation: geophysical causes and consequences*

(CUP). Three works included as an indulgence, although excellent in their presentation of information on optical phenomena in the sky, are R. H. Eather (1980) *Majestic lights* (AGU), which describes the aurora for general audiences, R. Greenler (reprinted 1989) *Rainbows, halos and glories* (CUP) and M. Gadsen and W. Schröder, *Noctilucent clouds* (Springer). The phenomenon of ball lightning, which has so far defied explanation, is covered in J. D. Barry (1980) *Ball lightning and bead lightning* (Plenum), with extensive coverage over 300 years. When discussing geophysics it is impossible to ignore the contribution of Akasofu, and the Chapman conference proceedings contains an excellent review article on the magnetosphere (S.-I. Akasofu (ed.) (1980) *Dynamics of the magnetosphere, substorms and related plasma processes* (Reidel)) and his work on solar-terrestrial relations (S.-I. Akasofu (1972) *Solar-terrestrial physics* (Clarendon)), although a little dated is still a classic. Linking the two great fluid media of the Earth, atmosphere and ocean, is a comprehensive work, A. E. Gill (1982) *Atmosphere – ocean dynamics* (Academic), and within the ocean area a wealth of information may be found in C. Emiliani (ed.) (1981) *The sea*, vol. 7, *The oceanic lithosphere* (Wiley).

As so much work has come from modern developments in space technology which have contributed to advances mentioned above, it is appropriate to conclude with several books in this area. M. Williams (1990) *The communications satellite* (Hilger) presents a comprehensive and clearly written account of all aspects of communications satellite design. The Open University has long been noted for the excellence of their texts and this can be seen in J. J. M. Leinders (1989) *Remote sensing* (2 vols, Open University) which was developed in association with the Department of Earth Sciences at the O.U. and there are further useful introductory texts by A. P. Cracknell and L. Hayes (1991) *Introduction to remote sensing* (Taylor and Francis) and a paperback by W. G. Rees (1990) *Physical principles of remote sensing* (CUP). A further introductory text is S. A. Drury (1990) *A guide to remote sensing* (OUP) which is a basic guide to remote sensing, setting out scientific principles and discussing applications. F. T. Ulaby *et al.* (1981–1983) *Microwave remote sensing – active and passive* (Addison-Wesley) is a unified treatment of the more specialized area of microwave remote sensing written for undergraduates and specialists, and E. C. Barrett and D. W. Martin (1981) *The use of satellite data in rainfall monitoring* (Academic) looks at satellite data for rainfall. D. King-Hele and others (1990) *The RAE table of earth satellites 1957–1987* (Royal Aircraft Establishment) is a comprehensive listing of all known satellites.

CHAPTER THIRTEEN

History of physics*

W. D. HACKMANN

Over the past few decades a revolution has taken place in the history of science in general, and in that of physics in particular. Previously, the common treatment was in terms of the triumphant march of scientific progress brought about by genius: the Galileos, Newtons and Einsteins of the scientific firmament. But a new generation of historians has appeared (not welcomed by those scientists interested in the pedagogic value of the cautionary historical tale or by those seeking unambiguous answers), who have firmly placed the history of physics in the discipline of historical research. The more complex picture of scientific progress that has emerged is not always appreciated by those physicists not particularly sympathetic to the kind of questions that may occupy the historian, but who simply want to trace the evolution of a concept or experiment that has borne fruit. For the professional historian of the subject, the physicist's pre-occupation is only one of the aspects that can be studied; he may even have trouble in defining what is historically fruitful.

An area that has grown dramatically since Thomas S. Kuhn (1962) *The structure of scientific revolutions* (U. of Chicago Pr., 2nd enlarged ed., 1970) is the sociology of science. Here we encounter a heady mix of what perhaps ought to be termed 'historico-socio-logical' analysis, often with a philosophical tinge as the various factions debate the validity of their own viewpoint or methodology. This is not the place to go into specifics, such as comparing the anthropological approaches in history, or 'ethnomethodology', as discussed in K.D. Knor-Cetina and M. Mulhay (eds.) (1983) *Science observed. Perspectives on the social study of science* (Sage), or in

---

* based on the first edition chapter written by A. R. Dorling

Bruno Latour (1979; 2nd edn., 1986) *Laboratory life. The social construction of scientific facts* (Sage, 2nd edn., Princeton U. Pr.), social constructivism, as used in A. Pickering (1984) *Constructing quarks. A sociological history of particle physics* (Edinburgh U. Pr.), or sociological relativism and reflexivity as in M. Ashmore (1989) *The reflexive thesis. Wrighting (sic) sociology of scientific knowledge* (U. of Chicago Pr.). The language in these works has the tendency to become as jargon-laden ('theory-laden' is the polite term) as in the most abstruse of physics textbooks. These various sociological approaches have percolated down into most of the recent histories of physics written by historians, but not in those by physicists or engineers, as for instance, in their volumes contributed for the IEE (Institution of Electrical Engineers) History of Technology Series, under the general editorship of Brian Bowers of the Science Museum in London, published by Peregrinus.

Perhaps it is best for the moment to return to more familiar ground and to consider the by now old chestnut of what constitutes 'pure' as opposed to 'applied' research. This debate has certainly gone on much longer than might be assumed. C. Babbage (1830) *Reflections on the decline of science in England, and on some of its causes* (repr. Irish U. Pr., 1972) argued that the decline of science could be halted if the Government employed scientists and instituted research grants. Babbage was one of those who suggested establishing the British Association for the Advancement of Science, whose impact on Victorian culture has been chronicled in J. B. Morrell and A. Thackray (1981) *Gentlemen of science. The early years of the British Association for the Advancement of Science* (OUP), and in R. M. MacLeod and Peter Collins (1981) *The parliament of science. The British Association for the Advancement of Science 1831–1981* (Science Reviews, Norwood). The funding of physics in Britain during the final decades of the nineteenth century is touched upon by several authors in G. L'E. Turner (ed.) (1976) *The patronage of science in the nineteenth century* (Noordhoff); by R. M. MacLeod (1972) 'Resources of science in Victorian England: the endowment of the science movement, 1868–1900', in *Science and society 1600–1900*, edited by P. Mathias, pp. 111–166 (CUP 1976); in R. M. MacLeod (ed.) (1988) *Government and expertise. Specialist administrators and professionals, 1860–1919* (CUP) and in Peter Alter (1987) *The reluctant patron. Science and the state in Britain 1850–1920*, translated from the German by Angela Davis (Berg), which concentrates on the period 1900 to 1920.

The distinction between pure and applied science became especially fuzzy in government-sponsored research (for instance for the military), and in industrial research. The two World Wars, in particular, accelerated the co-operation between physicists, governments and the military. The crucial rôle of science and technology in the First World War is described in Guy Hartcup (1988) *The war of invention. Scientific developments, 1914–18* (Brassey's), in which the author brings together recent historical research. Tim Travers (1992) *How the war was won. Command and technology in the British army on the Western Front 1917–1918* (Routledge) has a rather confusing title for the non-military historian, as he does not deal with the development of war technologies but with how the generals of the British Expeditionary Force coped tactically with the emerging modern mechanical warfare (tanks, machine guns and airplanes) in the last phases of this terrible war. The origins of naval physical research in the First World War have been examined in R. M. MacLeod and E. K. Andrews (1971) 'Scientific advice in the war at sea, 1915–17: the Board of Invention and Research', in *Journal of Contemporary History*, vol. 6, 3–40. This Board supervized the Royal Navy's work in underwater acoustics, or asdics. Indeed, two of the earliest successes of military physics research were asdics or sonar which began in World War I, and radar started in the interwar period. For the 'official' history of sonar see W. D. Hackmann (1984) *Seek and strike. Sonar, anti-submarine warfare and the Royal Navy 1914–1954* (HMSO). A detailed account of the early days of radar is given in S. S. Swords (1986) *Technical history of the beginnings of radar* (Peregrinus). The British contribution to radar has been described by several of those involved in the pioneering days. A dated classic is R. A. W. Watt (1957) *Three steps to victory* (Odhams). More recent accounts are G. Hartcup (1970) *Challenge of war* (David and Charles); E. G. Bowen (1987) *Radar days* (Hilger), which has a fascinating story about how the Boot and Randall cavity magnetron was introduced into America; E. B. Callick (1990) *Metres to microwaves. British development of active components for radar systems 1937 to 1944* (Peregrinus); and H.D. Howse (1993) *Radar at sea. The Royal Navy in World War 2* (Macmillan), which is the first detailed history of radar research in the Royal Navy sanctioned by the surviving naval scientists. The official American story is told in the two-volume history by Henry E. Guerlac (1987) *Radar in World War II* (Tomash and American Institute of Physics). This volume is in the series 'The History of Modern Physics 1800–1950' which reprints 'classical' texts on modern physics. Twelve volumes have been published to date, including Max Jammer (1989) *The conceptual development of quantum mechanics*, previously published by McGraw-Hill in 1966. Electronics have had a tremendous impact on warfare. Some of the early American developments have been described in L. S. Howeth (1963) *History of communications. Electronics in the United States Navy* (US Gov. Pr. Office), and the British, in particular as they related to the Royal Navy, in A. R. Hezlet (1975) *The electron and sea power* (Peter Davies). Colourful accounts of electronic warfare are given in A. W. Price (1967) *Instruments of darkness. The history of electronic warfare* (Kimber; rev. edn., Macdonald and Jane's, 1977) and R. V. Jones (1978) *Most secret war* (Hamilton). The history of the atomic bomb and subsequent atomic energy policy have also received much attention. The British wartime and post-war developments up to 1952 have been dealt with in M. M. Gowing (1964) *Britain and atomic energy 1939–45* (Macmillan); (1974) *Independence and deterrence. Britain and atomic energy, 1945–52*, 2 vols. (Macmillan) (assisted by L. Arnold), while the official American account is given in R. G. Hewlett and O. E. Anderson (1962) *The new world, 1939–46*, vol. 1 of *A history of the United States atomic energy commission* (Penn. St. U. Pr.). The 'myth of the German atomic bomb' is discussed in Mark Walker (1989 and 1993) *German National Socialism and the quest for nuclear power 1939–1949* (CUP), which touches on the wider issues of the relationship between physics research and the state. R. G. Hewlett and E. Duncan (1974) *Nuclear navy 1946–62* (U. of Chicago Pr.) deals with the introduction of this technology into the US Navy. The strategic implications of this kind of research and development are drawing increasing attention, but fall outside the scope of this chapter.

D. Greenberg (1967) *The politics of pure science. An inquiry into the relationship between science and government in the United States* (New Am. Lib.); also as *The politics of American science* (Penguin, 1969) is a highly critical account of the politics of American 'big' science battling for funding, while D. J. Kevles (1978) *The physicists. The history of a scientific community in modern America* (Knopf) is a historian's view of the political and economic factors affecting progress in physics. The latter contains a comprehensive 'essay on sources', permitting the reader to delve further into this topic. The same subject has been broadly covered more from a physicist's viewpoint in J. Ziman (1976) *The force of knowledge. The scientific dimension of society* (CUP). Ziman's book is intended primarily for science undergraduates, who may find it thought provoking.

Of special interest is the impact of technological firms, such as IBM, and government agencies. The histories of some of these establishments have been written. Recent examples are M. D. Fagen (ed.) (1975–1978) *A history of engineering and science in the Bell system*, 2 vols (Bell Labs.), which is a comprehensive history of the research at the Bell Telephone Laboratories, to which a third volume, edited by G. E. Schindler, was added in 1982; Robert Clayton and Joan Algar (1989) *The GEC Research Laboratories 1919–1984* (Peregrinus); Leonard S. Reich (1985) *The making of American industrial research. Science and business at GE and Bell, 1876–1926* (CUP); W. B. Carlson (1991) *Innovation as a social process. Elihu Thomson and the rise of General Electric, 1870–1900* (CUP); J. F. Wilson (1988) *Ferranti and the British electrical industry, 1864–1930* (Manchester U. Pr.); and M. Campbell-Kelly (1990) *ICL. A business and technical history* (OUP). On a smaller scale, there is the history of one of the best known British instrument-making firms, M. J. G. Cattermole and A. F. Wolfe *Horace Darwin's shop. A history of the Cambridge Scientific Instrument Company 1878–1968* (Hilger).

Among the fruits of large-scale research are electronic broadcasting and communication, the semiconductor industry, and lasers. These topics have been dealt with recently in R. W. Burns (1986) *British television. The formative years* (Peregrinus); Gordon Bussey (1990) *Wireless the crucial decade 1924–34* (Peregrinus); K. Geddes and Gordon Bussey (1991) *The setmakers. A history of the radio and television industry* (BREMA); P. R. Morris (1990) *A history of the world semiconductor industry* (Peregrinus), which is rather an ambitious title for this slim book; Hans Queisser (1988) *The conquest of the microchip. Science and business in the silicon age* (Harvard U. Pr.); and J. L. Bromberg (1991) *The laser in America, 1950–1970* (MIT). A comprehensive study is M. Eckert and H. Schubert (1990) *Crystals, electrons, transistors. From scholar's study to industrial research*, translated by Thomas Hughes (American Institute of Physics). Narrowly focused historical monographs have also been written on specific electronic components, such as the two volumes by V. J. Phillips (1980) *Early radio wave detectors* (Peregrinus), and (1987) *Waveforms. A history of early oscillography* (Hilger), and P. Povey and R. A. J. Earl (1988) *Vintage telephones of the world* (Peregrinus). The two publishers specializing in such publications in this country are Peregrinus for the Institution of Electrical Engineers and Hilger for the Institute of Physics. The theme of the telephone is developed in C. Marvin (1988) *When old technologies were new. Thinking about electric communication in the late nineteenth century* (OUP). A thought-provoking book on the impact of improved communication (including electronic) on society is J. M. Beniger (1986) *The control revolution. Technological and economic origins of the information society* (Harvard U. Pr.).

The organization of science in twentieth-century Britain has been discussed in, among others, D. S. L. Cardwell (1957) *The organisation of science in England* (Heinemann, rev. edn., 1972), H. Melville (1962) *The Department of Scientific and Industrial Research* (Allen & Unwin), J. B. Poole and K. Andrews (1972) *The government of science in Britain* (Weidenfeld), I. M. Varcoe (1974) *Organising for science in Britain. A case study* (OUP) and P. Gummett (1980)

Scientists in Whitehall (MUP). E. Pyatt (1983) *The National Physical Laboratory. A history* (Hilger) lacks references and a bibliography. Government laboratories deserve further (and better) treatment. Large university (institutional) laboratories are also drawing the attention of historians, as places of research, their links with government institutions, and as communities of scientists. J. G. Crowther (1979) *The Cavendish Laboratory 1879–1974* (Macmillan) is a bulky history of this famous Cambridge Laboratory, written in the traditional mould. Another recent area of interest is 'big science' which is dealt with in the same way as institutional science generally. Notable histories are A. Hermann *et al.* with a contribution by L. Belloni (1987) *The history of CERN* (2 vols., North-Holland) and by J. L. Heilbron and R. W. Seidel (1990) *Lawrence and his laboratory. A history of the Lawrence Berkeley Laboratory* (U. of Cal. Pr.). For an analysis of the historical literature dealing with the interaction between science and politics, see M. M. Gowing (1983) 'The history of science, politics and political economy', in *Information sources in the history of science and medicine*, edited by P. Corsi and P. Weindling, pp. 99–115 (Butterworths).

Two other areas should be briefly discussed in this introduction: the philosophy of physics and instrumentation. Many of the great physicists such as Bohr, Mach and Einstein, were very interested in the philosophical implications of their work. Historical examples were used as case studies for philosophical analysis, as, for instance, by Mach in his book on mechanics, greatly extended by Duhem (1914), and on heat, both cited later. An analysis of Mach is J. Blackmore (ed.) *Ernst Mach – a deeper look. Documents and new perspectives* (Kluwer). Bohr's epistemological ideas (1958) never received the popularity they deserved. A useful book is A. Pais (1991) *Niels Bohr's times. In physics, philosophy and polity* (Clarendon). For an interesting account of the reaction of Einstein to his ideas, see M. Sachs (1989) *Einstein versus Bohr. The continuing controversies in physics* (Open Court). The implications of Einstein's work were obviously immense, but he also liked to popularize. He wrote with Leopold Infield (1938) a charming account of the evolution of physics entitled *The evolution of physics* (CUP, 2nd edn., reissued 1971). His popular *Essays in physics* has recently (1985) been published by the Philosophical Library. Two volumes have, so far, been edited of *The collected papers of Albert Einstein* by J. Stachel (1987) *Volume 1. The early years, 1897–1902* and (1989) *Volume 2. The Swiss years. Writings, 1900–1909* (Princeton U. Pr.). The ideas underlying Western science have also been fully explored in W. F. Bynum *et al.* (1981 and 1983) *Dictionary of the history of science* (Macmillan), and in R. C. Olby *et al.* (eds.) (1989) *Companion to the history of modern science* (Routledge). An old classic is

D. J. de Solla Price (1963) *Little science, big science* (Columbia U. Pr. several later editions).

The history of scientific instruments has been growing steadily in the last decade. Initially, the approach was primarily either anti-quarian or cultural, but more recently sociologists and philosophers of science, too, have become interested. Their focus is not on the instruments *per se*, but on how they are used by the 'scientific com-munity', and on how they interact with the complex process of experimentation. For many years astronomical, mathematical and surveying instruments received most of the attention. Several good survey books now exist of this early material, such as Anthony Turner (1987) *Early scientific instruments Europe 1400–1800* (Sotheby), which moves from the classical angle-measuring instru-ments used in astronomy and surveying to the earliest devices used in natural philosophy; J. A. Bennett (1987) *The divided circle. A history of instruments for astronomy, navigation and surveying* (Christie); and Allan Chapman (1990) *Dividing the circle. The development of critical angular measurement in astronomy 1500–1850* (Ellis Horwood), which is the first thorough study of how the degree scales were laid down on these devices. Optical instruments, too, have been much studied, especially the telescope and the micro-scope. Of the latter, the best source for further study is G. L'E. Turner (1989) *The great age of the microscope. The collection of the Royal Microscopical Society through 150 years* (Hilger).

Up until recently, the apparatus of physics received a pitifully small amount of genuine academic interest. An early classic remains Maurice Daumas (1954 French edn., 1972 English edn.) *Scientific instruments of the 17th and 18th centuries and their makers*, translated and edited by Mary Holbrook (Batsford). The electrostatic generator and measuring instruments have been studied in some detail (W. D. Hackmann (1978) *Electricity from glass. The history of the frictional electrical machine 1600–1850* (Sijt. Noord.); (1978) 'Eighteenth-century electrostatic measuring devices', in *Annali dell'Istituto e Museo di Storia della Scienza di Firenze*, 3, 3–58), and scattered articles have appeared on other laboratory apparatus in journals and symposium volumes. An impression of what could be found in a well-appointed eighteenth century laboratory can be gleaned from an illustrated catalogue G. L'E. Turner and T. H. Levere (1973) *Van Marum's scientific instru-ments in Teyler's Museum*, vol. 4 of *Martinus van Marum. Life and Work*, edited by E. Lefebvre and J. G. de Bruijn (previous volumes by R. J. Forbes), 6 vols (Noordhoff International for the Hol-landsche Maatschappij der Wetenschappen, 1969–1976). Part II of this volume, which is the catalogue by Turner, is also printed separately. G. L'E. Turner (1983) *Nineteenth-century scientific instruments* (Sotheby) is the only survey of scientific instruments of this period, and the lacunae in this volume indicate what still needs to be done. Magnetic and electrical instruments, primarily of the nineteenth century, have been catalogued in K. Lyall (1991) *Electrical and magnetic instruments. Catalogue 8 of the Whipple Museum* (Whipple, Cambridge), which has a useful bibliography. An interesting work on the evolution of laboratories which house these instruments is F. A. J. L. James (ed.) (1989) *The development of the laboratory. Essays on the place of experiment in industrial civilisation* (Macmillan). This book neatly links us to the growth in recent years in placing instruments and experiments in a social context, of which an excellent example is Steve Shapin and Simon Schaffer (1985 and 1989) *Leviathan and the air-pump* (Princeton U. Pr.); this book has an extensive bibliography to assist the reader with delving further into the subject. Of interest, too, are attempts to analyse either historically or philosophically the interaction between instrumentation, theory and progress in physics. A pioneering paper is W. D. Hackmann (1979) 'The relationship between concept and instrument design in eighteenth-century experimental science', in *Annals of Science*, 36, 205–224. The analysis has become much more sophisticated, and a summary of recent research is David Gooding, Trevor Pinch and Simon Schaffer (1989) *The uses of experiments. Studies in the natural sciences* (CUP). This book highlights several of the areas that are being usefully explored. The literature of the history of instrumentation has been surveyed in G. L'E. Turner (1983) 'Scientific instruments', in *Information sources in the history of science and medicine*, edited by P. Corsi and P. Weindling, pp. 243–258 (Butterworths). Turner, however, focuses on the development of the actual devices (as artefacts of applied science or technology) and not on the philosophical or social contexts, and is also now a few years out of date. Another useful guide is P. Carroll *et al.* (1991) *Guide to the history of science* (Philadelphia).

The aim of this chapter is to list some of the classics of physics, and those commentaries and secondary sources (concentrating on those in the English language) which make further study possible. Many of the classics have been reprinted, making these texts more generally available to those interested in following the thought processes and experiments of the pioneers, rather than just reading about them 'second hand'. Popular histories, especially, can be misleading and inaccurate. Foremost among the reprint firms dealing with physics is Dover, handled in the UK by Constable. Their editions are reasonably priced but they lack information about the original publications. Other reprint series are Pergamon's 'Selected Readings in Physics' (placing the material in the context of contemporary physics); the

'Cass Library of Science Classics'; the 'Collection History of Science', by the Editions Culture et Civilisation in Brussels; Johnson's 'The Source of Science', published in New York, and another American firm, Readex, is producing microfiche editions of the works of famous scientists, including such key physicists as Cavendish, Faraday, Foucault, Maxwell and Rutherford. There are many other reprint firms such as Kraus Reprint, a division of the Kraus-Thomson Organization of New York and Liechtenstein. A directory of these firms and current titles can be found in the annual *Guide to Reprints* published by Guide to Reprints, Inc. of Kent, Connecticut, USA. Details of reprints of classical scientific texts can be found in D. H. D. Roller and M. M. Goodman (1976) *The Catalogue of the history of science collections of the University of Oklahoma libraries* (Mansell). Microfiche and microfilm firms are taking an interest in this market. A useful publication to consult is the annual *Guide to Microforms in Print* (Meckler).

Natural philosophy

This may seem at first sight a strange section for this chapter, but the most common term for the physical sciences in the seventeenth and eighteenth centuries was 'natural philosophy'; a word which more adequately reflects the scientific preoccupations of those days than does the modern 'physics'. Two surveys of this subject (S. Schaffer (1980) 'Natural philosophy', in *The ferment of knowledge. Studies in the historiography of eighteenth-century science*, edited by G. S. Rousseau and R. S. Porter, pp. 55–91 (CUP); (1983) 'History of physical science', in *Information sources in the history of science and medicine*, edited by P. Corsi and P. Weindling, pp. 285–314 (Butterworths)) show how modern historians treat the development of classical physics and, incidentally, also contain detailed references to the secondary historical literature. The experimental aspects of eighteenth-century philosophy have been dealt with in J. L. Heilbron (1980) 'Experiments in natural philosophy', in *The ferment of knowledge. Studies in the historiography of eighteenth-century science*, edited by G. S. Rousseau and R. S. Porter, pp. 357–387 (CUP).

One of the most popular accounts of medieval physics is A. C. Crombie (1952) *Augustine to Galileo*, 2 vols (Heinemann; repr. in 1 vol., 1979). However, what the medieval scholar meant by 'experiment' is not so easy to define. Conventionally, Galileo is always taken as one of the harbingers of the scientific revolution. His work on statics and falling bodies is chiefly found in his *Dialogues concerning two new sciences* (1638), in which he argues convincingly in the guise of Sagredo against Aristotelian physics. Galileo's attempts to undermine Aristotle were continued by the Accademia del Cimento, founded in Florence for this purpose in 1657. Their experiments have been published in English in W. E. K. Middleton (1971) *The experimenters. A study of the Accademia del Cimento* (Johns Hopkins). Descartes' work in physics, such as his attempt to formulate the laws of motion, has been described in J. F. Scott (1952) *The scientific work of René Descartes* (Taylor and Francis), and more recently in S. Gaukroger (ed.) (1980) *Descartes. Philosophy, mathematics and physics* (Harvester Press; Barnes & Noble), whose own paper in this volume is particularly ambitious.

The impact of Newton's *Principia* (1687) and *Opticks* (1704) on the development of experimental physics has been stressed to such an extent that this has rather overshadowed the contributions of others such as Robert Boyle and Robert Hooke. Boyle's own works were collected by Thomas Birch (1772), available in a facsimile reproduction from Georg Olms Verlag. M. Epinasse (1956) *Robert Hooke* (Heinemann) is still the best scientific biography of this important contemporary of Newton, while his contributions to mechanics have been described in F. F. Centore (1970) *Robert Hooke's contributions to mechanics. A study in seventeenth century natural philosophy* (Martinus Nijhoff). For the most recent assessment, see M. Hunter and S. Schaffer (eds.) (1989) *Robert Hooke. New studies* (Boydell and Brewer). The surviving manuscripts of both Boyle and Hooke are receiving scholarly attention and will be published in this decade.

The only paperback reprint of Newton's *Principia* is that of Motte's original English translation, revised by Cajori in 1934 (U. of Cal. Pr., 1962), which is not entirely satisfactory. A scholarly introduction to this work is I. B. Cohen (1971) *Introduction to Newton's 'Principia'* (CUP). In I. B. Cohen (1981) *The Newtonian revolution* (CUP) the author has discussed Newton's dominant role in seventeenth-century science. This phase of science was referred to as the 'Scientific Revolution', but recent scholarship dislikes this term as it gives the impression of a dramatic change from late-Medieval science, and a more acceptable term has become the 'Scientific Movement'. Useful assessments on this topic are M.C. Jacob (1990) *The Newtonian and the English Revolution 1689–1720*, Classics in the History and Philosophy of Science, 7 (Gordon and Breach); I. B. Cohen (1987) *The Newtonian revolution* (Burndy); D. C Lindberg and R. S. Westman (eds.) (1990) *Reappraisals of the scientific revolution* (CUP); and Larry Stewart (1992) *The rise of public science. Rhetoric, technology, and natural philosophy in Newtonian Britain, 1660–1750* (CUP). R. S. Westfall (1971) *Force in Newton's physics. The science of dynamics in the seventeenth century* (Macdonald;

American Elsevier) and (1980) *Never at rest. A biography of Isaac Newton* (CUP) are also worthy of perusal in this context, as is D. G. King-Hele and A. R. Hall (eds.) (1988) *Newton's Principia and its legacy*, proceedings of a Royal Society Discussion Meeting held on 30 June 1987 (Royal Society). The publishers Blackwell have started the 'Blackwell Science Biographies' series under the General Editorship of David Knight. The well known scholar of the seventeenth century, A. R. Hall, has written for this series *Isaac Newton. Adventurer in thought* (1992), a hefty volume of some 400 pages. Other physicists (broadly speaking) in this series are *Humphry Davy. Science and power* (1992), by David Knight, and biographies of Galileo Galilei and André Marie Ampère (forthcoming). The 1730 edition of Newton's *Opticks* (in English) has been reprinted with a foreword by Einstein, an introduction by Whittaker and preface by Cohen (Dover, 1952). A challenging interpretation of the inconvertibility of light and matter raised by Newton's Query 30 in the *Opticks* is P. Rowlands (1990) *Newton and the concept of mass-energy* (Liverpool U. Pr.). The best introduction which demonstrates the great diversity and importance of Newton's work is J. Fauvel *et al.* (eds.) (1988 and 1989) *Let Newton be! A new perspective on his life and works* (OUP). In the late eighteenth century, natural philosophy began to split up into distinct fields of research, and these will now be dealt with separately.

Mechanics and relativity

Successors of Newton's 'rational mechanics', such as Laplace, have been dealt with in I. Todhunter (1873) *A history of the mathematical theories of attraction and the figure of the earth from the time of Newton to that of Laplace*, 2 vols. (Macmillan) (rep. Dover, 1962). C. A. Truesdell (1968) *Essays in the history of mechanics* (Springer) is a compilation of this author's essays on the history of mechanics which began as lectures. This volume lacks detailed references but has some interesting insights. He has excluded his three main historical treatises on fluid and continuum mechanics, which have been published instead in an edition of the collected works of Euler (Füssli, 1960, 1954, 1955). M. Jammer's books *Concepts of force. A study in the foundation of mechanics* (Harvard U. Pr., 1957); *Concepts of mass in classical and modern physics* (Harvard U. Pr., 1961) and *Concepts of space. The history of theories of space in physics* (Harvard U. Pr., 1954) are lucid accounts of the development of the concepts of force, mass and space. These are particularly useful for the undergraduate, as is the critical analysis of the concepts of absolute space and time, by E. Mach (1883), translated as *The science of mechanics* (Open Court, 1960).

The most thorough early history of fluid mechanics is I. Todhunter and K. Pearson (1886–1893) *A history of the theory of elasticity and the strength of materials from Galileo to the present time*, 2 vols. in three (CUP), while the more recent history is covered rather superficially in G. A. Tokaty (1971) *A history and philosophy of fluid mechanics* (Foulis). H. Rouse and S. Ince (1957) *History of hydraulics* (Inst. Hyd. Res.) (repr. Dover, 1963) is also not very satisfactory. Acoustics has fared little better as far as its history is concerned. The classic textbook on sound for many years was that by Rayleigh (1877). The naval physicist A. B. Wood incorporated in his textbook (1930) the First World War's research on underwater acoustics as a means of detecting submarines. It has remained an important topic ever since. W. H. Bragg was also involved in this research at that time, which he described in (1920) *The world of sound* (Bell, 1920) (rep. Dover, 1968), a very popular book for a time, based on lectures given at the Royal Institution. The sourcebook (R. B. Lindsay (ed.) (1973) *Acoustics. Historical and philosophical development* (Hutchinson Ross)) contains a good selection of the classic papers on sound and also a synoptic historical overview, while a detailed narrative for the early period is E. V. Hunt (1978) *Origins in acoustics. The science of sound from antiquity to the age of Newton* (Yale U. Pr.). Hunt was one of the pioneers of underwater acoustics in America during the Second World War and coined the acronym 'sonar'. His book (1954) *Electroacoustics. The analysis of transduction and its historical background* (Harvard U. Pr.) is an authoritative account of the early work on this subject. Alas, he died before the completion of his history of sound for the post-Newton period.

The literature on Einstein and relativity is immense. A very readable book on the development of field theory is W. Berkson (1974) *Field of force. The development of a world view from Faraday to Einstein* (Routledge) which, although it has no bibliography, has extensive references. A. I. Miller (1981) *Albert Einstein's special theory of relativity. Emergence (1905) and early interpretation (1905–11)* (Addison-Wesley) is a detailed account of the emergence and early interpretation of Einstein's special theory of relativity and has a useful bibliography, while J. Mehra (1974) *Einstein, Hilbert, and the theory of gravitation. Historical origins of general relativity* (Reidel) is a slim volume covering the crucial period 1907–1919, when the foundations of the modern theory of gravitation were laid in the general theory of relativity. Hilbert formulated his field equations quite independently from Einstein. A lively debate ensued, as Einstein disliked Hilbert's axiomatic method. The development of both the special and the general theory of relativity can be followed from Einstein's *The principle of relativity* (1922), which includes papers by Lorentz and Minkowski, and also from the clear expositions in C. W. Kilmister *Special theory of relativity* (Pergamon, 1970); and *General theory of relativity* (Pergamon, 1973). A useful short historical account is G. J. Whitrow (1973) *Time, gravitation and the universe. The evolution of relativistic theories* (Imperial College), which was this author's inaugural lecture given at Imperial College, while J. C. Graves (1971) *The conceptual foundations of contemporary relativity theory* (MIT) gives a rather more complex philosophical treatment, starting with the concepts of space of Plato, Descartes and Newton and ending with post-Einsteinian 'geometrodynamics'. Useful surveys are S. W. Hawking (the present holder of Newton's chair at Cambridge) and W. Israel (eds.) (1987) *Three hundred years of gravitation* (CUP), and C. Ray (1987) *The evolution of relativity* (Hilger). Einstein spent the last 30 years of his life in the vain hunt for a theory that would unite gravitation and electromagnetic phenomena. For such attempts, see Abdus Salam (1990) *Unification of fundamental forces. The first of the 1988 Dirac Memorial Lectures* (CUP).

Light, electricity and magnetism

The new research into magnetism and electricity was heralded by Gilbert (1600). He elucidated the magnetic properties of the lodestone by means of experiments, and devoted one chapter to what we now call electrostatics, simply to distinguish between magnetic and electrical phenomena. During the next two centuries little advance was made in magnetism, while that of electricity developed at a pace although, until Poisson in 1811, the work remained almost entirely experimental. The best survey of this period is undoubtedly J. L. Heilbron (1980) *Electricity in the 17th and 18th centuries. A study of early modern physics* (U. of Cal. Pr.). Franklin has always been singled out because of his 'single fluid' concept of electricity. The Harvard University Press edition (1941) of his *Experiments and observations on electricity* of 1751 has an extensive introduction on the significance of Franklin's work by I. B. Cohen, who in another book, *Franklin and Newton. An inquiry into speculative Newtonian experimental science and Franklin's work in electricity as an example thereof* (APS; Harvard U. Pr., 2nd edn., 1966), has compared the scientific methods of Franklin and Newton. For a decade, Cohen was the foremost Franklinian scholar and his main papers on this subject have been gathered in I. B. Cohen (1990) *Benjamin Franklin's science* (Harvard U. Pr.).

One of the first to attempt a rigorous mathematical treatment of electricity and magnetism was Aepinus (1759). For an English translation of this book with an introductory monograph, see R. W. Home (ed.) (1979) *F. U. T. Aepinus, essay on the theory of electricity and magnetism* (Princeton U. Pr.). Aepinus spearheaded the 'mathematization' of physics which became the dominant force in the nineteenth century.

Volta's electrical researches have been collected in a seven-volume Italian edition (1918–1929), of which there is not yet an English translation. Coulomb's *Mémoires* (1777–1805), in which he describes the experiments with his torsion balance, were published by Gauthier-Villars in 1884; his seminal researches in electricity and magnetism have been analysed in C. S. Gillmor (1971) *Coulomb and the evolution of physics and engineering in eighteenth century France* (Princeton U. Pr.). The debate between Alessandro Volta and the anatomist Luigi Galvani, which led to the discovery of the electrochemical battery in 1800, has been well brought out in M. Pera (1992) *The ambiguous frog. The Galvani-Volta controversy on animal electricity*, translated by J. Mandelbaum (Princeton U. Pr.). Pera shows that their controversy derived from two basic, irreducible interpretations of nature based on their training: Galavani saw the frog phenomenon in terms of a biological organ, and Volta in terms of a physical apparatus. The phenomena associated with the battery (the first producer of current electricity) led to new speculations and was followed by the work of Faraday and Henry on electromagnetic induction. Oersted's famous paper of 1820 has been translated in B. Dibner (1961) *Oersted and the discovery of electromagnetism* (Burndy), and is included in R. A. R. Tricker (1965) *Early electrodynamics. The first law of circulation* (Pergamon), which also contains unsatisfactorily translated extracts of Ampère's *Théorie mathématique des phénomènes électrodynamiques* (1827). Ohm's 1827 monograph on electrical resistance was translated from the German by William Francis (1891).

Faraday's multi-volume *Experimental researches in electricity* (1839–1855) was reprinted by Dover in 1965, and the work is described in R. A. R. Tricker (1967) *The contributions of Faraday and Maxwell to electrical science* (Pergamon) and in L. P. Williams (1965) *Michael Faraday* (Simon and Schuster), a biography of Faraday which has not found universal favour with historians of physics. The recent biography by Geoffrey Cantor (1991) *Michael Faraday. Sandemanian and scientist* (Macmillan) is in the new mould of the history of science in that it deals with Faraday's scientific work in a social context, in this case with reference to the strict religious sect to which he belonged. His work is also reassessed in D. Gooding and F. J. L. James (eds.) (1985, reprinted with corrections

1989) *Faraday rediscovered. Essays on the life and work of Michael Faraday, 1791–1867* (Macmillan). Frank James is also the editor of (1991) *The correspondence of Michael Faraday. Volume 1: 1811 to 1831* (Peregrinus); five volumes are planned in all. Nathan Reingold has been editing the papers of the American Faraday, Joseph Henry, for almost two decades. Of the 15 planned volumes of *The papers of Joseph Henry* (1972–), five have been published (Smithsonian Inst. Pr.). A rather rambling reference work, but in spite of this a marvellous volume to dip into, is P. F. Mottelay (1922) *Bibliographical history of electricity and magnetism chronologically arranged* (Griffin), which terminates in 1821. To make the reader feel even less secure, he has arranged the entries by date, sometimes erroneously in the light of more recent research. In the mid-1840s Franz Ernst Neumann devised the first mathematical theory of electrical induction, the process of converting mechanical into electrical energy. His research in precision physics is described in K. Olesko (1991) *Physics as a calling. Discipline and practice in the Konigsberg seminar for physics* (Cornell U.Pr.).

The historical literature on the development of optics is less voluminous. Robert Hooke put forward an undulatory theory of light in his *Micrographia* (1665), further developed by Christiaan Huygens (1690), which was very successful for a time but ultimately failed, because it could not explain polarization phenomena, owing to Huygens' assumption of longitudinal instead of transverse vibrations. Newton's corpuscular theory, bearing some resemblance to the modern theory of light quanta, was described in his *Opticks*. For a good synopsis of these theories, see A. I. Sabra (1967) *Theories of light from Descartes to Newton* (Oldbourne; CUP, 1981). Young, Arago and Fresnel established the transverse-wave theory of light in the early nineteenth century. Of these, the collected works of Young and Fresnel have been reproduced by Johnson. How they overcame the adherents of the corpuscular theory is described in V. Ronchi (1970) *The nature of light. An historical survey*, translated with additional material from the Italian (Harvard U. Pr.). This period has also been dealt with in a not entirely satisfactory manner in E. Mach (1921) translated as *The principles of physical optics. An historical and philosophical treatment* (Methuen, 1926; Dover, c. 1953).

Doppler's discovery of 1842 and its application to the theory of light has been described in T. P. Gill (1965) *The Doppler effect. An introduction to the theory of the effect* (Academic). Doppler's papers (1842–1847) have been reprinted in vol. 161 (1907) of W. Ostwald (1889–1938) *Ostwald's Klassiker der exakten Wissenschaften*, 244 vols. (Engelmann). In the late nineteenth century, the wave-theory of light was reinforced by the spectroscopic studies of Fraunhöfer, Bunsen and Kirchhoff, as described in W. McGucken (1969)

Nineteenth-century spectroscopy (Johns Hopkins), who, rather sadly, has terminated his narrative with the discovery of the electron by Thomson in 1897. Luckily, this has been partly rectified in C. L. Maier (1981) *The rôle of spectroscopy in the acceptance of the internally structured atom, 1860–1920* (Arno) the author's 1964 PhD. thesis How the velocity of light determinations by Fizeau and Foucault, and the quest for a background aether by Michelson and Morley, eventually led to Einstein's space-time theory is described in L. S. Swenson (1972) *The ethereal aether. A history of the Michelson-Morley-Miller aether-drift experiments 1880–1930* (U. of Texas Pr.). This book has a very good bibliography, as has the volume by G. N. Cantor and M. J. S. Hodge (eds.) (1981) *Conceptions of ether. Studies in the history of ether theories 1740–1900* (CUP), which is especially good on the nineteenth century. A recent book that must also be consulted is J. Z. Buchwald (1989) *The rise of the wave theory of light. Optical theory and experiments in the early nineteenth century* (U. of Chicago Pr.).

To return to the development of electricity and magnetism, Dover has reprinted (1954) the third edition (1891) of Maxwell's *Treatise on electricity and magnetism*, in which he achieved the synthesis of light, electricity and magnetism. The present form of Maxwell's 'Equations' was first formulated by Hertz (collected papers 1892), and discussed in J. Henry (1986) *James Clerk Maxwell and the theory of the electromagnetic field* (Hilger), and in J. G. O'Hara and W. Pricha (1987) *Hertz and the Maxwellians. A study and documentation of the discovery of electromagnetic wave radiation, 1873–1894* (Peregrinus). Maxwell never rejected the aether concept; his ideas on this matter were published in 1876. He was also very interested in the history of electricity and felt honour-bound, as the first Cavendish Professor of Physics at Cambridge, to edit the unpublished electrical papers of his famous eighteenth-century predecessor, Henry Cavendish, who had anticipated some of the discoveries of Faraday and others, this work appeared after Maxwell's death (1879). Bruce Hunt (1991) *The Maxwellians* (Cornell U. Pr.), is mainly on Oliver Lodge, George Fitzgerald and Oliver Heaviside. Maxwell's papers are also receiving scholarly attention: volume 1 of *The scientific letters and papers of James Clerk Maxwell*, covering the years 1846–1862, edited by P. M. Harman, was published in 1990 (CUP). Essential reading is Jed Z. Buchwald (1985) *From Maxwell to microphysics. Aspects of electromagnetic theory in the last quarter of the nineteenth century* (Chicago U. Pr.). Two other key members of this nineteenth-century group of British physicists were Kelvin and G. G. Stokes; their temperaments were very different. J. J. Thomson (1936) in *Recollections and reflections* (Bell) describes a memorable occasion when Kelvin and Stokes had an argument about the existence of atoms when they met up at the Cavendish Laboratory in Cambridge. On their relationship, see D.B. Wilson (1987) *Kelvin and Stokes. A comparative study in Victorian physics* (Hilger), and by the same author (1990) *The correspondence between Sir George Gabriel Stokes and Sir William Thomson, Baron Kelvin of Largs*, volume 1 *1846–1869* and volume 2 *1870–1901* (CUP). In many respects Kelvin is the link between classical and modern physics. His views are expressed in R. Kargon and P. Achinstein (eds.) (1987) *Kelvin's Baltimore lectures and modern theoretical physics. History and philosophical perspectives* (MIT Pr.). An excellent insight into the physics in Cambridge at this time is in P. Harman (ed.) (1985) *Wranglers and physicists. Studies on Cambridge physics in the nineteenth century* (Manchester U. Pr.).

Lorentz described his electromagnetic theory in 1909, and a very fine reference work on the history of electromagnetism is E. T. Whittaker (1910) *A history of the theories of the aether and electricity from the age of Descartes to the close of the 19th century* (Dublin U. Pr.), with a new edition of two volumes (Nelson, 1951–1953; and Dover reprint 1989). A rather more unorthodox view has been presented in A. O'Rahilly (1938) *Electromagnetics. A discussion of fundamentals* (Longmans; Cork U. Pr.), reprinted as *Electromagnetic theory*; while M. B. Hesse (1961) *Forces and fields. The concept of action at a distance in the history of physics* (Nelson) is a wide-ranging survey from the Greeks to quantum mechanics. For the more recent developments of relativistic electromagnetic theory, see the historical sections of F. Rohrlich (1965) *Classical charged particles* (Addison-Wesley). A useful survey is C. Jungnickel and R. McGormach (1986) *Intellectual mastery of nature. Theoretical physics from Ohm to Einstein* (Chicago U. Pr.). Two surveys of the technical history of electrical engineering, suitable also for the non-specialist reader, are B. Bowers (1982 and 1991) *A history of electric light and power* (Peregrinus), and W. A. Atherton (1984) *From compass to computer. A history of electrical and electronics engineering* (San Francisco Pr.).

Heat, thermodynamics and statistical mechanics

D. McKie and N. H. de V. Heathcote (1935) *The discovery of specific and latent heats* (Arnold) deserves to be better known, as it is a useful historical introduction. Fourier (1822) attempted to formulate a theory of heat, based on his study of thermal conduction. Carnot's 'Second Law of Thermodynamics', which preceded the 'First Law' by 20 years, was introduced in 1824. Included in the Dover edition (1960) is a paper by Clausius reconciling the two laws. A new

French edition of Carnot's *Réflexions* (1824) was published in 1978 with an introduction by Robert Fox; an English edition, with a new translation, also by Fox, was published in 1986 (Manchester U.Pr.). D. S. L. Cardwell (1971) *From Watt to Clausius. The rise of thermodynamics in the early industrial age* (Heinemann) deals with the early development of thermodynamics, emphasizing the rôle of industry. He has also recently written (1989) *James Joule. A biography* (Manchester U. Pr.). The discovery of the 'First Law' belongs partly to the German physician, J. R. Mayer, but more particularly to Joule, who has deservedly received a fair amount of attention from the historians of science, and one of the more recent books on this is H. J. Steffens (1979) *James Prescott Joule and the concept of energy* (Science History; Dawson). The vagaries of the development of classical thermodynamics from 1822 to 1854 have been described in C. A. Truesdell (1980) *The tragicomical history of thermodynamics 1822–1854* (Springer), which includes a comprehensive list of the primary sources in chronological order. E. Mach (1896) *Die Principien der Wärmelehre. Historisch-kritisch entwickelt* (Barth) has been translated into English (1986) as *Principles of the theory of heat, historically and critically elucidated*, edited by Brian McGuinness (Reidel).

An important contributor to the history of statistical mechanics is S. G. Brush. Key source material has been reprinted in his book *Kinetic theory*, 3 vols. (Pergamon, 1965–1972); the first volume, entitled *The nature of gases and heat*, has selections from classical thermodynamics, starting with Robert Boyle; the second volume, entitled *Irreversible processes*, covers statistical mechanics from Maxwell to Boltzmann; and the third volume deals with the Chapman-Enskog solution of the transport equation for moderately dense gases. Brush's main historical papers have been reprinted in S. G. Brush (1976) *The kind of motion we call heat. A history of the kinetic theory of gases in the 19th century*, 2 vols (North-Holland). The first volume entitled *Physics and the atomist*, and the second *Statistical physics and irreversible processes*, make a useful guide to the source material published in his earlier work, and a very comprehensive bibliography for the period 1801–1900 is included in the second volume. Brush has also translated and edited Boltzmann's *Lectures on gas theory* (1896–1898).

Bumstead and Van Name have edited the collected papers of the pioneer of chemical thermodynamics and statistical mechanics, J. W. Gibbs (1906; Dover, 1961). His papers on thermodynamics are found in the first volume. Gibbs' *Elementary principles in statistical mechanics* (1902) was reprinted by Dover in 1960. A thorough analysis of Boltzmann and Gibbs can be found in P. Ehrenfest and T. Ehrenfest (1959) *The conceptual foundations of the statistical approach in mechanics*, translated from the German (Cornell U. Pr.). The 'Third Law' was formulated by Nernst in 1905–1906, and his seminal treatise has been reprinted by Dover (1969). For a thought-provoking biography, see K. A. G. Mendelssohn (1973) *The world of Walther Nernst. The rise and fall of German science* (Macmillan; U. of Pittsburg Pr.), who has also written a popular history of low-temperature physics in 1966 entitled *The quest for absolute zero. The meaning of low temperature physics* (Weidenfeld; 2nd edn, 1977, with SI Units, Taylor & Francis). R. Jancel (1969) *Foundations of classical and quantum statistical mechanics*, translated from the French (Pergamon) concentrates on the more recent history, especially on the statistical interpretation of the 'Second Law'.

Quantum theory

Planck introduced the concept of quanta of energy in 1900. This attempt at quantifying the new atomic phenomena before the introduction of wave mechanics in the mid-1920s by Heisenberg (1930) and Schrödinger, became known as the 'old quantum theory'. The genesis of this theory has been covered in both A. Herrmann (1971) *The genesis of quantum theory (1899–1913)* (English translation MIT) and T. S. Kuhn (1978) *Black body theory and the quantum discontinuity, 1894–1912* (OUP), which includes an extensive bibliography. A selection of the papers by the leading exponents (Planck, Einstein, Rutherford, Bohr, Pauli) are gathered in D. ter Haar (1967) *The old quantum theory* (Pergamon). An insight into what has been achieved by the early 1920s can be gleaned from Lorentz's 1922 lecture series (1927); while in M. Jammer (1966) *The conceptual development of quantum mechanics* (McGraw-Hill; reprinted Tomash and American Institute of Physics, 1989) the story is taken up to 1926 in a detailed mathematical account with extensive footnotes. In a later work (1974), *The philosophy of quantum mechanics. The interpretation of quantum mechanics in historical perspective* (Wiley-Interscience), Jammer has attempted an historical analysis of the logical foundations and philosophy of quantum mechanics. A more recent work on this subject is P. Suppes (ed.) (1980) *Studies in the foundations of quantum physics* (Philos. Sc. Assn). For the quantum physical interpretation of cosmology, see J. Gribbin (1986) *In search of the big bang. Quantum physics and cosmology* (Heinemann).

Louis de Broglie's popular non-mathematical exposition (1954) contains few references, but it is interesting to read the account of one of the pioneers of wave mechanics. The genesis of this concept is already to be found in his 1924 thesis (Masson, 1963), a portion of which has been translated in G. Ludwig (1968) *Wave mechanics* (Pergamon), a sourcebook which also contains material from Schrödinger, Heisenberg, Born and Jordan. Another, somewhat overlapping, sourcebook is B. L. Van der Waerden (ed.) (1967) *Sources of quantum mechanics* (Dover), in which 17 papers have been reproduced covering the period 1917–1926. The fiftieth anniversary of de Broglie's prediction of the wave nature of the electron is celebrated in W. C. Price *et al.* (eds.) (1973) *Wave mechanics. The first fifty years* (Butterworths). Most contributors have concentrated, after a brief retrospective review, on the present-day impact of de Broglie's discovery, but among the few historical chapters is one by de Broglie on the beginnings of wave mechanics, and another by J. C. Slater on the crucial period 1924–1926. Slater's scientific autobiography (1975) *Solid-state and molecular theory. A scientific biography* (Wiley) is also worthy of study in this context. Stern's molecular beam experiments have been described by I. Estermann (1959) *Recent research in molecular beams* (Academic), who became one of his collaborators. The second paper in this collection, by Nierenberg, deals with the more recent research at Berkeley.

Heisenberg gave a lecture series on the wave-particle duality and his matrix mechanics in 1929 and these were published in 1930. His work has been celebrated in W. C. Price and S. C. Seymour (1977) *The uncertainty principle and foundations to quantum mechanics. A fifty-year survey* (Wiley), which is yet another commemorative collection of essays reviewing the philosophical implications and the past, current, and potential future developments of quantum mechanics. S. I. Tomonga's *Quantum mechanics*, 2 vols., translated from the Japanese (North-Holland, 1962–1966) is a textbook along historical lines. The first volume is concerned with the old, and the second with the new, quantum theory. A valuable feature is the extensive account of Schrödinger's classical wave theory, but unfortunately references are lacking. W. Yourgrau and A. van der Merwe (1979) *Perspectives in quantum theory* (Dover) consists of a collection of scientific and philosophical papers by Born, de Broglie, Wigner, Bondi and Popper (and others). The most exhaustive history is undoubtedly J. Mehra and H. Rechenberg (1982) *The historical development of quantum theory*, 5 vols. (Springer), (three more volumes are planned). Rechenberg was Heisenberg's last doctoral student. The volumes published so far deal with the quantum theory of Planck, Einstein, Bohr and Sommerfeld; with the discovery of quantum mechanics in 1925; with matrix mechanics during the period 1925–1926; with the reception of the new quantum mechanics during the same period; and with the rise of wave mechanics in 1926. Projected volumes will deal with the completion of quantum mechanics during 1925–1965, and lastly with the interpretation and epistemology of quantum mechanics from 1926 to the present day. The contents are based on discussions over many years with most of the architects of quantum mechanics, and on archival material from the Niels Bohr Institute in Copenhagen and elsewhere. A valuable feature is the numerous excellent footnotes.

An indispensable tool for the historian of quantum mechanics is the detailed inventory of source material in T. S. Kuhn *et al.* (eds.) (1967) *Sources for history of quantum mechanics. An inventory and report* (APS). This project, which was under the auspices of the American Philosophical Society and the American Institute of Physics, and was funded by the National Science Foundation, consisted of locating the unpublished papers, notebooks and letters of more than 250 physicists involved in the development of quantum mechanics. Project members also recorded interviews with many prominent physicists and located other oral records.

Atomic, nuclear, particle and solid-state physics

The nineteenth-century position is summarized in P. M. Harmann (1982) *Energy, force and matter the conceptual development of nineteenth-century physics* (CUP). B. F. J. Schonland (1968) *The atomists (1805–1933)* (Clarendon) is a popular account tracing the concept of the atom from the beginning of the nineteenth century to the mathematical models proposed in the 1930s. A readable introduction to the infancy of atomic physics can be found in A. Keller (1983) *The infancy of atomic physics. Hercules in his cradle* (Clarendon). The Rutherford-Soddy collaboration during 1901–1903 on radioactive decay has been described, among others, in T. J. Trenn (1977) *The self-splitting atom. The history of the Rutherford-Soddy collaboration* (Taylor and Francis), which not only cites their main publications, but also contains a useful 'Who's who' of the scientists active at that time. Soddy's annual reports on advances in radioactivity to the Chemical Society in London from 1904 to 1920 have been reproduced in T. J. Trenn (ed.) (1975) *Radioactivity and atomic theory. Presenting facsimile reproductions of the annual progress reports on radioactivity, 1904–1920, to the Chemical Society* (Taylor and Francis) (part II consists of the reports of F. Soddy). Soddy's lucid style not only makes these reports a pleasure to read, but they also give a unique overview of the progress made during this period. Trenn's historical introduction and commentaries are helpful.

Ernst Mach died in 1916 still not believing in the real existence of the atom. H. A. Boorse and L. Motz (1966) *The world of the atom* (Basic) is a comprehensive anthology with commentaries by the editors of papers illustrating atomic theory from Lucretius to Bethe. The period from Greek atomism to X-rays is covered in the first volume, and from early twentieth-century atomic theory to high energy and particle physics in the second. The selection gives the flavour of this development, but a severe limitation is the lack of a bibliography. Their more recent book (with J.H Weaver) (1989) is *The atomic scientists. A biographical history* (Wiley). C. Weiner and E. Hart (eds.) (1972) *Exploring the history of nuclear physics* (AIP), R. H. Stuewer (ed.) (1979) *Nuclear physics in retrospect. Proceedings of a symposium on the 1930s* (U. of Minn. Pr.) and W. R. Shea (ed.) (1983) *Otto Hahn and the rise of nuclear physics* (Reidel) are all useful histories of nuclear physics. The first two books are the products of symposia at which pioneers of nuclear physics and historians were present. Such symposia have become quite popular as a means of gathering historical information. Weiner and Hart have assembled the transcripts of the discussion by the participants attempting to identify which areas of nuclear physics required historical exploration. The impression is of rather rambling musings, but there are some intriguing insights. Stuewer's volume deals squarely with nuclear physics in the 1930s and is more relevant. Among the participants were Bethe, Wigner and Peierls. A physicist's memory is no less fallible than that of ordinary mortals and, therefore, such oral histories should be treated with caution. Rudolph Peierls's (1985) reminiscences in *Bird of passage. Recollections of a physicist* (Princeton U. Pr.), are nevertheless very interesting.

Shea's book deals with Otto Hahn's impact on nuclear physics. The papers have been written by established historians of modern physics, including Stuewer, Weart and Trenn, and they are well referenced. Two earlier books should perhaps also be mentioned. T. T. Beyer (1949) *Foundations of nuclear physics. Facsimile of thirteen fundamental papers* (Dover) contains 13 papers covering the foundations of nuclear physics, but the bibliography, although extensive, is badly organized. Selected papers on quantum electrodynamics, starting with Dirac in 1927 and stopping in the mid-1950s, have been reproduced in J. Schwinger (ed.) (1958) *Selected papers on quantum electrodynamics* (Dover).

In 1980, a symposium was held at the Fermi National Accelerator Laboratory in Batavia, Illinois, USA, on the development of particle physics during the decade before the Second World War, and pioneers in the field offered papers and personal recollections. This material has been put together in L. M. Brown and L. Hoddeson (eds.) (1983) *The birth of particle physics* (CUP), and includes an historical introduction on the birth of particle physics in which the editors argue that this subject evolved out of cosmic ray and nuclear physics research during the period 1930–1950, while the theoretical structure was provided by relativistic quantum field theory. Among the papers is one by Weisskopf on the development of quantum electrodynamics, another by Dirac on the origins of the quantum field theory, while others deal with early research on cosmic rays.

Dirac was also one of the participants at a two-week school on the history of physics held at Varenna in 1972 under the auspices of the Italian Physical Society's International School of Physics 'Enrico Fermi'. These papers have been reproduced in C. Weiner (ed.) (1977) *History of twentieth-century physics. Proceedings of the International School of Physics 'Enrico Fermi', course LVII* (Varenna, 1972) (Academic), which is well referenced. Dirac submitted two papers, one giving personal recollections and the other on Ehrenhaft, the sub-electron and the quark. Klein wrote on the beginning of quantum theory, Heilbron on atomic physics from 1900 to 1922 and Casimir on solid-state physics. Personal impressions of the development of particle physics by one of the pioneers can be found in E. Segré (1976) *From X-rays to quarks. Modern physics and their discoveries*, translated from the Italian (1980) (Freedman). R. H. Stuewer (1975) *The Compton effect. Turning point in physics* (Science History) is an excellent account describing the importance of Compton's 1923 discovery, which demonstrated that X-rays could act as particles as predicted in quantum theory.

Almost the entire subject of weak interaction theory is based on Fermi. His 1934 paper has been reproduced with others dating from the 1950s and 1960s in P. K. Kabir (ed.) (1963) *The development of weak interaction theory* (Gordon and Breach). This volume, apart from a short introduction, has no commentary or bibliography. Strong interaction theory is dealt with in M. Gell-Mann and Y. Ne'eman (1964) *The eightfold way* (Benjamin), which covers the early 1960s when SU(3) became the central organizing principle of particle physics. The next attempt was to find larger symmetries such as SU(6) and SU(12). Thirty-two key papers for the period 1937–1965, the majority on SU(6), have been gathered in F. J. Dyson (1966) *Symmetry groups in nuclear and particle physics* (Benjamin). Recent useful histories of particle physics are Y. Ne'eman and Y. Kirsh (1986) *The particle hunters* (CUP); L. M. Brown, M. Dresden and L. Hoddeson (eds.) (1989) *Pions to quarks. Particle physics in the 1950s* (CUP); B. Foster and P. H. Fowler (eds.) (1988) *40 years of particle physics* (Hilger); and R.N. Cahn and G. Goldhaber (1989 and 1991) *The experimental foundation of particle physics* (CUP). Pickering's sociological study in this context has already been referred to. The impact of the three high-intensity proton accelerators built in the early 1970s (LÁMP at Los Alamos, PSI near Zurich and TRIUMF in Vancouver) are dealt with in T. O. Ericson, V. W. Hughes and D. E. Nagle (1991) *The Meson factories* (U. of Cal. Pr.).

Solid state physics has also entered the historical domain, in particular studies concerned with semiconductors and solid-state electronic devices. For a short historical review, see H. B. G. Casimir (1977) 'Development of solid-state physics', in *History of twentieth-century physics*, edited by C. Weiner, pp. 158–169 (Academic). For a detailed account by one of the pioneers, see N. F. Mott (ed.) (1980) *The beginnings of solid-state physics* (Royal Society), which includes reminiscences by Bardeen, Bethe, Braun, Frölich, Peierls and Ziman. The most recent book is L. Hoddeson , E. Braun, J. Teichmann and S. Weart (1992) *Out of the crystal maze. Chapters from the history of solid-state physics* (OUP). The influence of Arnold Sommerfeld is assessed in H. Welker (1969) 'Impact of Sommerfeld's work on solid-state research and technology', in *Physics of the one- and two-electron atoms*, edited by F. Bopp and H. Kleinpoppen, pp. 32–43 (North-Holland). The ninth Solvay conference of 1951 on solid-state physics is included in J. Mehra (1975) *The Solvay conferences on physics. Aspects of the development of physics since 1911* (Reidel), which deals with these conferences from their inception in 1911 until 1973. In his historical introduction, the author shows the impact these conferences had on the progress of modern physics, especially in the early years. He next reviews the most important topics that formed the substance of the Solvay reports, such as the structure and properties of atomic nuclei (1933), elementary particles (1948) and the fundamental problems associated with particle physics (1967) and quantum field theory (1961), to name just a few.

Social histories of laboratories and the sociology of physics

C. P. Snow (1959) *The two cultures and the scientific revolution* (CUP), and his (1964) *The two cultures and a second look* (CUP) generated a debate on the potential cultural differences between science and the arts. One way to resolve this issue was by studying science sociologically as pioneered by Kuhn (1962). He used social factors to explain the nature of scientific change embodied in his term 'paradigm'. More recently, an anthology of his philosophical papers appeared in T. S. Kuhn (1977) *The essential tension. Selected studies in scientific tradition and change* (U. of Chicago Pr.). Two of these, 'Mathematical versus experimental traditions in the development of physical science' and 'The function of measurement in modern physical science', might be of special interest to the physicist. His ideas, drawn largely from the history of physics, have influenced a wide range of disciplines such as philosophy, economics, the social sciences and theology, judging from an anthology compiled in G. Gutting (1980) *Paradigm and revolutions. Applications and reappraisals of Thomas Kuhn's philosophy of science* (U. of Notre Dame Pr.). Kuhn's paradigm concept appears to have achieved the same popular appeal as Heisenberg's uncertainty principle did in the 1930s, but he ignores it in his history of black-body radiation (1978). For an assessment of his work, see B. Barnes (1982) *T. S. Kuhn and social science* (Macmillan), whose earlier book (1974) *Scientific knowledge and sociological theory* (Routledge and Paul) gives a good introduction to this topic. Many scholars are active in this field. Well known studies are B. Latour (1986) *Science in action* (Open U. Pr.; Harvard U. Pr.) and P. K. Feyerabend (1987) *Farewell to reason* (Verso).

As we have seen in this survey, the sociological interpretative framework has become one of the most influential in the history of physics in recent years. Two key sociological studies of experimental physics not mentioned elsewhere are H. Collins (1985) *Changing orders. Replication and induction in scientific practice* (Sage), which deals with the difficulties of replicating the operation of TEA lasers and detecting gravity waves, and D. Gooding (1990) *Experiments and the making of meaning* (Kluwer), which is a philosophical and sociological study of Faraday's laboratory notebooks. This is a difficult book which can only be assessed adequately when its methodology is applied to the work of other physicists. A number of books in which laboratories are treated as communities of scientists have already been cited. Experiments, too, can be analysed in terms of communal acts. A very interesting book studying high energy physics in this context is Peter Galison (1987) *How experiments end* (Chicago U. Pr.). Issues such as what counts as a scientific discovery are discussed in Barnes's 1982 book already cited, in A. Brannigan (1981) *The social basis of scientific discoveries* (CUP), and in L. Fleck (1979) *The genesis and development of a scientific fact* (Chicago U. Pr.); and on the rôle of the scientific language in G.N. Gilbert and M. Mulkay (1984) *Opening Pandora's box. A sociological analysis of scientists' discourse* (CUP), and L. Jordonova (ed.) (1986) *Languages of nature. Critical essays on science and literature* (Rutgers U. Pr.; new ed. by Free Association Books). Another issue that is receiving increasing attention in recent years is the use of illustrations in scientific communications both to other scientists and to the public at large. Such scientific iconography can be very powerful. A useful book, but steeped in the jargon of sociology, is M.

Lynch and S. Woolgar (1990) *Representation in scientific practice* (MIT Pr.), especially the chapter by Greg Myers (which is also one of the best written).

Reminiscinences, biographies, bibliographies and periodicals

Reminiscences as a form of oral history are quite popular in certain quarters but are treated with suspicion by historians. Several such reminiscences have already been listed in the course of this survey. One suspects that they are an easy source of income for publishers as the books can be put together relatively easily. Nevertheless, it is interesting to get an insight into how the pioneers of modern physics see the evolution of their subject, and of their career. Hilger has published for the Institute of Physics several books in this *genre*: Rajkumar Williamson (ed.) (1987) *The making of physicists* (Hilger); J. Roche (ed.) *Physicists look back. Studies in the history of physics* (Hilger); and B. N. Kursunoglu and E. P. Wigner (1987) *Reminiscences about a great physicist. Paul Adrien Maurice Dirac* (CUP). H. S. Kragh (1990) *Dirac. A scientific biography* (CUP) should be noted here.

Biographies, scientific or otherwise, are a very popular form of recreating the past. They come in many different shapes and sizes, this art form being extremely susceptible to the abilities and preoccupations of the author. Scientific contents, too, can vary tremendously. The more 'old-fashioned' kinds such as D. Brewster (1858) *Memoirs of the life, writings and discoveries of Sir Isaac Newton*, 2 vols (repr. Johnson Repr.) tend to have a reasonably high scientific content (though this is no guarantee that the interpretation is correct), while others stress the social context and may look at the subject's psychological make-up. An example of this genre is E. Manuel (1968) *A portrait of Isaac Newton* (Harvard U. Pr.). More traditional biographies are D. Wilson (1983) *Rutherford. Simple genius* (Hodder and Stoughton), and P. Rowlands (1990) *Oliver Lodge and the Liverpool Physical Society* (Liverpool U. Pr.). An excellent example of the new school of biographies is the massive tome by C. W. Smith and M. N. Wise (1989) *Energy and empire. A biographical study of Lord Kelvin* (CUP), which almost tries too hard to place Kelvin in a social context. It is instructive to compare the approach of P. Tunbridge (1992) *Lord Kelvin. His influence on electrical measurements and units* (Peregrinus), which narrowly focusses on Kelvin's involvement in the standardization of electrical measurements, and makes no reference at all to the Smith and Wise 800-page blockbuster. Scientists are beginning to copy politicians in their penchant for autobiographies, which can give valuable insights into particular developments. It is perhaps not necessary to point out that, again like politicians, scientists do not have a monopoly of the 'truth', whatever this commodity may mean in historical terms.

An accessible source of brief biographies of nineteenth- and early twentieth-century scientists (including their main publications) was begun in J. C. Poggendorff (1858–) *Biographisch-literarisches Handwörterbuch zur Geschichte der exacten Wissenschaften* (Barth; varies; see P. Corsi (1983) for full details), but consult first C. C. Gillispie (ed.) (1970–1980) *Dictionary of scientific biography*, 16 vols. (American Council of Learned Societies; Scribners). The entries in these volumes, written by many specialists, are a valuable aid to further research, as they list the subject's main publications, the locations of his manuscripts (when known) and relevant secondary literature; supplement volumes are being planned. Popular single-volume biographical compendia are A. G. Debus (1968) *World who's who in science. A biographical dictionary of notable scientists from antiquity to the present* (Marquis); T. I. W. Williams (1969) *A biographical dictionary of scientists* (A & C Black; 3rd enlarged edn., 1982); I. Asimov (1972) *Asimov's biographical encyclopedia of science and technology* (David and Charles); and I. David and J. and M. Millar (1989) *Chambers concise dictionary of scientists* (Chambers and CUP). Renowned British scientists are included in the *Dictionary of national biography* (1882–), and in the *DNB missing persons* volume (1993). Detailed life histories of Fellows of the Royal Society appear in *Obituary notices of Fellows of the Royal Society of London* (1901–1954). (Before 1901, and between 1904 and 1932 these notices formed part of the *Proceedings of the Royal Society* and after 1955 as *Biographical memoirs of Fellows of the Royal Society of London*). Useful compendia which include well-known contemporary American physicists are (1979) *American men and women of science. Physical science and biological sciences*, 8 vols. (Bowker, 14th edn.) and *McGraw-Hill modern men of science* (1966–1968), 2 vols. (McGraw-Hill). For a guide to biographical dictionaries, see R. Slocum (1967) *Biographical dictionaries and related works* (Gale).

Only a few of the most relevant bibliographies of scientific works can be listed here; for a more comprehensive discussion, see P. Corsi (1983) 'Guide to bibliographical sources', in *Information sources in the history of science and medicine*, edited by P. Corsi and P. Weindling, pp. 137–156 (Butterworths). Poggendorff (1858–) has already been mentioned. Another source for the nineteenth- and early twentieth-century periodical literature is the Royal Society of London (1867–1925) *Catalogue of scientific papers, 1800–1900*, 19 vols. (HMSO and (from vol. 9) CUP), and its successor, the

International catalogue of scientific literature 1901–1904 (1902–1918) 28 vols. (The Royal Society for the International Council). A bibliography of eighteenth-century papers published by scientific societies was produced by J. Reuss (1801–1821) *Repertorium commentationum a societatibus litterariis editarum secundum disciplinarum ordinem*, 16 vols. (repr. B. Franklin, 1962). Useful modern surveys are F. Russo *Eléments de bibliographie de l'histoire des sciences et des techniques* (lst edn. 1954;, 2nd rev. edn., Hermann, 1969) entitled *Histoire des sciences et des techniques: bibliographie*; J. L. Thornton and R. I. J. Tully (1971) *Scientific books, libraries and collectors. A study of bibliography and the book trade in relation to science*, 3rd edn. (Library Assoc. rep. 1975) (supplement to 3rd edn. covering 1969–1975 (1978)), which also lists reprints such as those by Dover, and D. M. Knight (1975) *Sources for the history of science, 1660–1914* (The Sources of History Ltd, Hodder & Stoughton; now distributed by CUP), which is particularly good on the nineteenth century. Two notable specialized bibliographies dealing with the early history of electricity and magnetism are A. J. Frost (1880) *A catalogue of books and papers relating to electricity, magnetism and the electric telegraph, etc., including the Ronalds Library* (Soc. Tel. Eng.) and W. D. Weaver (1909) *Catalogue of the Wheeler gift of books, pamphlets and periodicals in the library of the American Institute of Electrical Engineers. With introduction, description and critical notes by Brother Potamian*, 2 vols. (AIEE), based on individual collections.

The two most important general listings are the *Isis Critical Bibliography* (1913–), produced yearly by the American history of science journal of that name, and the 'Histoire des sciences et des techniques' section of the *Bulletin Signalétique* (Centre National de la Recherche Scientifique (1947–) *Bulletin Analytique: Philosophie* (1947–1955), several changes until *Bulletin Signalétique 522 Histoire des sciences et des techniques* (1969–)). The latter has undergone a number of reorganizations over the years and the present section heading dates from 1961. This section has been printed as a separate volume since 1969. The annual *Isis* bibliographies, up to 1975, have been reprocessed by M. Whitrow (1972–1982) *Isis cumulative bibliography. A bibliography of the history of science formed from Isis critical bibliographies, 1–90, 1913–65*, 5 vols., and by J. Nice (1980–) *91–100, 1965–75* (Mansell), which is, without a doubt, the most comprehensive listing in the history of science. J. L. Heilbron and B. R. Wheaton (eds.) (1981) *Literature on the history of physics in the 20th century*, Berkeley Papers in the History of Science, 5 (Berkeley: OHST, University of California) should be consulted for the literature on twentieth-century physics. Another useful volume in the same series is B. R. Wheaton and J. L.

Heilbron (1982) *An inventory of published letters to and from physicists, 1900–1950,* Berkeley Papers in History of Science, 6 (Berkeley: OHST, University of California) and these volumes are really intended for the serious historical researcher. Heilbron and Wheaton also organize the 'Inventory of sources for history of twentieth-century physics' (ISHTCP) at the Office for History of Science and Technology of the University of California at Berkeley. The inventory to date records the whereabouts of some 750 000 letters scattered in almost a thousand collections. R.C. Hovis and H. Kragh have produced a bibliography of the history of elementary particles, covering some 600 titles, in 'Resource letter HEEPP-1: history of elementary-particles', in *American Journal of Physics,* 59 (1991) 779–807, which also refers to 37 other 'resource letters' in this journal, covering subjects from magnetic monopoles to particle accelerators.

The compilation in S. G. Brush (1972) *Resources for the history of physics* (U. Pr. New Eng.) is primarily intended for the physics teacher and has two sections: a 'Guide to books and audiovisual material' and a 'Guide to original works of importance and their translation into other languages', listed by scientist in alphabetical order. Brush has also compiled 'Resource letter HP-1: history of physics', in *American Journal of Physics,* 55, 683–691. Another useful volume is R.W. Home (1984) *The history of classical physics. A selected annotated bibliography* (Garland). A very comprehensive survey of bibliographies which includes the history of science is the five-volume work, T. Besterman (1965–1966) *A world bibliography of bibliographies, and of bibliographical catalogues, calendars, abstracts, digests, and the like,* 5 vols. (Soc. Bibl., 4th edn.). The two-volume supplement by A. F. Toomey (1977) *A world bibliography of bibliographies. 1964–1974. A list of works represented by Library of Congress printed cards. A decennial supplement to Theodore Besterman, A world bibliography of bibliographies,* 2 vols. (Rowman) is a bibliography intended for the history of the exact sciences and technology. Another useful reference work is S. A. Jayawardene (1982) *Handlist of reference books for the historian of science,* Science Museum Library Occasional Publications, 2 (Science Museum), which is based on the holdings of the Science Museum Library in London, the premier library on this subject in Britain.

Articles dealing with the history of physics are spread through a wide range of journals and only the most widely circulated English-language ones are listed here. For a more detailed discussion, consult P. Weindling (1983) 'Periodical literature and societies', in *Information sources in the history of science and medicine,* edited by P. Corsi and P. Weindling, pp. 157–171 (Butterworths). Sadly lacking is a directory of current history of science journals. A general review journal is *History of Science* (1962–), but it does not provide a systematic coverage. Important current journals are:

Isis (1912–).
Annals of Science (1936–).
Notes and Records of the Royal Society (1938–).
Archive for History of Exact Sciences (1960–).
British Journal for the History of Science (1962–), preceded by the *Bulletin* (1949–1958).
Japanese Studies in the History of Science (1962–), changed to *Historia Scientiarum* (1980–).
Historical Studies in the Physical Sciences (1969–1979), changed to *Historical Studies in the Physical and Biological Sciences* (1980–).
Science in Context (1987–).
Studies in the History and Philosophy of Science (1970–), which from 1993 includes a new section entitled *Studies in the History and Philosophy of Modern Physics*.
The Natural Philosopher (1963–1964), which ceased after only three volumes.

Other less well-known journals are *Janus* (1896, but present form started in 1957–), *Centaurus* (1950–), *Physis* (1959–) and *Annali dell'Istituto e Museo di Storia della Scienza* (1976), renamed *Nuncius, Annali di Storia della Scienza* (1986–). The American journal *Technology and Culture* (1959–) contains many useful book reviews, and accounts of modern discoveries can be found in *Adventures in Experimental Physics* (1972–). Obviously, this survey has barely scratched the surface of what is available in the history of physics, but hopefully it has made the readers aware of the abundance of material that awaits them.

Bibliography

Ampère, A.-M. (1827) *Théorie mathématique des phénoménes électrodynamiques, uniquement déduite de l'expérience* (Paris: rep. Librairie scientifique Albert Blanchard, 1958).
Bohr, N. H. D. (1958) *Atomic physics and human knowledge* (Wiley).
Bohr, N. H. D. (1972–) *Collected scientific works* (North-Holland).
Boltzmann, L. (1896–1898) *Voriesungen über Gastheorie*, translated as *Lectures on gas theory* (U. of Cal. Pr., 1964).
Boyle, R. (1772) *The works of the honourable Robert Boyle*. 2nd edn., 6 vols. (London: facsimile reprint by Georg Olms Verlag).
Carnot, S. (1824) *Réflexions sur la puissance motrice du feu*, translated as *Reflections on the motive power of fire* (Dover, 1960); new French edition with introduction and commentary by R. Fox (Vrin, 1979), and English edition, with new translation, also by Fox (Manchester U. Pr, 1986).

Coulomb, C. A. (1884) *Mémoires de Coulomb* (Gauthier-Villars).

de Broglie, L. (1924) *Recherches sur la théorie des quanta* (repr. Masson, 1963).

de Broglie, L. (1954) *The revolution in physics. A non-mathematical survey of quanta*, translated from the French (Routledge).

Doppler, J. C. (1907) 'Abhandlungen von Christian Doppler' edited by H. A. Lorentz. In *Ostwald's Klassiker der exakten Wissenschaften*, no. 161, Leipzig.

Duhem, P. (1914) *La théorie physique – son objet, sa structure*, translated as *The aim and structure of physical theory* (Atheneum, 1962).

Ehrenfest, P. (1959) *Collected scientific papers* (North-Holland).

Einstein, A. *et al.* (1922) *Das Relativitätsprinzip*, translated as *The principle of relativity* (Dover, 1952).

Euler, L. (1960, 1954, 1955) *Leonhardi Euleri opera omnia*. Series secunda. Vols 11/2 (Truesdell, 'The rational mechanics of flexible or elastic bodies, 1638–1788'), 12 (Truesdell, 'Rational fluid mechanics, 1687–1765'), and 13 (Truesdell, 'Introduction to commentationes mechanicae'). (Orell Füssli).

Faraday, M. (1839–1855) *Experimental researches in electricity* (repr. in 2 vols, Dover, 1965).

Faraday, M. (1859) *Experimental researches in chemistry and physics* (Taylor & Francis, 1991).

Fermi, E. (1962–1965) *Collected papers*, 2 vols. (U. of Chicago Pr.).

Fourier, J.-B. (1822) *Théorie analytique de la chaleur*, translated as *The analytical theory of heat* (Dover, 1955).

Franklin, B. (1769) *Experiments and observations on electricity* (repr. Harvard U. Pr., 1941).

Fresnel, A. J. (1866–1870), *Oeuvres complètes d'Augustin Fresnel*, 3 vols., edited by H. de Senaramont, E. Verdet and L. Fresnel, (repr. Johnson Repr.). Also repr. Johnson Repr. in 3 vols.

Galilei, Galileo (1610) *Siderius Nuncius or The sidereal messenger*, translated and edited by A. van Helden (Chicago U. Pr., 1989).

Galilei, Galileo (1638) *Dialogues concerning two new sciences*, translated by H. Crew and A. de Salvio (Dover, 1954).

Gibbs, J. W. (1902) *Elementary principles in statistical mechanics* (repr. Dover, 1960).

Gibbs, J. W. (1906) *The scientific papers of J. Willard Gibbs*, 2 vols., edited by H. A. Bumstead and R. G. Van Name (repr. Dover, 1961).

Gilbert, W. (1600) *De magnete*. Reissue of the P. F. Mottelay translation (Dover, 1958).

Hauksbee, Francis (1719) *Physico-mechanical experiments on various subjects*, 2nd ed. Introduction by D. H. Roller (repr. Johnson Repr.).

Heisenberg, W. C. (1930) *Die physikalische, Prinzipien der Quantentheorie*, translated as *The physical principles of the quantum theory* (U. of Chicago Press, 1930; Dover, 1949).

Hertz, H. (1892) *Untersuchungen über die Ausbreitung der elektrischen Kraft*, translated as *Electric waves* (Macmillan, 1900; Dover, 1962).

Hooke, R. (1665) *Micrographia*. Readily available in reprint editions. (Dover, 1961).

Huygens, C. (1690) *Traité de la lumière*, translated as *Treatise on light* (Macmillan, 1912; Dover, 1962).

Lorentz, H. A. (1909) *The theory of electrons and its application to the phenomena of light and radiant heat* 2nd edn., 1916 (repr. Dover, *c.* 1952).

Lorentz, H. A. (1927) *Problems of modern physics* (repr. Dover, 1967).

Lorentz, H. A. (1934–1939) *Collected papers*, 9 vols. (Martinus Nijhoff).

Lorenz, H. (1902) *Lehrbuch der technischen Physik*, 4 vols. Introduction by G. Oravas (repr. Johnson Repr.).

Maxwell, J. C. (1873) *A treatise on electricity and magnetism*, 2 vols. (Clarendon, 3rd edn., 1891; repr., Dover, 1954).

Maxwell, J. C. (1876) *Matter and motion*. Repr. 1920 with notes by Sir Joseph Larmor (Dover, 1952).

Maxwell, J. C. (ed.) (1879) *The electrical researches of the honourable Henry Cavendish* (CUP: 2nd edn., 2 vols, 1921). Facsimile reprint by Frank Cass.

Maxwell, J. C. (1890) *The scientific papers of James Clerk Maxwell*, edited by W. D. Niven, 2 vols. (CUP) (repr. in 1 vol., Dover, 1952).

Nernst, H. W. (1918) *Die theoretischen und experimenteilen Grundlagen des neuen Warmesatzes*, translated as *The new heat theorem. Its foundations in theory and experiment* (Methuen, 1926) (repr. Dover, 1969).

Newton, I. (1687) *Philosophiae naturalis principia mathematica*, translated by Andrew Motte (1729) as *Mathematical principles of natural philosophy* (U. of Cal. Pr., 1962).

Newton, I. (1704) *Opticks* (3rd edn., 1730 repr. Dover, 1952).

Ohm, G. S. (1827) *Die galvanische Kette mathematisch berarbeitet*, translated as *The galvanic circuit investigated mathematically* (1891). Van Nostrand Science Series, no. 102. Facsimile reprint by Editions Culture et Civilisation, 1969.

Pauli, W. (1964) *Collected scientific papers*, 2 vols. (Interscience).

Planck, M. (1958) *Physikalische Abhandlungen und Vorträge*, 3 vols. (Friedr. Vieweg Sohn).

Planck, M. (1989) *The theory of heat radiation*. Introduction by A. A. Needell. Translated by Morton Masius (AIP).

Power, H. and Hall, M.B. (1664) *Experimental philosophy*, 3 bks. (repr. Johnson Repr.).

Priestley, J. (1767) *The history and present state of electricity* 3rd edn., 1775 (repr. Johnson Repr., 1966).

Rayleigh, J. W. S. (1878) *The theory of sound* 2nd edn., 2 vols. (Macmillan, 1894–1896) (repr. Dover, 1956).

Rutherford, E. (1962–1965) *The collected papers of Lord Rutherford of Nelson*, 3 vols. (Pergamon).

Societé d'Arcueil (1807) *Memoires de physiques et de chemie*, 3 vols. Introduction by M. P. Crosland (repr. Johnson Repr.).

Sommerfeld, A. (1968) *Gesammelte Schriften*, 4 vols. (Vieweg).

Volta, A. G. A. A. (1918–1929) *Le opere*, 7 vols. (repr. Johnson Reprint).

Wood, A. B. (1930) *A textbook of sound. Being an account of the physics of vibrations with special reference to recent theoretical and technical developments* (Bell).

Young, T. (1855) *Miscellaneous works of the late Thomas Young* 3 vols., edited by G. Peacock and J. Leitch (repr. Johnson Reprint).

CHAPTER FOURTEEN

Mechanics and acoustics*

E. C. FINCH

Mechanics

Mechanics is defined as that branch of science which describes and predicts the conditions of rest or motion of bodies and deals with the forces that produce such conditions. It is a long-established subject of study and is fundamental to physics, astronomy and engineering, and indeed also to the development of applied mathematics.

It is not always easy to determine where the literature of mechanics ends and that of engineering or mathematics begins. An attempt has been made to concentrate on material of particular interest to physicists. It is obviously impossible in the space available here to survey all the literature of mechanics, so apart from references to the classics of the literature, much of the attention is given to publications appearing later than about 1970. A few conference proceedings are also mentioned where it seems appropriate, but the selection is not particularly systematic or comprehensive.

Many of the most important developments in mechanics took place in previous centuries, starting with the work of Newton, and much of the material presented here is divided into subsections corresponding roughly to the traditional subdivisions of the subject. However, one new and rapidly expanding area is nonlinear dynamics and chaos, and a subsection is also devoted to the literature in this field. In addition, there are contributions on two or three other 'newer' areas (at least in nomenclature), such as tribology and biomechanics.

---

* based on the first edition chapter written by R. H. de Vere

The closely related topic of acoustics, being concerned with mechanical disturbances, is treated in a second section of this chapter.

General references

Theoretical treatments of all branches of the subject can be found in the relevant volumes of *Encyclopedia of physics* (see Chapter 3). Detailed references are given in the subsections concerned.

Not surprisingly for such a fundamental subject there is no shortage of texts at varying introductory levels, such as (in alphabetical order by author):

Barford, N. C. (1973) *Mechanics* (Wiley).
Cowan, B. P. (1984) *Classical mechanics* (Chapman and Hall).
French, A. J. and Ebison, M. (1986) *Introduction to classical mechanics* (Van Nostrand Reinhold). A text in the MIT Physics series.
Kibble, T. W. B. (1985) *Classical mechanics* (Longman, 3rd edn.).
Lunn, M. (1991) *A first course in mechanics* (OUP).
Smith, P. and Smith, R. C. (1990) *Mechanics* (Wiley, 2nd edn.). Incorporates computer-based approaches to problems.
Synge, J. L. and Griffith, B. A. (1959) *Principles of mechanics* (McGraw-Hill, 3rd edn.). A well-known older textbook.

A selection of more advanced books includes the following:

Abraham, R. and Marsden, J. E. (1978) *Foundations of mechanics* (Addison-Wesley, 2nd edn.). An advanced mathematical exposition for those able to use the language of differential topology.
Goldstein, H. (1980) *Classical mechanics* (Addison-Wesley, 2nd edn.). A well-known intermediate text on particle and rigid body dynamics. The excellent critical bibliographies give much more detail about each book cited than can appear here in the space available.
Lanczos, C. (1970) *The variational principles of mechanics* (U. of Toronto Pr., 4th edn., reprinted by Dover, 1986). An advanced yet leisurely survey of the basics of mechanics. A summary is provided after each chapter section.
Landau, L. D. and Lifshitz, E. M. (1976) *Mechanics* (*Course of theoretical physics*, vol. 1) (Pergamon, 3rd edn.). An outstanding graduate textbook, succinctly written.
Symon, K. R. (1971) *Mechanics* (Addison-Wesley, 3rd edn.). An important intermediate-level book.

Two books concentrating on the topic of stability are:

Leipholz, H. (1987) *Stability theory: an introduction to the stability of dynamic systems and rigid bodies* (Wiley, 2nd edn.). An advanced text.

Pippard, B. (1985) *Response and stability: an introduction to the physical theory* (CUP). Covers the response of systems (mechanical and otherwise) in equilibrium to perturbing forces.

The periodical literature in mechanics includes the following commoner journals. Foreign-language journals and those with a large engineering content are omitted:

Acta Mechanica (Springer, 1965–), 20 issues annually.
Archive for Rational Mechanics and Analysis (Springer, 1957–), 20 issues annually. Concerned with mathematical mechanics, especially pure analysis in applied contexts.
Computer Methods in Applied Mechanics and Engineering (North-Holland, 1972–). 24 issues annually.
Journal of Applied Mathematics and Mechanics (Pergamon, 1958–), bi-monthly. Cover-to-cover translation of *Prikladnaia Matematika i Mekhanika*.
Journal of Applied Mechanics and Technical Physics (Consultants, 1966–), bi-monthly. Cover-to-cover translation of *Zhurnal Prikladnoi Mekhaniki i Tekhnicheskol Fiziki*.
Mechanics Research Communications (Pergamon, 1974–), bi-monthly. A rapid communications journal for the basic and applied mechanics of fluids, solids, particles, continua, rigid bodies, mechanisms and systems. Includes a software survey section.
Quarterly Journal of Mechanics and Applied Mathematics (Clarendon, 1948–).

Reviews of various topics, published at irregular intervals, include *Advances in Applied Mechanics* (Academic, 1948–) and *Studies in Applied Mechanics* (Elsevier, 1979–)

The following abstracting service specializes in mechanics: *Applied Mechanics Reviews* (ASME, 1948–), monthly. Over 10 000 abstracts each year obtained from a wide range of sources. Also includes review articles, bibliographic surveys, book, database and software reviews, and conference proceedings. With annual indices.

A useful multi-lingual dictionary specializing in mechanics is: Troskolanski, A. T. (ed.) (1962,1967) *Vocabulary of mechanics in five languages: English, German, French, Polish and Russian*, 2 vols. (Pergamon and Wyd. Nauk.-Tech.). Terms are defined in English.

For many years an important set of publications has been the courses and lectures given at CISM (International Centre for Mechanical Sciences), Udine, Italy. From 1969, Springer published several volumes annually. The latest volume announced is no. 315: Kluwick, A. (ed.), *Nonlinear waves in real fluids* (1991).

Mention should also be made of other national and international organizations, such as the International Union of Theoretical and Applied Mechanics (IUTAM), which exist to promote and encourage the development of mechanics through publishing and other activities.

Dynamics

Dynamics is concerned with the motions of systems with reference to the forces required to produce such motions. (Some books using the term 'mechanics' in the title refer mostly just to dynamics, and not, for example, to statics as well, where bodies are at rest and forces are in equilibrium).

Early classics on the subject which are still frequently referred to include:

Routh E. J. (1897, 1905) *A treatise on the dynamics of a system of rigid bodies*, 2 vols. (Macmillan, 6th edn.).

Whittaker, E. T. (1937, 1988) *A treatise on the analytical dynamics of particles and rigid bodies* (CUP, 4th edn.). A standard work, although famous for its almost total lack of figures.

There is an expert exposition entitled *Classical dynamics* by J. L. Synge in *Encyclopedia of physics*, vol. III/1, pp. 1–225 (1960).

Among more modern books are the following:

Chorlton, F. (1983) *Textbook of dynamics* (Ellis Horwood, 2nd edn.). A mathematical text at undergraduate level.

Greenwood, D. T. (1988) *Principles of dynamics* (Prentice-Hall, 2nd edn.). A fairly advanced account of the subject.

Griffiths, J. B. (1985) *The theory of classical dynamics* (CUP). An undergraduate text emphasizing fundamental concepts.

Marion, J. B. and Thornton, S. T. (1988) *Classical dynamics of particles and systems* (Harcourt-Brace Jovanovich, 3rd edn.). Intermediate level, with emphasis on the physics concerned. Includes a computer 'workbook'.

Pars, L. A. (1965, 1979) *A treatise on analytical dynamics* (Heinemann, Oxbow). A scholarly theoretical account.

Rosenberg, R. M. (1977) *Analytic dynamics of discrete systems* (Plenum). Intermediate level.

Sposito, G. (1976) *An introduction to classical dynamics* (Wiley). Gives the physical basis of dynamics at an intermediate level.

Gyroscopes, as well as appearing in some of the books already cited, are the subject of specialized treatment in Arnold, R. N. and

Maunder, L. (1961) *Gyrodynamics and its engineering applications* (Academic).

Nonlinear dynamics and chaos

Within the last two or three decades there has been an explosion of work in dynamical systems exhibiting nonlinear behaviour. Such systems can be mechanical, like the damped driven pendulum. The ideas are, however, equally applicable in fields as diverse as fluid turbulence, meteorology, feedback control, defect formation in solids, laser physics, chemical reactions, population dynamics in biology and so on. The books in the area often reflect this, and thus interpret the word 'dynamics' in a wide sense beyond the strictly mechanical.

A selection of possible books neither too mathematical nor too engineering in style includes:

Baker, G. L. and Gollub, J. P. (1990) *Chaotic dynamics: an introduction* (CUP). A short undergraduate-level text presenting a wide range of concepts.

Bergé, P., Pomeau, Y. and Vidal, C. (1984) *Order within chaos: towards a deterministic approach to turbulence* (Wiley). A well-received introduction suitable for physicists.

Guckenheimer, J. and Holmes, P. (1983) *Nonlinear oscillations, dynamical systems, and bifurcations of vector fields* (*Applied Mathematical Sciences*, vol. 42) (Springer). Designed for scientists and engineers as a 'user's guide' to the mathematical background.

Jackson, E. A. (1989, 1990) *Perspectives of non-linear dynamics*, 2 vols. (CUP). A substantial text covering a broad spectrum of systems.

Moon, F. C. (1987) *Chaotic vibrations: an introduction for applied scientists and engineers* (Wiley). A good general review at graduate level of the whole field of chaotic dynamics.

Sagdeev, R. Z., Usikov, D. A. and Zaslavsky, G. M. (1988) *Nonlinear physics: from the pendulum to turbulence and chaos* (Harwood). A thorough survey of all aspects of nonlinear theory. Accompanying diskettes are available.

Schuster, H. G. (1988) *Deterministic chaos* (VCH, 2nd edn.). An introductory account.

Two journals of interest which each cover a wide range of applications are *Dynamics and Stability of Systems* (OUP, 1986–), quarterly; and *Nonlinearity* (IOP and London Mathematical Society, 1988–), quarterly.

Relativity

Many books exist which cover the fields of special and general relativity at all levels, ranging from accounts written for the general public through to advanced rigorous mathematical treatises. Some texts focussing mainly or entirely on special relativity include:

French, A. P. (1968) *Special relativity* (Nelson). A good, much-used book.

Rindler, W. (1991) *Introduction to special relativity* (Clarendon, 2nd edn.). Well-received, with a conceptual and mathematical emphasis. Includes a chapter on the relativistic mechanics of continua.

Rosser, W. G. V. (1992) *Introductory special relativity* (Taylor and Francis).

Synge, J. L. (1964) *Relativity: the special theory* (North-Holland, 2nd edn.).

Turner, R. (1984) *Relativity physics* (Chapman and Hall).

A selection of texts which concentrate on general relativity is given below:

Hughston, L. P. and Tod, K. P. (1991) *An introduction to general relativity* (CUP). A concise text.

Kenyon, I. R. (1990) *General relativity* (OUP). Describes the concepts and formalism of the subject together with experimental results. Contains an informative bibliography.

Martin, J. L. (1988) *General relativity: a guide to its consequences for gravity and cosmology* (Wiley). An introduction aimed at undergraduates.

Schutz, B. F. (1985) *A first course in general relativity* (CUP). Sets out to start from a minimum of pre-requisites.

Synge, J. L. (1960) *Relativity: the general theory* (North-Holland).

A well-received graduate text, which has an unusual coverage of applications in space, nuclear and electrical engineering as well as in physics, is:

Van Bladel, J. (1984) *Relativity and engineering* (Springer Series in Electrophysics, vol. 15) (Springer).

One comparatively new physics journal which includes articles bearing on the ideas of general relativity is:

Classical and Quantum Gravity (IOP, 1984–), monthly.

Since the development of relativity was essentially the work of just one person, mention must be made here of the splendid biography

Subtle is the Lord: the science and the life of Albert Einstein by A. Pais (Clarendon, 1982).

Continuum mechanics

Continuum mechanics is a term coined to cover the underlying principles of mechanics applicable to both the solid and the fluid states. Since 1945 it has become much more of a distinct subject in its own right. It is a subdivision of continuum physics, which is concerned in addition with phenomena such as classical electromagnetism and heat flow. The term is also used to describe literature whose contents cover both solid and fluid mechanics.

Extensive articles by C. A. Truesdell and R. A. Toupin and by C. A. Truesdell and W. Noll appear in *Encyclopedia of physics*, vol. III/1, pp. 226–858 (1960) and vol. III/3 (1965), respectively.

Books on continuum mechanics include:

Chadwick, P. (1976) *Continuum mechanics: concise theory and problems* (Allen & Unwin). An introductory graduate text.
Eringen, A. C. (ed.) (1971–1976) *Continuum physics*, 4 vols. (Academic).
Gurtin, M. E. (1981) *An introduction to continuum mechanics* (Academic). A theoretical graduate-level treatment, with a useful list of references.
Lai, W. M. *et al.* (1978) *Introduction to continuum mechanics* (Pergamon, rev. edn.). Takes the subject up to beginning-graduate level.
Ling, F. F. (1973) *Surface mechanics* (Wiley). A specialist advanced theoretical survey of surface behaviour from the continuum-mechanics viewpoint.
Sedov, L. I. (1971–1972) *A course in continuum mechanics*, 4 vols. (Wolters-Noordhoff).
Spencer, A. J. M. (1980) *Continuum mechanics* (Longmans). A useful short elementary introduction.

The proceedings of a recent conference, of relevance to several other subsections of this chapter as well as this one, appear in:

Graham, G. A. C. and Malik, S. K. (eds.) (1989) *Continuum mechanics and its applications* (Hemisphere).

Research articles and reviews in continuum mechanics occur in the reviews and periodical literature listed above in the 'General references' subsection.

Solid mechanics

Solid mechanics is the study of the statics and dynamics of deformable solids. It includes *elasticity*, which is concerned with substances which regain their original shape when relieved from stress, *plasticity*, which involves instantaneous (that is, time-dependent), permanent deformation of solids, and *rheology*, the time-dependent flow of quasi-solid materials. The behaviour of liquids and gases under stress is dealt with in the subsection on 'Fluid mechanics'.

Advanced expositions on elasticity appear in *Encyclopedia of physics*, vol. VI (*Elasticity and plasticity*) and vol. VIa (*Mechanics of solids*).

Among classic texts must be mentioned:

Landau, L. D., Lifshitz, E. M., Kosevich, A. M. and Pitaevskii, L. P. (1986) *Theory of elasticity* (*Course of Theoretical Physics*, vol. 7) (Pergamon, 3rd edn.).
Love, A. E. H. (1927) *A treatise on the mathematical theory of elasticity* (CUP, 4th. edn.).

Other books on elasticity include:

Reismann, H. and Pawlik, P. S. (1980) *Elasticity: theory and applications* (Wiley).
Sokolnikoff, I. S. (1956) *Mathematical theory of elasticity* (McGraw-Hill). A highly-regarded older text.
Timoshenko, S. P. and Goodier, J. N. (1970) *Theory of elasticity* (McGraw-Hill, 3rd edn.).

Plasticity forms the subject of books such as:

Chakrabarty, J. (1987) *Theory of plasticity* (McGraw-Hill).
Hill, R. (1950, reprinted 1983) *The mathematical theory of plasticity* (Clarendon).

General texts which can be referred to are:

Hearn, E. J. (1985) *Mechanics of materials: an introduction to the mechanics of elastic and plastic deformation of solids and structural components* (Pergamon, 2nd edn.).
Shames, I. H. (1990) *Introduction to solid mechanics* (Prentice Hall, 2nd edn.). A rigorous introduction.

A major older work on rheology is:

Eirich, F. R. (ed.) (1956–1969) *Rheology: theory and applications* (Academic, 5 vols).

More recent works include:

Barnes, H. A., Hutton, J. F. and Walters, K. (1989) *An introduction to rheology* (Elsevier). Aimed especially at graduate newcomers to the field.

Ferguson, J. and Kemblowski, Z. (1991) *Applied fluid rheology* (Elsevier).

Tanner, R. I. (1988) *Engineering rheology* (Clarendon, revised edition). Focusses especially on polymer melts and solutions.

Journals covering solid mechanics, or some part of it, include:

Acta Mechanica Solida Sinica (HUST/Pergamon, 1988–), quarterly. Covers solid state mechanics research in China.

European Journal of Mechanics: A/Solids (Gauthier-Villars, 1982–), bi-monthly. (Formerly *Journal de Méchanique (Théorique et Appliquée)*, founded in 1961).

International Journal of Non-Linear Mechanics (Pergamon, 1966–, bi-monthly). Covers both theoretical and experimental research into the non-linear mechanics of solids and fluids. Includes a software survey section.

International Journal of Plasticity (Pergamon, 1985–), eight issues annually.

International Journal of Solids and Structures (Pergamon, 1965–), semi-monthly. Publishes applied research into the mechanics of solids and structures.

Journal of the Mechanics and Physics of Solids (Pergamon, 1952–), eight issues annually.

Rheologica Acta (Steinkopff, 1958–), bi-monthly.

Rheology Abstracts (Pergamon, for the British Society of Rheology, 1958–), quarterly. Surveys the world's literature, and includes annual subject and author indices.

Solid Mechanics Archives (OUP, 1976–), irregular. Covers advances and trends in solid mechanics research.

The online database RHEOLOGY, produced by BAM, Germany, contains about 46 000 abstracts of publications on the deformation and flow of matter. It covers the period 1976–87 but is not updated on a regular basis.

Fracture mechanics

The study of the fracture of materials under stress is obviously of vital importance to engineering applications. Books covering the underlying physics include:

Ewalds, H. L. and Wanhill, R. J. H. (1989) *Fracture mechanics* (Arnold, corrected edn.). Covers both basic concepts and applications in actual materials.

Freund, L. B. (1990) *Dynamic fracture mechanics* (CUP). With a substantial bibliography. Here 'dynamic' refers to situations in which the inertia of the material during fracture becomes significant.

Kanninen, M. F. and Popelar, C. H. (1985) *Advanced fracture mechanics* (Clarendon). Covers not only the linear elastic regions but also the dynamic and elasto-plastic regimes.

A journal for the field is:

Engineering Fracture Mechanics (Pergamon, 1968–), 18 issues annually.

Oscillations, vibrations and waves

Oscillations, vibrations and waves occur in many branches of science other than mechanics, and this is reflected in the wide-ranging subject matter of some of the books listed. There are many introductory texts, including:

French, A. P. (1971) *Vibrations and waves* (Chapman and Hall).

Gough, W., Richards, J. P. G. and Williams, R. P. (1983) *Vibrations and waves* (Ellis Horwood).

Hussey, M. (1983) *Fundamentals of mechanical vibrations* (Macmillan).

Main, I. G. (1984) *Vibrations and waves in physics* (CUP, 2nd edn.).

Pain, H. J. (1983) *The physics of vibrations and waves* (Wiley, 3rd edn.).

Among other recent books are the following:

Hagedorn, P. (1988) *Non-linear oscillations* (Clarendon, 2nd edn.). A graduate-level treatment, with extensive references, of this important topic.

Ingard, K. U. (1988) *Fundamentals of waves and oscillations* (CUP). An extensive text at advanced undergraduate level.

Pippard, A. B. (1989) *The physics of vibration* (CUP). A well-received book covering the physical principles underlying a wide range of phenomena. An omnibus edition of earlier texts on classical and quantum vibrations.

Schmidt, G. and Tondl, A. (1986) *Non-linear vibrations* (CUP). Graduate level.

Thomson, W. T. (1981) *Theory of vibration with application* (Prentice-Hall, 2nd edn.; also Allen & Unwin). A substantial textbook. The versions from the two publishers are similar but not identical; for example, that from Allen & Unwin does not use SI units.

Some works on wave propagation of different types are listed here. Other books which are more specifically on the propagation of elastic waves in media are discussed in the section below on 'Acoustics'.

Baldock, G. R. and Bridgeman, T. (1981) *The mathematical theory of wave motion* (Ellis Horwood). An intermediate-level treatment of various types of wave motion.

Bland, D. R. (1988) *Wave theory and applications* (Clarendon). Undergraduate level. Includes elastic and non-linear waves in a solid mechanics context.

Brekhovskikh, L. M. (1980) *Waves in layered media* (Academic, 2nd edn.). A substantial exposition of the theory of propagation of elastic and electromagnetic waves in layered media.

Lighthill, M. J. (1978) *Waves in fluids* (CUP). A theoretical account, with reference to applications, of sound and other waves in liquids and gases. The extensive and critical bibliography is particularly useful for material up to that time.

Taniuti, T. and Nishihara, K. (1983) *Nonlinear waves* (Pitman).

Whitham, G. B. (1974) *Linear and non-linear waves* (Wiley-Interscience). A standard advanced work.

Solitons, or solitary waves as they are sometimes called, form the subject of:

Drazin, P. G. and Johnson, R. S. (1989) *Solitons: an introduction* (CUP). A clearly written introduction to the theory.

Lamb, G. L. (1980) *Elements of soliton theory* (Wiley).

Two journals designed specifically to cover wave phenomena are:

Wave Motion (Elsevier, 1979–), irregular, but approximately bi-monthly.

Waves in Random Media (IOP, 1991–), quarterly.

Tribology

Tribology, a modern term, is the science of solid surfaces in contact and in relative motion. It is concerned with friction, lubrication and wear, and is thus of considerable technological importance as well as scientific interest.

An early book treating the subject at an advanced level is:

Bowden, F. P. and Tabor, D. (1950, reprinted 1986) *The friction and lubrication of solids* (Clarendon).

A few of the many newer books are:

Arnell, R. D., Davies, P. B., Halling, J. and Whomes, T. L. (1991) *Tribology: principles and design applications* (Macmillan). An introductory text.

Booser, E. R. (ed.) (1983, 1984) *Handbook of lubrication (Theory and practice of tribology),* 2 vols. (CRC). An extensive guide for practitioners.

Quinn, T. F. J. (1991) *Physical analysis for tribology* (CUP). Concentrates on the methods for analysing wear in mechanical systems.

Sarkar, A. D. (1980) *Friction and wear* (Academic). Another introductory book.

Information on specific tribological terms, listed alphabetically, is given in:

Kajdas, C., Harvey, S. S. K. and Wilusz, E. (1990) *Encyclopedia of tribology* (Elsevier).

Journals include:
Tribology International (Butterworth-Heinemann, 1968–), bimonthly. Covers the practice and technology of tribology.
Wear (Elsevier, 1958–), fortnightly. Covers the science and technology of friction, lubrication and wear.

Two thousand abstracts in tribology appear each year in the following: *Tribology and Corrosion Abstracts* (Elsevier, 1991–), monthly. Incorporates the earlier abstracting journal *Tribos*, first published in 1968 by BHRA.

The online database TRIBOLOGY INDEX, produced by BAM, Germany, contains over 70 000 abstracts covering the period from 1972 onwards. It is updated quarterly.

Biomechanics

An important and interesting area of application of mechanics is biological systems. Human, plant and cell mechanics, animal locomotion and biological structures and materials all form part of the subject. The following three books illustrate the literature available:

Alexander, R. McN. (1983) *Animal mechanics* (Blackwell, 2nd edn.). A wide survey of mechanics applied to zoology.

Alexander, R. McN. (1988) *Elastic mechanisms in animal movement* (CUP). A short, very readable undergraduate text (see, for example, the mathematical model of a worm).

Nyborg, W. L. (1975) *Intermediate biophysical mechanics* (Cummings). A substantial text which covers most areas of the subject.

The main journal for the field is:
Journal of Biomechanics (Pergamon, 1968–), monthly.

Fluid mechanics

Fluid mechanics deals with the forces and pressures exerted on a fluid at rest (fluid statics) and with the forces on a fluid and the resulting motions (fluid dynamics). By 'fluid' is meant matter in the gaseous, liquid or plasma state. In practice, of course, books on 'fluid mechanics' often concentrate largely on fluid dynamics, which is by far the more substantial part of the topic.

The literature of the subject is very extensive and only a small selection can be given. Some books on fluid mechanics are of an applied mathematical character, perhaps dealing only with classical problems and their analytical solutions (irrotational and laminar flows, for example). Others concentrate heavily on engineering applications, like flows in ducts and open channels (often using imperial rather than SI units). An attempt is made here to give an impression of what may be particularly interesting for physicists.

GENERAL REFERENCES

The 'classics' of the literature include:

Batchelor, G. K. (1967) *An introduction to fluid dynamics* (CUP). A substantial theoretical introduction.
Lamb, H. (1932) *Hydrodynamics* (CUP, 6th edn.).
Landau, L. D. and Lifshitz, E. M. (1989) *Fluid mechanics* (Course of Theoretical Physics, vol. 6) (Pergamon, corrected 2nd edn.).
Prandtl, L. (1952) *Essentials of fluid dynamics* (Blackie). Primarily experimental in character.

More recent texts include:

Acheson, D. J. (1990) *Elementary fluid dynamics* (Clarendon). A wide-ranging and relatively concise text which has been well received.
Fox, R. W. and McDonald, A. T. (1985) *Introduction to fluid mechanics* (Wiley, 3rd edn.). Emphasizes physical concepts and methods of analysis.
Kundu, P. K. (1990) *Fluid mechanics* (Academic). An intermediate level text, with an instructor's manual available.
Lighthill, J. (1989) *An informal approach to theoretical fluid mechanics* (Clarendon, corrected reprint). A succinct and illuminating introduction which covers both practical and theoretical aspects.
McCormack, P. D. and Crane, L. (1973) *Physical fluid dynamics* (Academic). Written particularly for physicists. Includes a chapter on superfluids.

Paterson, A. R. (1984) *A first course in fluid dynamics* (CUP). A very readable account.

Vennard, J. K. and Street, R. L. (1982) *Elementary fluid mechanics* (Wiley, 6th edn.). The latest edition of a long-established student textbook. Emphasizes physical concepts rather than mathematical manipulation.

Books which are primarily restricted to the physics of incompressible flow include:

Panton, R. L. (1984) *Incompressible flow* (Wiley). A substantial text.

Tritton, D. J. (1988) *Physical fluid dynamics* (Clarendon, 2nd edn.). A good presentation at undergraduate level, with a useful bibliography.

Williams, J. and Elder, S. A. (1988) *Fluid physics for oceanographers and physicists: an introduction to incompressible flow* (Pergamon).

Boundary layer flow, of considerable importance to fluid mechanics, is covered in:

Schlicting, H. (1979) *Boundary layer theory* (McGraw-Hill, 7th edn.).

Theoretical treatments of all aspects of fluid mechanics appear in *Encyclopedia of physics*, vols. VIII and IX (1959–1963).

For reference texts the following should be noted:

Cheremisinoff, N. P. (ed.) (1986) *Encyclopedia of fluid mechanics*, vol. 1, *Flow phenomena and measurement* (Gulf). There are also six further volumes dealing with more applied topics.

Parker, S. P. (ed.) (1988) *Fluid mechanics source book* (McGraw-Hill). In the Science Reference Series.

The division of fluid dynamics within the American Physical Society holds a large meeting each year. The abstracts of the invited lectures and other contributions (at present totalling nearly 700 each meeting) are published in advance in the *Bulletin of the American Physical Society*.

A recent symposium of interest, in which several specialists reviewed various aspects of the field, appeared in:

Coles, D. (ed.) (1988) *Perspectives in fluid mechanics* (Lecture Notes in Physics 320) (Springer).

Journals in fluid mechanics which are of interest to physicists include:

European Journal of Mechanics: B/Fluids (Gauthier-Villars, 1982–), bi-monthly.

Experiments in Fluids: Experimental Methods and their Applications to Fluid Flow (Springer, 1983–), quarterly.
Fluid Mechanics. Soviet Research (Scripta, 1972–), bi-monthly. English translations of selected Russian papers.
Journal of Fluid Mechanics (CUP, 1956–), monthly.
Physicochemical Hydrodynamics (Pergamon, 1980–), bi-monthly.
Physics of Fluids. A: Fluid Dynamics and *B: Plasma Physics* (AIP, 1989–), each monthly. Previously appeared from 1958 in only one part.

Reviews appear in:

Annual Review of Fluid Mechanics (Annual Reviews, 1969–).

The literature is abstracted by the usual services in physics. Some entries of interest can also be found in specialist publications such as: *Fluid Abstracts: Civil Engineering* and *Fluid Abstracts: Process Engineering* (Elsevier, 1991–), each monthly. These replace earlier abstracting journals from BHRA.

The online database FLUIDEX, produced by Elsevier, contains over 200 000 abstracts of publications on fluids and fluid mechanics. It covers the period from 1973 onwards and is updated monthly.

AERODYNAMICS, GAS DYNAMICS AND COMPRESSIBLE FLOW

These topics all involve in different ways the flow of gases. A few of the more modern texts include:

Anderson, J. D. (1991) *Fundamentals of aerodynamics* (McGraw-Hill, 2nd edn.) A textbook which covers most aspects.
Clancy, L. D. (1975, reprinted 1986) *Aerodynamics* (Pitman, Longman). A good introduction to both low-speed (incompressible flow) and high-speed (compressible flow) phenomena.
Kuethe, A. M. and Chow, Chuen-Yen (1986) *Foundations of aerodynamics: bases of aerodynamic design* (Wiley, 4th edn.). An intermediate-level account of all aspects.
Ramm, H. J. (1990) *Fluid dynamics for the study of transonic flow* (Clarendon). Covers aerodynamics at speeds near that of sound.
Schreier, S. (1982) *Compressible flow* (Wiley). A wide-ranging graduate-level book.
Zucrow, M. J. and Hoffman, J. D. (1977) *Gas dynamics*, 2 vols. (Wiley).

Applications of low-speed aerodynamics to atmospheric physics, meteorology and insect and bird flight are the subject of:

Scorer, R. S. (1978) *Environmental aerodynamics* (Ellis Horwood).
Ward-Smith, A. J. (1985) *Biophysical aerodynamics and the natural environment* (Wiley). A mostly descriptive text.

High-speed gas flow is exhaustively treated from an aerodynamic viewpoint in:

Von Kármán, T. (Chairman of editors) (1955) *High speed aero-* '
dynamics and jet propulsion, 12 vols. (OUP).

The proceedings of the latest (i.e. the 17th) in a series of symposia stretching back to 1958 appear in:

Beylich, A. E. (ed.) (1991) *Rarefied gas dynamics* (VCH).

Many articles on aerodynamics appear in *AIAA Journal* (AIAA, 1963–), monthly.

NON-NEWTONIAN FLUIDS

Non-Newtonian fluids are specially covered in the following books and journal:

Böhme, G. (1987) *Non-Newtonian fluid mechanics* (North-Holland).
Schowalter, W. R. (1978) *Mechanics of non-Newtonian fluids* (Pergamon). An introductory graduate text.
Journal of Non-Newtonian Fluid Mechanics (Elsevier, 1976–), irregular.

TURBULENCE AND MIXING

Turbulence, a subject of much theoretical and practical interest, is responsible for often considerable energy losses but it also can be used for mixing and heat transfer. An early classic text is Batchelor, G. K. (1953, reprinted 1982) *The theory of homogeneous turbulence* (CUP).
Some more recent books include:

Frost, W. and Moulden, T. H. (eds.) (1977) *Handbook of turbulence*, vol. 1 (Plenum).
Hinze, J. O. (1975) *Turbulence* (McGraw-Hill, 2nd edn.). A substantial theoretical and experimental treatment, with many references.
Landahl, M. T. and Mollo-Christensen, E. (1986) *Turbulence and random processes in fluid mechanics* (CUP). A graduate introduction stressing the dynamics of processes which help to create and maintain turbulent flows.
McComb, W. D. (1991) *The physics of fluid turbulence* (Clarendon, corrected edn.). A graduate-level text.
Ottino, J. M. (1989) *The kinematics of mixing: stretching, chaos and transport* (CUP). Merges ideas from fluid mechanics with ones

from the theories of dynamical systems and chaos. Contains an informative bibliography.

Flow stability and the processes accompanying the transition to turbulence are the subject of:

Chandrasekhar, S. (1961) *Hydrodynamic and hydromagnetic stability* (Clarendon, reprinted by Dover, 1981).
Drazin, P. G. and Reid, W. H. (1981) *Hydrodynamic stability* (CUP).
Swinney, H. L. and Gollub, J. P. (eds.) (1981) *Hydrodynamic instabilities and the transition to turbulence* (Topics in Applied Physics, vol. 45) (Springer). Contains nine review articles.

'Traditional' versus 'modern' approaches to turbulence are surveyed by several specialists in the following recent workshop proceedings:

Lumley, J. L. (ed.) (1989) *Whither turbulence? Turbulence at the crossroads* (Lecture Notes In Physics 357) (Springer).

FLOW VISUALIZATION

The techniques of flow visualization, with examples of the beautiful photographs obtainable, are covered in:

Nakayama, Y. (Chairman of editorial committee) (1988) *Visualized flow* (Pergamon). Compiled by the Japan Society of Mechanical Engineers. A handbook which includes over 200 photographs.
Yang, Wen-Jei (ed.) (1989) *Handbook of flow visualization* (Hemisphere). Describes the various methods for visualization and their applications.

The most recent proceedings in a series of conferences which began in 1979 appear in:

Řezníček, R. (ed.) (1990) *Flow visualization V* (Hemisphere).

MULTIPHASE FLOW

Multiphase flow is concerned with both gas-liquid and solid-liquid flow, and occurs in areas of application as diverse as fluidized beds and carbonated drinks. Two books are suggested:

Hetsroni, G. (ed.) (1982) *Handbook of multiphase systems* (Hemisphere). An exhaustive multi-author tome on all aspects of multiphase flow.
Soo, S. L. (1990) *Multiphase fluid dynamics* (Science Press/Gower).

A journal which covers the subject is *International Journal of Multiphase Flow* (Pergamon, 1973–), bi-monthly.

FLOW MEASUREMENT

The following books and journal specialize in the subject of flow measurement from a practical viewpoint:

Ower, E. and Pankhurst, R. C. (1977) *The measurement of air flow* (Pergamon, 5th edn.).
Scott, R. W. W. (ed.) (1982) *Developments in flow measurement-1* (Applied Science). Covers both liquids and gases.
Flow Measurement and Instrumentation (Butterworth-Heinemann, 1989–), quarterly.

NUMERICAL METHODS AND COMPUTER TECHNIQUES

The importance nowadays of numerical methods and computer techniques, enabling previously intractable problems in fluid mechanics to be tackled, is matched by a rapid rise in the amount of literature produced on the subject. A few of the more recent books are:

Abbott, M. B. and Basco, D. R. (1989) *Computational fluid dynamics: an introduction for engineers* (Longman). Should also be useful for physicists.
Fletcher, C. A. J. (1988) *Computational techniques for fluid dynamics*, 2 vols. (Springer Series in Computational Physics) (Springer). The two volumes are titled *Fundamental and general techniques* and *Specific techniques for different flow categories*.
Holt, M. (1984) *Numerical methods in fluid dynamics* (Springer Series in Computational Physics) (Springer, 2nd edn). A graduate text.
Peyret, R. and Taylor, T. D. (1983) *Computational methods for fluid flow* (Springer Series in Computational Physics) (Springer). A well-received graduate-level account.

A short introductory course, with accompanying diskette, on simple methods of numerically modelling fluid flows is presented in:

Ninomiya, H. and Onishi, K. (1991) *Flow analysis using a PC* (Computational Mechanics/CRC).

Conference proceedings of interest include the following:

Morton, K. W. (ed.) (1990) *Twelfth international conference on numerical methods in fluid dynamics* (Lecture Notes in Physics 371) (Springer). Earlier proceedings back to 1971 appear in previous issues of the Lecture Notes.
Taylor, C. *et al.* (eds.) (1991) *Numerical methods in laminar and*

turbulent flow (Pineridge). This forms the proceedings of the seventh conference in a series which started in 1978.

Wesseling, P. (ed.) (1990) *Proceedings of the eighth GAMM – conference on numerical methods in fluid mechanics* (Notes on Numerical Fluid Mechanics, vol. 29) (Vieweg). These appear in a series of volumes beginning in 1978 which includes reviews and monographs as well as conference proceedings.

Periodicals specially introduced for this field include:

Annual Review of Numerical Fluid Mechanics and Heat Transfer (Hemisphere, 1987–).
Computers and Fluids (Pergamon, 1973–), quarterly.
International Journal for Numerical Methods in Fluids (Wiley, 1981–), semi-monthly.

MAGNETOHYDRODYNAMICS

Magnetohydrodynamics or plasma fluid dynamics (the flow of conducting media, particularly in the presence of magnetic fields), are covered by books such as:

Cowling, T. G. (1976) *Magnetohydrodynamics* (Hilger). An introductory text which includes astronomical and geophysical applications in particular.
Dragoş, L. (1975) *Magnetofluid dynamics* (Abacus). An advanced theoretical treatment.
Freidberg, J. P. (1987) *Ideal magnetohydrodynamics* (Plenum). Develops a single fluid description of ideal (i.e. long wavelength, low frequency) macroscopic plasma behaviour.
Woods, L. C. (1987) *Principles of magnetoplasma dynamics* (Clarendon).

Specialist journals on the subject include:

Journal of Plasma Physics (CUP, 1967–), bi-monthly.
Plasma Physics and Controlled Fusion (IOP and Pergamon, 1959–), 13 issues annually.

Articles of interest also appear in *Physics of Fluids*, as already cited. Where appropriate both parts of the journal should be consulted, despite their subtitles. A more general discussion of the literature on plasmas will be found in Chapter 9.

MISCELLANEOUS REFERENCES

To conclude this subsection on fluid mechanics, a selection of books on miscellaneous topics is given.

Benedict, R. P. (1980) *Fundamentals of pipe flow* (Wiley). Covers ideal and real (i.e. with turbulence and viscous losses) incompressible and compressible fluid flow in pipes.

Dullien, F. A. L. (1979) *Porous media – fluid transport and pore structure* (Academic).

Putterman, S. J. (1974) *Superfluid hydrodynamics* (North-Holland). The fluid mechanics of liquid helium.

Trevena, D. H. (1987) *Cavitation and tension in liquids* (Hilger). A short monograph covering the rupture of liquids.

Yih, Chia-Shun (1980) *Stratified flows* (Academic). Deals with flows of fluids of variable density or entropy in a gravitational field.

Young, F. R. (1989) *Cavitation* (McGraw-Hill). Introduces the research into bubble and cavity formation and behaviour in liquids.

Zel'dovich, Y. B. and Raizer, Y. P. (1966,1967) *Physics of shock waves and high temperature hydrodynamic phenomena*, 2 vols. (Academic). This is an area closely associated with topics such as plasmas and high-speed gas flow.

Acoustics

Acoustics is the science of sound, including its production, transmission and effects. It deals with all types of elastic stress waves and vibrations in all media in the audible frequency range, above it (ultrasonics) and below it (infrasonics). This definition is, of course, far wider than some popular conceptions of the meaning of the word. Since a mechanical disturbance is involved, there are strong connections with the material covered in the previous section on mechanics, especially with waves.

The following subsections cover most of the different interests in acoustics. As in the previous section, quantum effects, such as phonons, are not considered. Books specifically on seismic waves and on the electrical and human production and reception of sound are also not treated.

General references

The cornerstone of the modern literature on the subject is:

Strutt, J. W., 3rd Baron Rayleigh (1894, 1896) *The theory of sound*, 2 vols. (Macmillan, 2nd edn., reprinted by Dover, 1945). Of lasting importance and significance.

Some modern introductions to the subject include:

Dowling, A. P. and Ffowcs Williams, J. E. (1983) *Sound and sources of sound* (Wiley). Includes material on the noise of turbulent flow.
Hall, D. E. (1987) *Basic acoustics* (Harper and Row, also Wiley). Covers audible sound in air.
Sen, S. N. (1990) *Acoustics: waves and oscillations* (Wiley).

A selection of other general books in the field is given below:

Kinsler, L. E., Frey, A. R., Coppens, A. B. and Sanders, J. V. (1982) *Fundamentals of acoustics* (Wiley, 3rd edn.).
Morse, P. M. and Ingard, K. U. (1968) *Theoretical acoustics* (McGraw-Hill, reprinted by Princeton, 1986).
Pierce, A. D. (1981) *Acoustics. An introduction to its physical principles and applications* (McGraw-Hill).
Rossing, T. D. (1990) *The science of sound* (Addison-Wesley, 2nd edn.). A comprehensive survey.
Skudrzyk, E. (1971) *The foundations of acoustics: basic mathematics and basic acoustics* (Springer). An exhaustive treatment.
Temkin, S. (1981) *Elements of acoustics* (Wiley). Emphasizes particularly the connections with fluid mechanics.

A series of extensive articles is to be found in *Encyclopedia of physics*, vol. XI/1 and 2 (1961, 1962) on acoustics, and vol. VIa/3 and 4 (1973, 1974) on waves in solids.
A very useful reference work is:

Parker, S. (ed.) (1988) *Acoustics source book* (McGraw-Hill). In the Science Reference Series.

Recent developments are reviewed by various authors in:

Mason, W. P. and Thurston, R. N. (eds.) *Physical acoustics: principles and methods* (Academic). Nineteen volumes have appeared between 1964 and 1990 in this series. (Following W. P. Mason's death, volume 19 was edited by R. N. Thurston and A. D. Pierce.)

The following books concentrate on solids:

Achenbach, J. D. (1973) *Wave propagation in elastic solids* (North-Holland).
Auld, B. A. (1973) *Acoustic fields and waves in solids*, 2 vols. (Wiley).
Eringen, A. C. and Şuhubi, E. S. (1974, 1975) *Elastodynamics*, 2 vols. (Academic). An advanced theoretical treatise.
Hudson, J. A. (1980) *The excitation and propagation of elastic waves* (CUP).
Miklowitz, J. (1978) *The theory of elastic waves and waveguides* (North-Holland).
Pollard, H. F. (1977) *Sound waves in solids* (Pion).

Below are listed some of the more important general journals in acoustics:

Acoustics Bulletin (Institute of Acoustics, 1976–), bi-monthly. The 'house' journal for members of the Institute.
Acoustics Letters (Parjon, 1977–), monthly.
Acustica (Hirzel, 1951–), monthly. Articles in English, French and German, with summaries in all three languages.
Journal of the Acoustical Society of America (ASA/AIP, 1929–), monthly. With cumulative indices and a current-awareness service.
Journal of Sound and Vibration (Academic, 1964–), semimonthly.

Acoustics is covered by the general abstracting services for physics and by:

Acoustics Abstracts (Multi-Sci. Publ., 1967–), two issues monthly. Contains about 3 000 entries each year, with annual indices.
Surface Wave Abstracts (Multi-Sci. Publ. 1971–), quarterly. Contains over 200 entries each year, with annual indices. See also the subsection below, and the section on 'Acoustic phenomena and electronics'.

A multi-lingual dictionary for the field, with particular emphasis on applied and technical aspects, is:

Stephens, R. W. B. (ed.) (1975) *Sound* (Crosby). Contains several thousand definitions (in English); terms are translated into and from seven other languages.

Several organizations publish reports, arrange conferences and so on. Two important bodies are the Institute of Acoustics and the Acoustical Society of America.
Two special areas of interest in acoustics are mentioned here; others are described in the following subsections. Flow acoustics, which deals with the sound generated by aerodynamic forces or motions originating in a fluid flow, is the subject of:

Goldstein, M. E. (1976) *Aeroacoustics* (McGraw-Hill).

Musical acoustics, or the interaction of physics and music, is treated in many books, and all levels of treatment from semi-popular to advanced can be found. Three recent books of interest are:

Benade, A. H. (1990) *Fundamentals of musical acoustics* (Dover, revised edition of the 1976 OUP text). A substantial reference work.
Johnston, I. (1989) *Measured tones* (Hilger). Accurately described as 'an entertaining exploration of the underlying physical principles of music'.

Rigden, J. S. (1985) *Physics and the sound of music* (Wiley, 2nd edn.).

Architectural acoustics

Architectural acoustics deals with the provision of a satisfactory acoustical environment, and particularly with the satisfactory propagation of sound within a confined space. It has close associations with noise, which is dealt with in the next subsection. Some books which are partly or wholly on the subject include:

Ando, Y. (1985) *Concert hall acoustics* (Springer Series in Electrophysics, vol. 17) (Springer). An important mathematical monograph.
Beranek, L. L. (1962) *Music, acoustics and architecture* (Wiley). A largely descriptive text, with detailed reviews of the acoustics up to that time of 54 concert halls across the world.
Cremer, L. and Müller, H. A. (1982) *Principles and applications of room acoustics*, 2 vols. (Applied Science). A good, thorough account.
Kuttruff, H. (1979) *Room acoustics* (Applied Science, 2nd edn.).
Lawrence, A. (1989) *Acoustics and the built environment* (Elsevier). A practical guide from a scientific viewpoint.
Parkin, P. H. *et al.* (1979) *Acoustics, noise and buildings* (Faber, 4th edn.).
Porges, G. (1977) *Applied acoustics* (Arnold).

Noise and vibration

Noise, in the acoustical as opposed to electrical sense, is unwanted sound. This includes topics such as echoes and reverberations in buildings, vibrations caused by machinery, aerodynamic effects and so on. The literature selected below concentrates on its physics aspects, its origin, transmission, measurement and control, rather than on the environmental, social, medical and legal consequences.

Among many books which treat the subject comprehensively are:

Harris, C. M. (ed.) (1979) *Handbook of noise control* (McGraw-Hill, 2nd edn.).
Irwin, J. D. and Graf, E. R. (1979) *Industrial noise and vibration control* (Prentice-Hall).
White, R. G. and Walker, J. G. (eds.) (1982) *Noise and vibration* (Ellis Horwood). An invaluable multi-author book containing 31 contributions, which is based on the annual advanced lecture course given at Southampton University, UK.

Recently 'active' noise control systems have been developed using digital signal processing devices. These systems create a secondary acoustic field that destructively interferes with the unwanted noise. Thus this work is of at least equal interest to engineers, but it is certainly worth listing here two books which cover this field:

Nelson, P. A. and Elliott, S. J. (1992) *Active control of sound* (Academic).

Tokhi, M. O. and Leitch, R. R. (1992) *Active noise control* (Clarendon) which is the shorter of these two books.

Specialist journals in noise and vibration include:

Applied Acoustics (Elsevier, 1968–), monthly.
Journal of Low Frequency Noise and Vibration (Multi-Sci. Publ., 1982–), quarterly.

An abstracting service specializing in the literature of noise is *Noise and Vibration Bulletin* (Multi-Sci. Publ., 1970–), monthly. Contains about 200 abridgements of articles each year.

Underwater acoustics

Underwater acoustics is of interest as a phenomenon in its own right and also because of its applications to oceanology, underwater communication, submarine detection and so on. Books on the subject include:

Brekhovskikh, L. and Lysanov, L. (1991) *Fundamentals of ocean acoustics* (Springer Series in Wave Phenomena, vol. 8) (Springer, 2nd edn.).

Clay, C. S. and Medwin, H. (1977) *Acoustical oceanography: principles and applications* (Wiley-Interscience).

DeSanto, J. A. (ed.) (1979) *Ocean acoustics* (Topics in Current Physics 8) (Springer).

Flatté, S. M. (ed.) (1979) *Sound transmission through a fluctuating ocean* (CUP).

Ross, D. (1976) *Mechanics of underwater noise* (Pergamon). Covers noise transmitted by structures and radiated into the sea.

Articles on the subject appear in the acoustics literature and in the literature of oceanology, such as:

Ocean Engineering (Pergamon, 1968–), bi-monthly.

Acoustic phenomena and electronics

Phenomena such as bulk and surface acoustic waves, piezoelectricity and the acousto-optic effect (the interaction in solids of acoustic and electromagnetic waves) have long been known, and they are now used in electronic devices, particularly at high frequencies. Books on the subject include:

Dieulesaint, E. and Royer, D. (1980) *Elastic waves in solids: applications to signal processing* (Wiley). A general book which covers the physics of all basic processes as well as of devices.
Ristic, V. M. (1983) *Principles of acoustic devices* (Wiley). Gives extensive coverage of the whole field.
Sapriel, J. (1979) *Acousto-optics* (Wiley-Interscience). Covers both principles and devices.

Ultrasonics

Ultrasonics is concerned with frequencies between about 20 kHz and the GHz region. Many scientific and technological applications arise from the use of ultrasonic radiation. Some of these have already been covered, as the principles of acoustics and ultrasonics are, of course, similar. Some general books are:

Beyer, R. T. and Letcher, S. V. (1969) *Physical ultrasonics* (Academic). A good introduction for physicists.
Ensminger, D. (1973) *Ultrasonics. The low and high intensity applications* (Dekker).
Kuttruff, H. (1991) *Ultrasonics: fundamentals and applications* (Elsevier). A wide-ranging introduction, with a useful bibliography.

The specialist periodical literature includes the following titles:

Russian Ultrasonics (Multi-Sci. Publ., 1971–), bi-monthly. Contains 60–70 translated papers per volume collected from a large number of Russian journals.
Ultrasonics (Butterworth-Heinemann, 1963–), bi-monthly.

The literature is abstracted by the usual services in physics and acoustics.

The most important industrial application of ultrasonics is its use in non-destructive testing (NDT). A relevant book is Krautkrämer, J. and H. *Ultrasonic testing of materials* (Springer, 2nd edn., 1977). Specialist journals include:

The British Journal of Non-Destructive Testing (British Institute of Non-Destructive Testing, 1959–), bi-monthly.
NDT International (Butterworth-Heinemann, 1976–), monthly.

A selection of books on imaging and other applications of ultrasonics (in, for example, medicine) is given below:

Briggs, A. (1992) *Acoustic microscopy* (Clarendon). A materials science text which describes the acoustic techniques now available for investigating microstructures in materials.

Greguss, P. (1980) *Ultrasonic imaging (seeing by sound)* (Focal). An important book, with extensive references.

Hildebrand, B. P. and Brenden, B. B. (1972) *An introduction to acoustical holography* (Plenum; also Hilger).

Rose, J. L. and Goldberg, B. B. (1979) *Basic physics in diagnostic ultrasound* (Wiley).

High-intensity ultrasound can be used to alter the properties of materials, and this is described in:

Pŭskár, A. (1982) *The use of high intensity ultrasonics* (North-Holland).

Finally, the interaction of acoustic and ultrasonic radiation with matter, including molecular effects, may be studied in:

Bhatia, A. B. (1967) *Ultrasonic absorption: an introduction to the theory of sound absorption and dispersion in gases, liquids and solids* (Clarendon, reprinted by Dover, 1985).

Matheson, A. J. (1971) *Molecular acoustics* (Wiley).

CHAPTER FIFTEEN

Nuclear and particle physics

H. BEHRENS

Introduction

Nuclear physics is that branch of physics which is concerned with the structure, properties and decays of atomic nuclei, including collisions and reactions between atomic nuclei such as have been investigated in the past few years, in particular in the field of heavy-ion physics. Questions concerning the atomic cloud of electrons surrounding the atomic nucleus are dealt with in Chapter 4. The literature on the different interactions within the atomic nucleus (strong interactions, electromagnetic interactions and weak interactions) is considered here.

Elementary-particle physics deals with the submicroscopic constituents of protons, neutrons and all other particles, the so-called quarks; the other particle group which is at present believed to be elementary, the leptons, is also included. As the alternative name 'High-energy physics' implies, this branch of physics treats all matters concerned with high energies, 'high' being now generally restricted to reactions at energies above about 1GeV.

Looking at the number of publications in the field of nuclear and particle physics we are confronted with figures of 7000–8000 in a each field (for 1989 and 1990). A comparison of these figures with the annual publications in physics as a whole is interesting, and shows that publications in nuclear physics account for about 6 per cent of all physics literature and those in elementary-particle physics also for about 6 per cent. The figures for nuclear physics mentioned above include publications on accelerators and nuclear instrumentation. Without these latter items the amount of publications in nuclear physics is less than that in elementary-particle physics (about 4 per cent).

'Publications', of course, are documents or written scripts of a wide variety. There are primary journal articles, letter articles, review articles, books, reports, theses, conference papers and pre-prints.

Primary publication in journals and books

In the following we will consider first the most important journals. For nuclear physics they are listed in Table 15.1(a). Looking at the third column we find the number of articles per year. Though quantity, of course, is only one indication of significance and reputation of a certain journal, the two named first, *Nuclear Physics A* and *Physical Review C*, not only come first on account of numbers but also because of their high reputation. The figures not only reveal something about the frequency of certain publications but also about the 'centre of mass' of the information in physics in primary nuclear physics articles.

Table 15.1(b) shows the corresponding list of journals for elementary-particle physics, and is self-explanatory. It will be noted that the first two journals in the ranking are *Physical Review D* and *Nuclear Physics B*.

The so-called 'letter' journals shown in Table 15.2 play a very important role, in nuclear physics as well as in elementary-particle physics. Letter journals have their origin in 'Letters to the Editor', originally published as an annex to the primary journals. In the course of time these letters became 'independent', and the letter journal was born. The two named first in the table are the ones best known. Perhaps one could even say that *Physical Review Letters* and *Physics Letters B* have the highest reputation among all publications in the field under consideration. Generally, one expects that the time span between a scientific paper being 'received' and its being published would be much shorter in the case of a letter journal than for the customary primary journals. The reality, however, is that this difference is not as great as one would think. The average time span for letter journals is 2–5 months and for the primary journals in Table 15.1 it lies between 3 and 9 months.

As far as the state of the art and surveys are concerned, review articles play a particularly important role in nuclear and in elementary-particle physics. For this purpose, some specialized publications have come into being, and these are listed in Table 15.3. Also included in this table are the review series, which are publications in book form containing chapters or articles written by different authors on different topics. A well-known example in the list, *Annual Review of Nuclear and Particle Science*, has appeared in its

TABLE 15.1 The most important journals for nuclear physics and elementary-particle physics

(a) Nuclear physics

Name	Number of volumes/ issues per year	Number of articles per year[a] (in nuclear physics)
Nuclear Physics A	15/60	~550
Physical Review C	2/12	~670
Zeitschrift für Physik A	4/16	~185
Soviet Journal of Nuclear Physics[b] (Yadernaya Fizika)	2/12	~220
Journal of Physics G	1/12	~130
Nuovo Cimento A	1/12	~90
Progress of Theoretical Physics	2/12	~40
Nuclear Instruments and Methods A	13/39	~660

(b) Elementary-particle physics

Name	Number of volumes/ issues per year	Number of articles per year[a] (in elementary-particle physics)
Physical Review D	2/24	~800
Nuclear Physics B	21/63	~480
Zeitschrift für Physik C	4/16	~260
Soviet Journal of Nuclear Physics[b] (Yadernaya Fizika)	2/12	~310
Nuovo Cimento A	1/12	~100
Journal of Mathematical Physics	1/12	~140
Progress of Theoretical Physics	2/12	~100
Annals of Physics (NY)	8/16	~50
Journal of Physics G	1/12	~50
Communications in Mathematical Physics	8/24	~80
International Journal of Modern Physics A	1/32	~35

[a] Based on the years 1989 and 1990.
[b] Now *Physics of Atomic Nuclei*.

TABLE 15.2 Letter journals

Name	Number of articles per year[a] in nuclear physics	Number of articles per year[a] in elemenetary-particle physics
Physical Review Letters	~130	~250
Physical Letters B	~300	~1100
JETP Letters	~20	~50
Modern Physics Letters A	~5	~50
Europhysics Letters	~20	~20

[a] Based on the years 1989 and 1990.

TABLE 15.3 Review journals and series

Name	Number of volumes/ issues per year	Remarks
Reviews of Modern Physics	4	
Physics Reports	15/52	
Fortschritte der Physik	12	
Comments on Nuclear and Particle Physics	1/6	Short articles (~5 pages)
Contemporary Physics	1/6	Short reviews
Annual Review of Nuclear and Particle Science	1	
Advances in Nuclear Physics	1	
Progress in Particle and Nuclear Physics	2	
Lecture Notes in Physics	~30	
Springer Tracts in Modern Physics (Ergebnisse der exakten Naturwissenschaften)	1–2	
Reports on Progress in Physics	12	

present form since 1978, and was formerly *Annual Review of Nuclear Science* between 1952 and 1977.

It should be emphasized that review articles are particularly important; besides allowing the novice quickly to gain an overview of a special field, they also represent a comprehensive collection of all material available, giving some idea of the subject and the state of the art. Unfortunately, not many review articles are written nowadays.

Let us now consider books (textbooks and monographs). As a rule, books provide more comprehensive and extensive cover than review articles. However, they have also particular disadvantages in the effort of writing and high costs of production. Consequently in many cases they are published only once; or if they are updated, then only after a long period of time. We list below, particularly for students, some recommended, recently published textbooks on the main topics in nuclear and elementary-particle physics.

As far as elementary-particle physics is concerned, we have restricted ourselves to the most recent texts. In this field the development has been so rapid that older works can be considered to be obsolete. We only have to mention 'gauge theory' to elucidate this point. Because of the drastic development in this field in particular, we have concentrated on books dealing with gauge formalism.

GENERAL INTRODUCTION TO NUCLEAR PHYSICS

Frauenfelder, H. and Henley, E. M. (1974) *Subatomic physics* (Prentice-Hall).

Segré, E. *Nuclei and particles* (Benjamin, 2nd edn., 1977).

Valentin, L. (1981) *Subatomic physics: nuclei and particles*, 2 vols. (North-Holland).

Cottingham, W.N. and Greenwood, D.A. (1986) *An introduction to nuclear physics* (CUP).

Jones, G. A. (1987) *The properties of nuclei* (Clarendon).

Jelley, N. A. (1990) *Fundamentals of nuclear physics* (CUP).

Williams, W. S. C. (1991) *Nuclear and particle physics* (Clarendon).

NUCLEAR STRUCTURE

Preston, M. A. and Bhaduri, R. K. (1975) *Structure of the nucleus* (Addison-Wesley).

deShalit, A. and Feshbach, H. (1974) *Theoretical nuclear physics, vol. 1 – Nuclear structure* (Wiley).

Bohr, A. and Mottelson, B. R. (1969) *Nuclear structure, vol. 1: Single-particle motion* (Benjamin).

Bohr, A. and Mottelson, B. R. (1975) *Nuclear structure, vol. 2: Nuclear deformations* (Benjamin).

Wilkinson, D. H. (ed.) (1969) *Isospin in nuclear physics* (North-Holland).

Brussaard, P. J. and Glaudemans, P. W. M. (1977) *Shell-model applications in nuclear spectroscopy* (North-Holland).

Lawson, R. D. (1980) *Theory of the nuclear shell model* (Clarendon).

Ring, P. and Schuck, P. (1980) *The nuclear many-body problem* (Springer).
Eisenberg, J. M. and Greiner, W. (1987) *Nuclear theory, vol. 1: Nuclear models* (North-Holland 3nd edn.).
Eisenberg, J. M. and Greiner, W. (1986) *Nuclear theory, vol. 3: Microscopic theory of nucleus* (North-Holland 2nd edn.).
Antonov, A. N., Hodgson, P. E. and Petkov, I. Zh. (1988) *Nucleon momentum and density distributions in nuclei* (Clarendon).
Casten, R. F. (1990) *Nuclear structure from a simple perspective* (OUP)
Heyde, K. L. G. (1990) *The nuclear shell model* (Springer).
Iachello, F. and van Isacker, P. (1991) *The interacting Boson-Fermion model* (CUP).

NUCLEAR REACTIONS

Gibson, W. M. (1980) *The physics of nuclear reactions* (Pergamon).
Glendenning, N. K. (1983) *Direct nuclear reactions* (Academic).
Satchler, G. R. (1983) *Direct nuclear reactions* (Clarendon).
Ericson, T. and Weise, W. (1988) *Pions and nuclei* (Clarendon).

SUPERHEAVY ELEMENTS

Bock, R. (ed.) *Heavy ions collisions*, vol. 1, 1979, vol. 2, 1980, vol. 3, 1982 (North-Holland).
Kumar, K. (1989) *Superheavy elements* (Hilger).

NUCLEAR FORCES

Rho, M. and Wilkinson, D. H. (eds.) (1979) *Mesons in nuclei*, vols. 1, 2, & 3 (North-Holland).

NEUTRON PHYSICS, FISSION

Michaudon, A. (ed.) (1981) *Neutron physics and nuclear data in science and technology. Vol. 1: Nuclear fission and neutron-induced fission cross sections* (Pergamon).
Cierjacks, S. (ed.) (1983) *Neutron physics and nuclear data in science and technology. Vol. 2: Neutron sources for basic physics and applications* (Pergamon).
Williams, W. G. (1988) *Polarized neutrons* (Clarendon).
Golub, R., Richardson, D. J. and Lamoreaux, S. K. (1991) *Ultra-cold neutrons* (Hilger).

BETA DECAY, WEAK INTERACTIONS, ELECTROMAGNETIC INTERACTIONS

Wu. C. S. and Moszkowski, S. A. (1966) *Beta decay* (Interscience).
Schopper, H. F. (1966) *Weak interactions and nuclear beta decay* (North-Holland).
Blin-Stoyle, R. J. (1973) *Fundamental interactions and the nucleus* (North-Holland).
Morita, M. (1973) *Beta decay and muon capture* (Benjamin)
Behrens, H. and Bühring, W. (1982) *Electron radial wave functions and nuclear beta decay* (Clarendon).
Eisenberg, J. M. and Greiner, W. *Nuclear theory, vol. 2: Excitation mechanism of the nucleus* (North-Holland, 2nd edn., 1988).
Boehm, F. and Vogel, P. (1992) *Physics of massive neutrinos*, 2nd edn. (CUP).
Holstein, B. R. (1989) *Weak interactions in nuclei* (Princeton).
Mohapatra, R. N. and P. B. Pal (1991) *Massive neutrinos in physics and astrophysics* (World Scientific).

GENERAL INTRODUCTION TO ELEMENTARY-PARTICLE PHYSICS

Segré, E. *Nuclei and particles* (Benjamin, 2nd edn., 1977).
Valentin, L. (1981) *Subatomic physics – nuclei and particles* 2 vols. (North-Holland).
Perkins, D. H. *Introduction to high energy physics* (Addison-Wesley 3rd edn., 1987).
Cooper, N. G. and West, G. B. (1988) *Particle physics. A Los Alamos primer* (CUP).
Bowler, M. G. (1990) *Femtophysics. A short course on particle physics* (Pergamon).
Hughes, I. S. *Elementary particles* (CUP, 3rd edn., 1991).
Coughlan, G. D. and Dodd, J. E. *The ideas of particle physics. An introduction for scientists* (CUP, 2nd edn., 1991).
Belokurov, V. V. and Shirkov, D.V. (1991) *The theory of particle interactions* (American Institute of Physics)
Williams, W. S. C. (1991) *Nuclear and particle physics* (Clarendon).
Martin, B. R. and Shaw G. (1992) *Particle physics* (Wiley).

QUARKS, GAUGE THEORIES

Flamm, D. and Schöberl, F. (1982) *Introduction to the quark model of elementary particles, vol. 1 – Quantum numbers, gauge theories and hadron spectroscopy* (Gordon and Breach).
Faddeev, L. D. and Slavnov, A. A. (1980) *Gauge fields. Introduction to quantum theory* (Benjamin/Cummings).
Leader, E. and Predazzi, E. (1982) *An introduction to gauge theories and the 'new physics'* (CUP).

Bailin, D. (1986) *Introduction to gauge field theory* (Hilger).
Aitchison, I. J. R. and Hey, A. J. G. *Gauge theories in particle physics. A practical introduction* (Hilger, 2nd edn., 1989).
Mosel, U. (1989) *Fields, symmetries and quarks* (Mc Graw-Hill).
Pokorski, S. (1990) *Gauge field theories* (CUP).
Guidry, M. (1991) *Gauge field theories. An introduction with applications* (Wiley).
Huang, K. *Quarks, leptons and gauge fields* (World Scientific, 2nd edn., 1992).

WEAK INTERACTIONS, GAUGE THEORIES

Okun, L. B. (1982) *Leptons and quarks* (North-Holland).
Commins, E. D. and Bucksbaum, P. H. (1983) *Weak interactions of leptons and quarks* (CUP).
Renton, P. (1990) *Electroweak interactions. An introduction to the physics of quarks and leptons* (CUP).
Grotz, K. and Klapdor, H.V. (1990) *The weak interaction in nuclear particle and astrophysics* (Hilger).
Winter, K. (ed.) (1991) *Neutrino physics* (CUP).

GENERAL INTRODUCTION TO THE THEORY

Pilkuhn, H. M. (1979) *Relativistic particle physics* (Springer).
Lee, T. D. (1981) *Particle physics and introduction to field theory* (Harwood).
Nachtmann, O. (1990) *Elementary particle physics. Concepts and phenomena* (Springer).
Mohapatra, R. N. (1992) *Unification and supersymmetry* (Springer).
Balachandran, A. P., Marmo, G., Skagerstam, B. S. and Stern, A. (1991) *Classical topology and quantum states* (World Scientific).
Smirnov, F. A. (1992) *Form factors in completely integrable models of quantum field theory* (World Scientific).
Grandy, W. T. (1991) *Relativistic quantum mechanics of leptons and fields* (Kluwer).

EXPERIMENTAL PARTICLE PHYSICS

Fernow, R. C. (1986) *Introduction to experimental particle physics* (CUP).
Sokolsky, P. (1989) *Introduction to ultra-high energy cosmic ray physics* (Addison-Wesley).

Two more types of publications, in addition to those discussed already, are worth mentioning: preprints and reports. Preprints are in most cases copies of manuscripts to be published in journals or in conference proceedings. They are mailed directly by the authors to scientists known to them and working in their field. Preprints guarantee a much more rapid circulation of new results than journal articles (the time span until publication of an article being some 3–9 months on average). Reports are published, for example, by research centres in special series and in limited numbers (say, a hundred). Sometimes theses also fall into this category. Reports (and theses) mostly describe experiments and theoretical derivations in more detail than is customary in primary journal articles. From this point of view, they certainly have their place on the publications scene. This is discussed more fully by John Chillag in Chapter 19.

Data compilations

General remarks

The question where to find which data is becoming important for all scientists, particularly for physicists. Not only is there the problem of finding data for physical quantities but also there is the question of the reliability of the data found. The latter is in no way as trivial as one might first imagine. One only has to consider the large number of different values for one specific physical quantity from different experiments, which in some cases even contradict each other. This is the reason that the recent establishment of data compilations and evaluations, both in printed and in machine-readable form, has gained so much in importance. By evaluations we do not mean a simple collection but an accumulation of recommended values following critical selection and assessment. In the following we deal briefly with the most important compilations of this kind.

Nuclear physics

NUCLEAR STRUCTURE AND DECAY DATA

Here we have international co-operation for the production of a machine-readable database which contains all nuclear structure and radioactive decay data, i.e. the data of all known isotopes. The name of this database is EVALUATED NUCLEAR STRUCTURE DATA FILE (ENSDF) (Table 15.4). Ten countries are involved in this, the work being shared in such a way that each institution is responsible for certain mass chains. In Table 15.4 figures concerning the numbers of various data sets are shown, a data set being defined as the sample of evaluated data for one isotope and for one special type of experiment (α-decay, (n,γ), (n,p), etc.). In addition, there is always one dataset of adopted energy levels for each isotope.

ENSDF is also published as NUCLEAR DATA SHEETS (Table 15.5),

TABLE 15.4 Contents of ENSDF (2385 Nuclides)

Type	Number of datasets
Energy levels (adopted)	2353
Decay	3035
beta decay, etc.	944
Reactions	5532
elastic scattering	172
inelastic scattering	930
Other (comments, etc.)	465
Sum	11385

which are printed directly from the database. The machine-readable database is available online via international telecommunication networks. 'Spinners' or 'hosts' providing access to the database are the National Nuclear Data Center, Brookhaven, the NEA Databank, Saclay and the IAEA, Vienna. For a description of these centres, see Behrens and Ebel (1982). The bibliographic database, NUCLEAR STRUCTURE REFERENCES (NSR), also belongs to ENSDF (see below).

Besides the printed version of ENSDF already mentioned, there are other data compilations which contain nuclear structure and decay data in general. These are listed in Table 15.5. In addition to these broad compilations there are a number of selected (or horizontal) compilations which cover specific properties such as atomic masses, nuclear moments, gamma rays, etc. There are hundreds of such compilations. In Table 15.6 we present a few which are of more general importance. Again, we have considered only the latest works. Besides these horizontal compilations there are a number of tables and handbooks which are needed for the evaluation of experiments. We have chosen a selection of the more important topics, and these are presented in Table 15.7. Of significance for a quick glance at properties (decay modes, half-lives, etc.) of all isotopes are the so-called wall charts, and the most recently published charts are listed in Table 15.8.

NUCLEAR REACTIONS (NON-NEUTRON)

When surveying the whole field of nuclear physics, nuclear reactions certainly represent a large part; evidence for this is the large number of publications in this area. Let us first consider one

TABLE 15.5 Most relevant printed compilations containing nuclear structure and decay data

Mass-range	Author(s) or editor(s)	Source	Publisher
3	D. R. Tilley et al.	*Nuclear Physics A*, **474**, 1 (1987)	North-Holland
4	D. R. Tilley et al.	*Nuclear Physics A*, **541**, 1 (1992)	North-Holland
5–10	F. Ajzenberg-Selove	*Nuclear Physics A*, **490**, 1 (1988)	North-Holland
11–12	F. Ajzenberg-Selove	*Nuclear Physics A*, **506**, 1 (1990)	North-Holland
13–15	F. Ajzenberg-Selove	*Nuclear Physics A*, **523**, 1 (1991)	North-Holland
16–17	F. Ajzenberg-Selove	*Nuclear Physics A*, **460**, 1 (1986)	North-Holland
18–20	F. Ajzenberg-Selove	*Nuclear Physics A*, **475**, 1 (1987)	North-Holland
21–44	P. M. Endt	*Nuclear Physics A*, **527**, 1 (1990)	North-Holland
45–263	International Cooperation (various different authors)	*Nuclear Data Sheets*	Academic
1–263	C. M. Lederer V. S. Shirley	*Table of Isotopes*	Wiley, 1978
	V. S. Shirley E. Browne R. B. Firestone	*Table of Radioactive Isotopes*	Wiley, 1986
Decay data dosimetry and radiology)	D. C. Kocher	*DOE/TIC-11026*	Technical Information Center, US Department of Energy, 1981

TABLE 15.6 Selected (horizontal) compilations for nuclear structure and decay data

Topic	Author(s)	Source	Publisher
Atomic masses	A. H. Wapstra G. Audi R. Hoekstra	*Atomic Data and Nuclear Data Tables,* **39**, 281 (1988) *Nuclear Physics,* **A432**, 1 (1985)	Academic North-Holland
Nuclear moments	Pramila Raghavan V. S. Shirley C. M. Lederer	*Atomic Data and Nuclear Data Tables,* **42**, 189 (1989) *Table of Isotopes*	Academic Wiley
Gamma rays (energies, intensities)	U. Reus W. Westmeier	*Atomic Data and Nuclear Data Tables,* **29**, 1 and 193 (1983)	Academic
Gamma rays (strengths) $A = 6$–44 $A = 45$–90 $A = 91$–150	P. M. Endt	*Atomic Data and Nuclear Data Tables,* **23**, 3 (1979) **23**, 547 (1979) **26**, 47 (1981)	Academic
Gamma rays (standards)	P. van Assche R. G. Helmer C. van der Leun A. L. Nichols	*Atomic Data and Nuclear Data Tables,* **24**, 39 (1979) *Nuclear Instruments and Methods,* **A286**, 467 (1990)	Academic North-Holland
Alpha decay	A. Rytz	*Atomic Data and Nuclear Data Tables,* **47**, 205 (1991)	Academic
Nuclear charge density distribution	H. de Vries C. W. Jager C. de Vries	*Atomic Data and Nuclear Data Tables,* **36**, 495 (1987)	Academic

aspect of nuclear reactions (in which the projectile is a charged particle), together with the photonuclear reactions (in which the projectile is a gamma quantum). Despite the flood of publications in this particular field, there have been few data compilations. This is particularly surprising, considering that the now highly 'fashionable' heavy-ion physics also comes under this heading. There are no longer machine-readable databases which are still updated regularly (perhaps with the exception of one previously mentioned data index in the bibliographic database NUCLEAR STRUCTURE REFERENCE). The machine-readable files (not offered online) were those by Kim *et al.* (1964) and the KACHAPAG file (Münzel *et al.*, 1979). The files are still available but they are not easy to handle. There are, however, some compilations in printed form, which are listed in Table 15.9. It should perhaps be noted that the printed version of the KACHAPAG file is listed in this table (*Physik Daten/Physics Data, no. 15*).

NEUTRON PHYSICS

There are special reasons why we deal with neutron physics (i.e. neutron reactions in which the projectile is a neutron) under a separate heading, namely that, in contrast to the last section, we have here the most comprehensive and the most sophisticated machine-readable databases as well as a large number of printed compilations. The reason for this difference is that there is substantial commercial need and use for these databases and compilations, for example for nuclear reactor construction and for shielding calculations. Hence it is not surprising that there has been long-standing co-operation in the collection of bibliographic information (CINDA) and in the collection of the corresponding experimental data (EXFOR). Four centres: (National Nuclear Data Center, Brookhaven, USA; NEA Data Bank, Saclay, France; Centr po Jadernym Dannym am Fiziko-Energeticheskij, Obninsk, Russia; and Nuclear Data Section, IAEA, Vienna, Austria) are jointly responsible; at each of these, the machine-readable databases CINDA and EXFOR are maintained. The contents of CINDA are regularly published in printed form and available from these centres. To obtain experimental data, it is necessary to gain access to the machine-readable EXFOR file (50 000 datasets) or to request specific data from one of the centres mentioned above. Much more important however, for the prospective users are the so-called evaluated neutron data files (libraries), i.e. the machine-readable databases containing evaluated data (see Table 15.10). These are very extensive and comprehensive databases, and serve mainly as a basis for major computations, i.e. they represent the input data for large computer calculations (e.g. for

TABLE 15.7 Tables for the evaluation of experiments

Topic	Author(s)	Source	Publisher
Internal conversion coefficients:			
$Z = 30–103$	R. S. Hager E. C. Seltzer	*Nuclear Data Tables*, **4**, 1 (1968)	Academic
$Z < 30$	I. M. Band M. B. Trzhaskovskaya M. A. Listengarden	*Atomic Data and Nuclear Data Tables*, **18**, 433 (1976)	Academic
$Z = 30–104$	F. Rösel H. M. Fries K. Alder H. C. Pauli	*Atomic Data and Nuclear Data Tables*, **21**, 291 (1978)	Academic
Beta decay (Fermi functions, Coulomb functions)	H. Behrens J. Jänecke	*Landolt–Börnstein, New Series,* Vol. I/4	Springer (1969)
Beta decay (Fermi functions, Coulomb functions)	B. S. Dzhelepov L. N. Zyrianova Y. P. N. Suslov	*Beta Processes*	Nauka, Leningrad (1972)

Beta decay (log f values)	N. B. Gove M. J. Martin	*Nuclear Data Tables*, **10**, 205 (1971)	Academic
Electron capture (capture probabilities, capture ratios, EC/β^+)	W. Bambynek H. Behrens M. H. Chen M. L. Fitzpatrick K. W. D. Ledingham H. Genz M. Mutterer R. L. Intemann	*Reviews of Modern Physics*, **49**, 77 (1977)	AIP
Fluorescence yields	M. O. Krause	*Journal of Physical and Chemical Reference Data*, **8**, 307 (1979)	ACS, AIP, and Natl Bur. Stand.
Auger transition probabilities	M. H. Chen	*Atomic Data and Nuclear Data Tables*, **24**, 13 (1979)	Academic
Clebsch–Gordan coefficients, Racah coefficients, $9j$-symbols, etc.	H. Appel	*Landolt–Börnstein*, New Series Vol. I/3	Springer
Fractional parentage coefficients	B. F. Bayman A. Lande S. Shlomo	*Nuclear Physics*, **77**, 1 (1965)	North-Holland
		Nuclear Physics A, **184**, 545 (1972)	North-Holland

TABLE 15.8 Wall charts of nuclides

Title	Author(s)	Publisher	Date
Chart of the Nuclides (US)	F. W. Walker J. R. Parrington F. Feiner	General Electric Co.	1989
Chart of the Nuclides (Japan)	T. Horiguchi T. Tachibana T. Tamura	Japanese Nuclear Data Committee and Nuclear Data Center Japan Atomic Energy Research Institute	1992
Karlsruher Nuklidkarte (Germany)	W. Seelmann-Eggebert G. Pfennig H. Münzel H. Klewe-Nebenius	Kernforschungszentrum Karlsruhe	1981

TABLE 15.9 Tables in the field of nuclear reactions (non-neutron)

Topic	Author(s)	Source	Publisher
Charged-particle nuclear reactions (integral cross-sections, bibliography)	N. E. Holden S. Ramavataran	*BNL–NCS–51771* Ed.1 (Suppl. 1, 1985) (Suppl. 2, 1986) (Suppl. 3, 1987) (Suppl. 4, 1988) (Suppl. 5, 1989)	Brookhaven National Laboratory
Charged-particle nuclear reactions (Q-values, excitations functions)	K. A. Keller J. Lange H. Münzel G. Pfennig	*Landolt–Börnstein*, New Series Vol. I/5 (pt. a, b, c)	Springer 1973, 1974
Charged-particle nuclear reactions (integral cross-sections, thick-target yields, product yields)	H. Münzel H. Klewe-Nebenius J. Lange G. Pfennig K. Hemberle	*Physik Daten/Physics Data* **15–1** (1979) **15–2** (1979) **15–3** (1982) **15–4** (1982) **15–5** (1982) **15**-Index (1982)	FIZ Karlsruhe

TABLE 15.9 (cont.)

Topic	Author(s)	Source	Publisher
Heavy-ion collisions (reaction parameters)	W. W. Wilcke J. R. Birkelund H. J. Wollersheim A. D. Hoover J. R. Huizinga W. U. Schröder L. E. Tubbs	*Atomic Data and Nuclear Data Tables*, **25**, 389 (1980)	Academic
Photonuclear reactions (data index)	E. G. Fuller H. Gerstenberg H. J. van der Molen T. C. Daun	*NBS Special Publications* 380 (1973) 380 (1982), Suppl. 2	Natl. Bur. Stand.
Photon mass attenuation coefficients (1 keV to 20 MeV)	J. H. Hubbel	*The International Journal of Applied Radiation and Isotopes*, **33**, 1269 (1982)	Pergamon
(100 eV to 100 keV)	E. B. Salomon J. H. Hubbel J. H. Scofield	*Atomic Data and Nuclear Data Tables*, **38**, 1 (1988)	Academic
Ranges of protons in matter (1 keV to 10 GeV)	J. F. Janni	*Atomic Data and Nuclear Data Tables*, **27**, 147 and 341 (1982)	Academic

Ranges of ions in matter (stopping powers)	F. Ziegler	U. Littmark	H. H. Andersen	*The Stopping and Ranges of Ions in Matter* Vol. 1 (1984) Vol. 2 (1977) Vol. 3 (1977) Vol. 4 (1977) Vol. 5 (1980) Vol. 6 (1980)	Pergamon
	F. Hubert	R. Bimbot	H. Gauvin	*Atomic Data and Nuclear Data Tables*, **46**, 1 (1990)	Academic

TABLE 15.10 Evaluated neutron data files

Topic	Name	Number of nuclides	Year of last update	Producer
Cross-sections, resonance parameters	ENDF/B–VI (Evaluated Nuclear Data File)	320	1989/90	CSEWG (Brookhaven, USA)
	JENDL–3 (Japanese Evaluated Nuclear Data Library)	324	1989	INDC and JAERI (Tokaimura, Japan)
	JEF–2 (Joint Evaluated File)	312	1990/91	NEA (Paris, France)
	BROND–2 (Russian Evaluated Neutron Data Library)	163	1991	IPPE (Obninsk, Russia)

reactor calculations). The files, as well as specific data from these files, are obtainable from the data and information centres mentioned above (see Behrens and Ebel, 1982).

The most important of all the bases mentioned is (or has been) ENDF/B. The format, too, is now something like a standard computer format for storing evaluated neutron data.

These machine-readable databases are of overriding importance. Nevertheless, some of the hundreds of printed compilations, small and large, are still of general interest, and a few are listed in Table 15.11. The best known of these, the Brookhaven National Laboratory report *BNL 325 – Neutron cross-sections* (the so-called 'Barn Book'), is now published by Academic.

Particle physics

PARTICLE PROPERTIES

The most important publication in this field is *Review of Particle Properties*. The Particle Data Group regularly compiles all data of relevant properties of stable and unstable elementary particles, such as value of masses, decay modes, branching ratios, etc. In the past this review was published once every two years, alternatively, either in *Review of Modern Physics* or in *Physics Letters*. The last edition appeared, however, in *Physical Review D* (see Hikasa *et al.*, 1992). It almost goes without saying that it is in everyday use in almost every high-energy laboratory, helped by the fact that there are also pocket-size versions (with reduced contents). The full listings of the *Review of Particle Properties* are also available in machine-readable form. Online access is possible at SLAC, CERN, Durham and Serpukhov.

PARTICLE REACTIONS

Unfortunately there is no corresponding group compiling comprehensive reaction data. However, there are some specific activities. First, there is the Durham – RAL HEP database group and the COMPAS group, which compile elementary particle-reaction data (experimental) and associated bibliographic information; these data are gained from published papers, preprints or directly from experimentalists. The last source is important, and for this reason the file has some kind of archive function. All data are stored in machine-readable form, controlled by a relatively sophisticated database management system (DBMS) and made accesible to the particle physicist by computer through SLAC, CERN, Durham and Serpukhov.

There are four main series of printed compilations of reaction data:

TABLE 15.11 Neutron data compilations in printed form

Topic	Author(s)	Source	Publisher
Neutron-induced reactions Resonance parameters, thermal cross-sections	S. F. Mughabghab R. R. Kinsey C. L. Dunford	*Neutron Cross Sections:* Vol. I Part A (1981) Part B (1984)	Academic
Neutron cross-section curves	V. McLane C. L. Dunford P. F. Rose	Vol. II (1988)	Academic
Fission product yields	B. F. Rider M. F. Meek E. A. C. Crouch	NEDO–12154–3C (1981) *Atomic Data and Nuclear Data Tables*, **19**, 419 (1977)	Vallecitos Nuclear Center General Electric, Co. Academic
Reactor neutron metrology	J. H. Baard W. L. Zijp H. P. Nolthenius	*Nuclear Data Guide for Reactor Neutron Metrology* (1989)	Kluwer
Activation cross sections	K. Okamoto (ed.)	*IAEA Technical Report Series No. 273* (1987)	IAEA, Vienna
Neutron scattering lengths	L. Koester H. Rauch E. Seymann	*Atomic Data and Nuclear Data Tables*, **49**, 65 (1991)	Academic

Landolt-Börnstein
CERN-HERA
Physik Daten/Physics Data
RAL-Reports.

A list of the more recent issues within these series is presented in Table 15.12. The compilations mentioned here contain not only tables of experimental or evaluated data but also details of the complete analyses (e.g. phase-shift analyses).

Two more collections of a slightly different kind are worth noting. First, there is the project *Current Experiments in Elementary Particle Physics*, and second, *Major Detectors in Elementary Particle Physics*. Both collections are published by the Lawrence Berkeley Laboratory (Wohl *et al.*, 1989, and Gidal *et al.*, 1985).

WEAK INTERACTIONS

At present a field theory of electro-weak interactions which is a renormalizable theory exists. It is often called the standard model of the weak interactions. A data compilation containing the corresponding facts and theoretical foundations has been published 1988 (Haidt and Pietschmann, 1988)

COUPLING CONSTANTS

Coupling constants are naturally of great significance in physics. Besides the so-called fundamental constants of the basic interactions of physics as a whole, there are also the coupling constants especially related to elementary-particle physics. For fundamental constants, evaluations are published regularly (approximately every 10 years; for the latest of these, see Cohen and Taylor, 1988). Constants particularly relevant for high-energy physics have been also evaluated and compiled. They can be found in Dumbrajs *et al.* (1983).

Bibliographic databases and abstracts journals

Bibliographies of the abstracts journal type inform the user of what is happening at present (current awareness) and what happened in the past (retrospective). They contain the bibliographic data of a publication, such as title, author, source (journal, name and number of report, etc.), affiliation, abstract, etc., and a subject analysis in the form of a classification code and keywords (descriptors). In the past, bibliographies published regularly – e.g. once every fortnight – stood in the fore, in the 1980s, however, the trend went in the direction of

TABLE 15.12 Selected compilations of data on particle reactions

Topic	Author(s)	Source	Publisher
Total reaction cross sections	A. Baldini V. Flamino W. G. Moorhead D. R. O. Morrison	*Landolt–Börnstein*, New Series Vol. I/12(a + b)	Springer (1988)
Pion photoproduction	D. Menze W. Pfeil R. Wilcke	*Physik Daten/Physics Data, 7–1* (1977)	FIZ Karlsruhe
Photoproduction of elementary particles	P. Joos W. Pfeil	*Landolt–Börnstein*, New Series Vol. I/8	Springer (1973)
Gamma–gamma–hadrons	R. G. Roberts M. R. Whalley	*Report RAL-86–058* (1986)	Rutherford
Scattering (elastic and charge exchange)	P. J. Carlson A. N. Diddens F. Mönnig G. Giacomelli H. Schopper	*Landolt–Börnstein*, New Series Vol. I/7	Springer (1973)
Pion-nucleon scattering (data + analysis)	G. Höhler	*Landolt–Börnstein*, New Series Vol. I/9b (Part 1 + Part 2)	Springer (1983)

Pion-induced reactions	V. Flaminio W. G. Moorhead D. R. O. Morrison N. Rivoire	*CERN-HERA 83-01* (1983)	CERN
Kaon-nucleon scattering, nucleon nucleon scattering	J. Bystricky P. Carlson C. Lechanoine F. Lehar F. Mönnig K. R. Schubert	*Landolt–Börnstein*, New Series Vol. I/9a	Springer (1980)
Kaon-induced reactions	V. Flaminio W. G. Moorhead D. R. O. Morrison	*CERN-HERA* 83-02 (1983)	CERN
Nucleon-nucleon scattering	M. K. Carter P. D. B. Collins M. R. Whalley	*Report RAL-86-002 (1986)*	Rutherford
Nucleon-nucleon scattering (analysis)	P. Kroll	*Physik Daten/Physics Data*, **22–1** (1981)	FIZ Karlsruhe
Structure functions in deep inelastic scattering	R. G. Roberts M. R. Whalley	*Journal of Physics*, **G17**, D1 (1991)	IOP
Proton (anti-proton) induced reactions	V. Flaminio W. G. Moorhead D. R. O. Morrison N. Rivoire	*CERN-HERA 84-01* (1984)	CERN

TABLE 15.12 (cont.)

Topic	Author(s)	Source	Publisher
Hadronic atoms	H. Poth	*Physik Daten/Physics Data*, **14–1** (1979)	FIZ Karlsruhe
$\gamma, \nu, \Lambda, \Sigma, \Xi$ and K_L^0 induced reactions	S. I. Alekhin et al.	*CERN-HERA 87-01* (1987)	CERN
Shielding	A. Fasso K. Goebel M. Höfert J. Ranft G. Stevenson	*Landolt–Börnstein*, New Series, Vol. I/11 (1990)	Springer

machine-readable databases. These databases are implemented at so-called 'Hosts' or 'Spinners', like STN International, DIALOG, etc. (For the corresponding addresses etc. see Marcaccio, 1992)). Access is possible via modern telecommunication networks. Every user in possession of a PC or a terminal has the possibility to search online in the chosen database. The reason for this development is that it brings advantages in so far as keywords, classification code and even free-text terms (of title and abstract) can be linked by some kind of Boolean algebra interactively in the computer. The result is that searches are much faster and more precise than a browse through the indexes of printed products. The online connection between user and host computer is not the only data distribution channel. Data can also be transfered via electronic media such as CD-ROMs, magnetic tapes, etc. Thus some of the databases under consideration are also available on CD-ROM or magnetic tapes. Then they can be loaded on the computer for in-house use. There are bibliographic databases which cover physics as a whole, like INSPEC-A, PHYS. In addition, there are some very specific databases dealing with nuclear or elementary particle physics only, like INIS, HEP, etc. Since there is still a place for printed abstracts journals despite the importance of machine-readable databases, the databases usually have their corresponding printed counterparts. All these, databases and printed products, are given in Table 15.13 together with a number of figures which elucidate volume and contents. The oldest specialized bibliography – or bibliographic database – has been *Nuclear Science Abstracts (NSA)*, founded 1948. NSA was discontinued in favour of INIS. However, for retrospective searches of publications prior to 1972, NSA is still of importance. All the databases mentioned in Table 15.13 are updated regularly. In the case of the major databases this updating is carried out every fortnight, in the case of those more specialized the intervals can be up to half a year. The updating is carried out either by a single institution (e.g. Institution of Electrical Engineers) or decentralized in international co-operation (e.g. INIS).

Conferences and proceedings

Conferences offer the unique opportunity for personal contacts and discussions. The main purpose of a conference is to keep the scientist up to date with the latest results and developments. These are presented either as *invited* or *contributed* papers. Invited papers usually give an excellent review or state of the art of the subject. Conference papers are usually published in printed form after the conference. The usual procedure nowadays is for the authors to deliver

TABLE 15.13 Bibliographic databases and abstracts journals for nuclear and particle physics

Scope	Name of the database	Name of printed version	Number of citations in the database	Producer	Date of earliest citation in the database
Physics in general	Inspec A	*Physics Abstracts*	2 500 000 in total 160 000 annually	Institution of Electrical Engineers (UK)	1969
	PHYS	*Physics Briefs*	1 700 000 in total 135 000 annually	FIZ Karlsruhe	1979
Nuclear research and technology	INIS	*INIS ATOMINDEX*	1 600 000 in total 90 000 annually	IAEA International cooperation (96 members)	

Nuclear physics (nuclear structure and decay)	Nuclear Structure References (NSR)[b]	*Nuclear Structure References* (published *Nuclear Data Sheets*)	120 000 in total 3700 annually	National Nuclear Data Center, Brookhaven (USA)	1910
Neutron physics (data index)	CINDA[a]	*CINDA*	243 000 in total 4000 annually	USA National Nuclear Data Center Russian Nuclear Data Center NEA Data Bank IAEA Nuclear Data Section	1935
Elementary-particle physics	HEP[b]	*High Energy Physics Index*	250 000 in total 20 000 annually	DESY (Germany) and SLAC (USA)	1969

[a] The number of citations means the number of indexed data, not the number of bibliographic units.
[b] No abstracts.

to the organizers a camera-ready copy of their manuscript. Invited papers are usually substantial, of a length equivalent to that of the customary journal article, whereas contributed papers are usually restricted to one or two pages in the proceedings. Conference papers do not generally have the same ranking as journal articles and they certainly do not replace them. The value of conference papers lies mainly in the rapid circulation of information which is achieved. However, the useful life of this kind of proceedings publication is, in general, short. Therefore it is of the utmost importance that proceedings should appear as soon as possible after the conference, which is not usually difficult with camera-ready manuscripts.

Nevertheless, it is not too unusual to find that proceedings are not published until one (or even two) years after the conference, and then their value is probably diminished. In the field of nuclear and high-energy physics there are several types of conferences.

First, we have the well-known large international conferences which are held at regular intervals (every two to four years), and at alternating venues. They are partly sponsored by or held under the auspices of IUPAP (see Chapter 2). Examples are:

International Conference on High Energy Physics
International Symposium on Lepton and Photon Interactions at High Energies
International Conference on Neutrino Physics and Astrophysics
International Symposium on Weak and Electromagnetic Interactions in Nuclei
International Conference on Nuclear Physics
International Conference on Atomic Masses and Fundamental Constants.

Substantial volumes of proceedings are always produced after these conferences.

Second, there are a number of conferences and workshops, also international, but more specialized; for instance, conferences on:

Quark matter
Photonuclear effect
Heavy ion collisions
Few body problems
Positron annihilation

Here, too, proceedings are published.

Third, there are numerous conferences of the various national physical societies, for example, the Spring Meeting of the American Physical Society (APS) and the Spring Meeting of the Deutsche Physikalische Gesellschaft (DPG). Contributions to these confer-

ences, as a rule, are published only in abstract form, for instance in the *Bulletin of the APS* or in the *Verhandlungen der DPG*.
In addition, there are schools (summer and winter). The main difference between a conference and a school is that the latter place an emphasis on teaching. Lectures of the most important schools are also published, and sometimes they present very good reviews. Schools are often named after the town they are held in. Well-known schools are:

Trieste (Italy)
Les Houches (France)
Erice (Italy)
Cargèse (France)
Varenna (Italy)
Schladming (Austria)

It should be noted that they do not always deal with elementary-particle physics or nuclear physics (see, for instance, Chapter 19).

Conference calendars published in well-known physics journals, such as *Physics Today*, and databases such as CONF available at the host STN International or CONF available at SLAC and CERN give regular information on various conference and summer schools. Reverting to the number of publications mentioned at the beginning of this chapter, proceedings represent an average of approx. 30 per cent of all publications in physics, which is higher than one would think.

References

Behrens, H. and Ebel, G. (1982) *Codata Bulletin*, 48.
Cohen, E. R. and Taylor, B. N. (1988) *Journal of Physical and Chemical Reference Data*, 17, 1795.
Dumbrajs, D., Koch, R., Pilkuhn. H. *et al*. (1983) *Nuclear Physics B*, 216, 277 (*Physik Daten/Physics Data*, 4.3).
Gidal, G., Armstrong, B. and Rittenberg, A. (1985) *Lawrence Berkeley Laboratory Report*, **LBL – 91**, Suppl. revised.
Haidt, D. and H. Pietschmann (1988) *Landolt-Börnstein, New Series vol.* I/10.
Hikasa, K. *et al*. (1992) *Physical Review D*, 45 (11), part II.
Kim, H. J., Milner, W. T. and McGowan, F. K. (1964) *Oak Ridge National Laboratory Report* **ORNL-CPX 1**, **ORNL-CPX 2**; (1966) Nuclear Data Tables, **A1**, 203; (1966) *Nuclear Data Tables*, **A2**, 1; (1967) *Nuclear Data Tables*, **A3**, 123.
Marcaccio, K.J. ed. (1992) *CUADRA Dictionary of Online Databases*, 13, 2 (Gale Research).
Münzel, H., Klewe-Nebenius, H., Lange, J. *et al*. (1979) *Physik Daten/Physics Data*, **15–1, 15–2**, (1982) *Physik Daten/Physics Data* **15–3, 15–4 15–5, 15**(Index).
Wohl, C. G., Armstrong, F. E., Rittenberg, A. *et al*. (1989) *Lawrence Berkeley Laboratory Report*, **BL-91** suppl.

CHAPTER SIXTEEN

Optoelectronics, quantum optics and spectroscopy

A. L. RODGERS

Introduction

Optoelectronics has been defined as the study of the interaction of optical radiation in the wavelength range $0.2\mu m$ to $20\mu m$ with matter and of the devices which depend upon these interactions. Examples are optical communications systems, fibre optics and waveguides, display devices, lasers and photodetectors.

Quantum electronics is a closely related subject which is concerned with the interactions involving quantum energy levels and resonance phenomena, particularly those applied to lasers and related phenomena such as non-linear optics, light modulation and detection, laser spectroscopy, quantum noise effects, etc.

Integrated optics is the study of optical devices that are based upon light transmission in planer waveguides which are dielectric structures that confine the propagating light to a very small region of the order of the optical wavelength. An example of a device working on this technique is the Integrated Optical Circuit (IOC) which is the optical analogy of the semiconductor type of integrated circuit. An IOC would be used in lasers, integrated lenses, switches, interferometers, modulators, detectors, etc., and forms part of signal processing and optical communications systems.

These three fields are closely related and although an attempt has been made within this chapter to separate communications and waveguides from optoelectronics devices, the reader is advised to expect considerable overlapping of subject material in the bibliography and is also referred to associated material in Chapter 6. Brief summaries to references are depicted where made available; otherwise author, title and publisher details are given.

It is very apparent that there have been substantial advances in all aspects of the subject and in no other area is this more evident than in the field of non-linear optics. This field is concerned with the interaction of optical radiation with materials either solid, liquid or gas, in which a non-linear optical response occurs, resulting in intensity dependent variation of characteristics. Although these non-linear effects were known before the advent of the laser it is the very high intensities of this device which have substantially expanded the range of materials and effects, allowing a considerable increase in scope of application. A particular example is the phenomenon of optical phase conjugation which has use in adaptive optics systems to correct automatically wavefront aberration, thus allowing for better quality laser beam output in one application and improving atmospheric laser beam propagation in another.

Quantum theory is the foundation of many of the devices referred to, and a section on quantum optics is included in the bibliography. Such theory is invaluable in understanding atomic and molecular spectra but may have limitations in explaining detailed phenomena, and the achievement of an all encompassing theory of relativistic quantum electrodynamics remains a future goal.

In a data retrieval exercise of this type it is inevitable that source information is missed and omissions are regretted. Full consultation with the particular publication is recommended and is essential in order to confirm interest before purchase.

Journals, series and conference proceedings

The reader is referred to the comprehensive lists of journals and series in Chapter 6, and attention is drawn to the remarks made in that chapter about conference proceedings.

Optoelectronics and quantum electronics

The subject of optoelectronics deals with devices combining optical and electrical parts, and has expanded enormously over the past few years as a result of the rapid expansion in the semi-conductor industry. See also references under 'Optical communications'.

Books and proceedings

QUANTUM OPTICS

The quantum theory of light is necessary to explain all photon interaction phenomena, including lasers. Knight, P. L. and Allen, P. L. (1983) *Concepts of quantum optics* (Pergamon) presents a tutorial account of the basis of quantum optics with a selection of association pedagogic papers; Loudon, R. *The quantum theory of light* (OUP, 2nd edn., 1979, Clarendon 1983) gives a comprehensive account in which illumination by both narrow and broad-band light is discussed. Simmons, J. W. and Guttman, M. J. (1970) *States waves and photons: a modern introduction to light* (Addison-Wesley) is a student text. Two further treatises in this field are Argarwal, G S. (1982) *Quantum optics* (Springer) and Haken, H. (1981) *Light. Vol. 1: Waves, photons and atoms* (North-Holland), which develops the interaction of light with matter from quantum theoretical aspects.

More recent books in the field are:

Knight, P. L. and Allen, L. (1983) *Concepts of quantum optics* (Pergamon) gives a basic approach suitable for final year undergraduates.

Pike, E. R. and Sarkar, S. (1986) *Frontiers in quantum optics* (Adam Hilger) is an advanced treatise covering squeezed states of light, quantum theory of optical bistability, quantum effects in parametric down-conversion of light and quantum fluctuations in the laser.

Pike, E. R. and Walther, H. (1988) *Photons and quantum fluctuations* (Hilger) is a series of advanced papers in quantum non-linear optics concerned with non-classical properties of radiation and applied to photon correlation techniques and dynamics of radiation-atom coupling.

Arecchi, F. T. and Harrison, R. G. (1987) *Instabilities and chaos in quantum optics* (Springer) is concerned with the use of the laser to study new concepts in deterministic chaos and other kinds of instabilities which have particular application in the construction of all optical computers.

Meystie, P. and Sargent III, M. (1990) *Elements of quantum optics* (Springer) grew out of two semester graduate courses in laser physics and quantum physics; requires a good understanding of basic electromagnetism as well as knowledge of quantum mechanics.

Bertolotti, M. and Pike, E. R. (eds.) (1991) *ECOOSA 90 quantum optics* (IOP) is the proceedings of a conference held by the Optics Division of the European Physical Society.

Akulin, V. M. and Karlov, N. V. (1991) *Intense resonant interactions in quantum electronics* (Springer).

Ehlotzky, F. (ed.) (1987) *Fundamentals of quantum optics II* (Springer) is the proceedings of the Third Meeting on Laser Phenomena held at Obergurgl, Austria in 1987.

Harvey, J. D. and Walls, D. F. (ed.) (1986) *Quantum optics IV* (Springer) is the proceedings of the Fourth International

Symposium, Hamilton, New Zealand, in 1986 and *Quantum optics V* the proceedings of the Fifth Symposium at the same location.

Andreev, A. V., Ernel'Yamov, V. I. and LL'inskii, Y. A. (1993) *Co-operative effects in optics* (IOP) is a systematic treatment of the modern theory of co-operative optical phenomena, giving a detailed explanation of their collective interactions.

OPTOELECTRONICS

Ross, D. S. (1979) *Optoelectronics devices and optical imaging techniques* (Macmillan) covers photon interaction with matter: the principles of light-emitting diodes, photoconductors, photo-diodes and phototransistors; noise in optoelectronic devices, solar cells, laser diodes and optical imaging techniques, including treatment of modulator transfer functions.

Other basic books are:

Herman, M. A. (ed.) (1980) *Semiconductor optoelectronics* (Wiley), dealing with physical principles and applications of opto-electronic devices and materials, and Wilson, J. and Hawkes, J. F. B. (1983) *Optoelectronics* (Prentice-Hall). Texas Instruments publish *Optoelectronics theory and practice* (1976) and *Opto-electronics data book*, giving details of optoelectronics components.

Bowden, C. M., Ciftan, M. and Robl, H. R. (1981) *Optical bistability* (Plenum) describes the non-linear effects which are driven by coherent radiation and are useful for optical switches in applica-tion to integrated optics communication system development.

A sourcebook on quantum electronics is *Principles of quantum electronics* by D. Marcuse (Academic, 1980).

Moss, T. S. and Stenholm, S. (eds.) (1983) Progress in Quantum Electronics, Vol. 7 (Pergamon) is part of a series based upon articles presenting both theoretical and experimental aspects of current research on the interaction of radiation with matter and is also reported in *International Review Journal,* by the same publisher.

Knight, P. L. (1983) *Quantum electronics and electro-optics* (Wiley) reports the proceedings of the Fifth National Conference on Quantum Electronics.

Camatini, E. (ed.) (1973) *Progress in electro-optics* and Drause, S. (1983) *Molecular electro-optics* are reports within the NATO Advanced Study Institute series.

Londe, L. D. (ed.) (1982) *Coherent properties of semiconductors under laser irradiation* is a volume in the NATO Advanced Study Institute series.

More recent books on this subject are:

Andonovic, I. and Uttamchandani, D. (1989) *Principles of modern optical systems* (Artech House) originates from a course run at Strathclyde University and covers basic principles and applications of optoelectronics. Subjects include semi-conductor sources, optical detection, optical fibres, networks, integrated optics, optical information processing, guided wave-signal processing, remote sensing.

Ghatak, A. and Thyagarajan, K. (1989) *Optical electronics* (CUP) is an advanced text intended for senior undergraduate and first year graduate students presenting the basic physical principles of optoelectronic effects and devices based on course workshops and summer schools.

Wilson J. and Hawkes, J. F. B. (1989) *Optoelectronics: an introduction* (Prentice-Hall, 2nd edn.) has been updated to include developments in optical communications and associated techniques. Most chapters are revised and a new chapter addresses fibre optic sensors and fibre light guides.

Nunley, W. and Bechtel, J. S. *Infrared optoelectronics, devices and applications* (Marcel Dekker, Opt. Eng. Series, vol. 12).

Hutcheson, L. O. (ed.) *Integrated optical circuits and components* (Marcel Dekker, Opt. Eng. Series, vol. 13).

Jones, K. A. (1987) *Introduction to optical electronics* (Harper and Row) is a readable textbook given at an instructional level and covering the principles of optical electronics and the devices which make use of them.

Yariv, A. (1985) *Optical electronics*, (Holt, Rinehart and Winston, 3rd edn.) is an advanced treatise covering electromagnetic theory, the propagation of optical beams in free space and in fibres, the interaction of optical radiation with atomic systems, the theory of optical resonators and laser oscillation, and examination of specific laser systems, including a new chapter on semi-conductor lasers, optical detection and treatment of noise, holography and the theory and applications of phase conjugate optics.

Yariv, A. (1989) *Quantum electronics* (Holt-Reinhart, 3rd edn.) covers similar ground to the author's *Optical electronics*.

Tamir, T. (1990) *Guided wave optoelectronics* (Springer, 2nd edn.).

Lugue, A. (ed.) (1989) *Solar cells and optics for photovoltaic concentration* (Hilger) is concerned with the application of semi-conductor device physics to the operation of solar cells for concentrated sunlight and associated optics.

Chaimowicz, J. C. A. (1989) *Lightwave technology: an introduction* (Butterworths) is aimed at the electro-optics engineer and covers the principles and operations of LEDs, lasers and laser beam engineering, optical fibres, p-n junctions, photodiodes, free space and fibre optic communications, insulated signal couplers, holography, Fourier transforms and integrated optics.

Waidelich, W. (ed.) (1989) *Laser optoelectronics in engineering* (Springer).

Saleh, B. E. and Teich, M. C. (1991) *Fundamentals of photonics* (Wiley) is an introductory textbook covering the interaction of optics and electronics and applications to lasers, optical fibres and semi-conductor optical devices.

McAuley, A. D. (1991) *Optical computer architecture* (Wiley) describes the basic mathematical concepts of optical systems and explores applications through the use of optical computing devices, associative memories, interconnections and optical logic.

Waidelich, W. and Waidelich, R. (eds.) (1988) *Laser optoelectronics in medicine* is the proceedings of the 7th International Conference (Springer) and the authors have also edited the 9th International Conference (Springer, 1990) under the same title.

Waidelich, W. and Waidelich, R. (eds.) (1987) *Laser optoelectronics in engineering* is the proceedings of the 8th International Conference (Springer).

Pratesi, R. (1991) *Optronic techniques in diagnostic and therapeutic medicine* (Plenum).

Martellucci, S. and Chester, A. N. (1991) *Opto-electronics for environmental science* (Plenum).

Defonzo, A. P. (1991) *Advances in interconnection and packaging* (SPIE, vol. 1389 and vol. 1390) is the proceedings of a conference held at Boston, USA, in 1991 and covers the micro-electronics and integrated optoelectronics in optical communication and computer systems.

Oron, M. and Shladov, I. (eds.) (1991) *Optical engineering* (SPIE, vol. 1442) is the proceedings of the 7th Meeting held at Tel-Aviv in 1990, covering microlithography, optical design, non-linear optics, optical thin films, infra-red systems, optoelectronics and applications in medicine.

Seery, B. D. (ed.) (1989) *Optomechanical design of laser transmitters and receivers* (SPIE, vol. 1044) is the proceedings of a conference held in Los Angeles in 1989 and covers laser diode modules, transmitter/receiver technology, advanced electro-optical technology diagnostics and materials.

Otrio, G. (ed.) (1989) *Optical space communication* (SPIE, vol. 1131) is the proceedings of a conference held in Paris in 1989 and covers the optoelectronic techniques and technologies underlying space communications.

Franz, G. (ed.) (1991) *Optical space communication II* (SPIE, vol.

1522) is the proceedings of a conference held in Munich in 1991 and develops these techniques further.

Prokhorov, A. M. and Ursu, I. (eds.) (1989) *Trends in quantum electronics* (SPIE, vol. 1033) is the proceedings of a conference held in Bucharest in 1988 and covers new developments in laser physics and technology, non-linear phenomena, surface studies with lasers, optical processing and diagnostics, optical wave-guides, laser spectroscopy and applications in medicine and biology.

Reichl, H. (ed.) (1990) *Microsystem technologies '90* (Springer) is the proceedings of the First International Conference on micro, electro, opto, and mechanic systems and components held in Berlin in 1990.

Midwinter, J. and Guo, Y. (1992) *Optoelectronics and lightwave technology* (Wiley) is a substantial review based from the view-point of lightwave telecommunication technology.

Arrathooon, R. (ed.) *Optical computing* (Marcel Dekker, Opt. Eng. Series, vol. 19).

Elsevier's *Dictionary of opto-electronics and electro-optics* (1986) is in four languages: English, German, French and Spanish.

Journals

Physics of Quantum Electronics (Addison-Wesley)
Journal of Optical and Quantum Electronics (Chapman and Hall)
Journal of Quantum Electronics (IEEE)
Quantum Optics (IOP)
Journal of European Optical Society, Part B: Quantum Optics (AIP)
Radiophysics and Quantum Electronics (Plenum)

Optical communications

The field of optical communications is a very rapidly expanding one. The optical waveband offers a large capacity, high data-rate, communication channel which can be used to meet the ever increasing demand for telecommunications space. Propagation can be either by direct transmission or by guiding the wave through a fibre optic. There are many books being published on communications in fibre optics technology and the associated subject of integrated optics. See also references under 'Optoelectronics and quantum electronics' in this chapter and also under 'Electromagnetic waves' in Chapter 6.

Books and proceedings

INTEGRATED OPTICS AND WAVE GUIDES

Integrated optics is concerned with the application of dielectric thin film and integrated electronics technology to optical circuits and devices, including wave guides for communications purposes.

Ostrowsky, D. B. (1978) *Fibre and integrated optics* is published by Plenum in the NATO Advanced Study Institute series, as well as *Integrated optics physics and applications*, by S. Martellucci and A. N. Chester (1982).

Other books are *Integrated optics* by D. Marcuse (IEEE, 1973), which includes thin-film waveguides and *Light transmission optics*, by the same author (Van Nostrand, 1972), covers optical fibres and optical wave-guides; Tamir, T. (ed.) (1979) *Integrated optics*, appearing in Topics in Applied Physics (Springer), covers the theory of optical waveguides and deals with techniques and applications. *Integrated optics: theory and application*, by R. G. Hunsparger (1982), is also published by Springer in the Series on Optical Sciences, vol. 33. *Integrated optics*, published by Peregrinus (1981) reports the 35 papers presented at the First European Conference on Integrated Optics and is referenced as IEE Conference Publication no. 201. The Second Conference is reported in IEE Conference Publication no. 227 (Peregrinus, 1983).

Adams, M. J. (1981) *An introduction to optical waveguides* (Wiley) is an introduction to the principles of waveguides, covering different types with applications.

Snyder, A. W. and Love, J. D. (1983) *Optical waveguide theory* (Methuen) is a comprehensive and unified treatment of optical waveguides.

Clarricoats, P. J. B. (ed.) (1975) *Optical fibre waveguides*, IEE Reprint Series No. 1 (Peregrinus), is a compendium of papers relevant to the theory of optical waveguides and design of optical fibre systems.

New directions in guided waves and coherent optics in the NATO Advanced Study Group Series (Plenum, 1983), covering fibre optics, integrated optics and coherent optics; and treating non-linear effects, soliton propagation and polarization conservation. Solitons are discussed in *Solitons* by P. G. Drazin (CUP, 1983). *Optical waveguide sciences*, by Huang Hung Chia and A. W. Snyder (eds.) (Nijhoff, 1983) discusses the integration of the educational, industrial and research aspects of this technology.

Additional publications in this field since 1984 are:

Najafi, S. I. (ed.) (1992) *Introduction to glass integrated optics* (Artech House).

Hunsperger, R. G. (1991) *Integrated optics: theory and technology*, (Springer, 3rd edn.).

Snyder, A. W. and Love, J. D. (1983) *Optical waveguide theory* (Chapman and Hall) is an in-depth self-contained treatment.

Lee, D. L. (1986) *Electromagnetic principles of integrated optics* (Wiley) covers basic theory, interaction at dielectric surface, waveguide geometrics and fabrication, mode coupling, Bragg scattering and optical fibres.

Solimeno, S., Crosignani, B. and Di Porto, P. (1986) *Guiding diffraction and confinement of optical radiation* (Academic) covers basic theory and ray optics, aperture diffraction and scattering from different geometrics of metallic and dielectric obstacles, optical resonators and Fabry-Perot interferometers, and propagation in optical fibres.

Marcuse, D. (1991) *Theory of dielectric optical waveguides* (Academic, 2nd edn.) covers dielectric slab waveguides, weakly guiding optical fibres, coupled mode theory and applications, coupled power theory and direction-coupling effects and methods.

Yeh, P. (1988) *Optical waves in layered media* (Wiley) is a mathematical analysis of this subject.

Lekner, J. (1987) *Theory of reflection of electromagnetic and particle waves* (Martinus Nijhoff) is another mathematical treatment.

Tsao, C. (1992) *Optical fibre waveguide analysis* (OUP) is a comprehensive but practical treatment intended for practising scientists and engineers.

Koshiha, M. (1990) *Optical wave guide analysis* (McGraw-Hill) is an introductory text for science and engineering students and researchers.

Tamir, T. (1988) *Guided wave optoelectronics* (Springer) covers integrated optics, optical wave guides, optoelectronic devices and semi-conductor lasers.

Tamir, T. (ed.) (1990) *Guided wave optoelectronics* (Springer, 2nd edn.) surveys recent advances including waveguides light-wave junctions and other thin-film components, lithium-niobate devices, guided mode semi-conductor lasers and integrated optic devices.

Das, P. K. (1991) *Optical signal processing. Fundamentals* (Springer) presents the background material necessary for an understanding of modern optical methods of signal processing.

Tada, K. and Hinton, H. S. (eds.) (1990) *Photonic switching* (Springer) covers the current status of research, achievements and trends in this field.

Gustafson, T. K. and Smith, D. W. (eds.) (1988) *Photonic switching I* (Springer) is the proceedings of the First Topical Meeting held in Nevada, 1987.

Tada, K. and Hinton, H. S. (eds.) (1991) *Photonic switching II* (Springer) is the proceedings of an International Meeting held in Kobe, Japan, in 1990.

Haug, H. and Banyai, L. (1989) *Optical switching in low-dimensional systems* (Plenum).

Vassallo, C. (1991) *Optical waveguide concepts* (Elsevier).

Begley, D. L. (ed.) (1991) *Free-space laser communications* (SPIE, vol. MS 30) gives selected papers on system performance and analysis, pointing acquisition and tracking, laser transmitters, detectors and receivers, communication channel characterization, and optical systems.

Begley, D. L. (ed.) (1990) *Free-space laser communication technologies III* (SPIE, vol. 1218) is the proceedings of a conference held in Los Angeles in 1990 and complements Begley's book cited above.

Robson, P. N. and Kendal P. N. (1991) *Rib waveguide theory by the spectral index method* (Wiley) is concerned with production of accurate propagation constants and mode profiles chiefly using Fourier transform methods.

Bendjaballah, C., Hirota, O and Reynard, S. (1991) *Quantum aspects of optical communications* (Springer) surveys the work of leading research groups in this area.

Franz, J. (ed.) (1991) *Optical space communication II* (SPIE, vol. 1522) is a proceedings of a conference held at Munich in 1991 covering systems and subsystems, pointing acquisition and tracking, lasers and transmitters, and coherent systems.

FIBRE OPTICS

Wolf, H. E. (ed.) (1979) *Handbook of fibre optics* (Granada) is an early review of the theory and applications.

Suematsu, Y. (1982) *Introduction to optical fibre communications* (Wiley) is an introductory book, as is *Optical fibre systems and their components* by A. P. R. Sharma and S. J. Holmes (Springer, 1981).

Howes, M. J. and Morgan, D. V. (1980) *Optical fibre communications* (Wiley) is another basic book, dealing with sources, components, fibres and systems.

Arnaud, J. A. (1976) *Beam and fibre optics* is a volume in the Quantum Electronics Series (Academic) and *Fundamentals of optical fibre communications* by M. K. Barnoski was also published by Academic in 1976.

Another book with a basic approach is *Optical fibres* by T. Okoshi (Academic, 1982), which describes the theoretical basis of transmission characteristics of optical fibres, and a similar book, by M. P. Lisita, *Fibre optics* was published by Wiley in 1973.

Lacey, E. A. (1982) *Fibre optics* is published by Prentice-Hall.

Midwinter, J. E. (1979) *Optical fibres for transmission* (Wiley) is a fundamental introduction to the subject introducing electromagnetic theory and dielectric wave guides, dealing also with coupling and dispersion effects in materials.

Marcuse, D. (1981) *Principles of optical fibre measurements* (Academic) deals in particular with the physical principles of refractive index profiling and transverse backscattering.

Kressel, H. (1990) *Semiconductor devices for optical communications*, Topics in Applied Physics no. 39 (Springer, 2nd edn.) and Suematsu, Y. (ed.) (1983) *Optical devices* (North-Holland) are concerned with system aspects, the latter discussing components, fibres, devices and systems with possible applications to the Japanese industrial and commercial scene.

Cherin, A. H. (1983) *Optical fibres for engineering and physics* (McGraw-Hill) is designed for courses in fibre optics communications and associated integrated optics for electrical engineers and covers both theoretical and practical aspects.

Another book covering similar ground for electrical engineers is *Optical fibre communications* by G. E. Keiser (McGraw-Hill, 1983).

Miller, S. E. and Chynoweth, A. G. (1979) *Optical fibre telecommunications* (Academic) gives a comprehensive in-depth coverage of waveguiding in fibres and provides information on dispersion effects and coupling. Another sourcebook is *Fundamentals of optics fibre communications* (Academic, 2nd edn., 1981) by M. K. Barnoski, which examines in tutorial form sources and detectors for fibre systems and multi-terminal networks.

Other books in this field are:

Sandbank, C. P. (ed.) (1980) *Optical fibre communications systems* (Wiley).

Personick, S. D. (1981) *Optical fibre transmission systems* (Plenum).

Gloge, D. (ed.) (1976) *Optical fiber technology* (IEEE).

Hodara, H. (1977) *Optical fibres* (AGARD).

Harger, R. O. (1977) *Optical communications theory* (Dowden).

The following which have been published since 1984:

Halley, P. (1987) *Fibre optics systems* (Wiley) is a review of the theory and applications at a practical level.

Briley, B. E. (1988) *An introduction to fibre optics system design*

(North-Holland) is aimed at the student and professional electrical or electronic engineer. It addresses the principles and technology of this subject with exercises to promote understanding.

Palais, J. C. (1988) *Fibre optic communications* (Prentice-Hall) is an introductory book based upon short courses. Minimum mathematics is involved.

Kao, C. K. (1988) *Optical fibre* (Peregrinus) is based upon a series of lectures at graduate level, dealing with the relevant physics and chemistry of fibre optic systems and confined to the theoretical basis of the fibre itself.

Katsuyama, T. and Matsumura, H. (1989) *Infra-red optical fibres* (Hilger) is a state-of-the-art review on infra-red optical fibres.

Ungar, S. (1989) *Fibre optics: theory and applications* (Wiley) is primarily concerned with industrial applications, viz., fabrication, interfacing and uses; with the aim of acquainting engineers with the technology and associated optics and electronics. A reasonable mathematical background is required.

Adams, M. J. and Henning, L. D. (1990) *Optical fibres and sources for communications* (Plenum) aims to provide a relatively speedy introduction to the basic principles of optical fibres and semiconductor sources, sensors and detectors for optical fibre communication systems, the systems themselves and applications.

Islam, M. (1992) *Ultrafast fibre switching devices and systems* (CUP) covers fundamental issues and engineering applications, of interest to graduates and researchers working in non-linear guided wave optics, photonic switching, optical data processing and high speed electronics.

Edwards, T. (1989) *Fibre optic systems* (Wiley) looks at the products and networking applications.

Udd, E. (1991) *Fibre optic sensors* (Wiley) is a comprehensive introduction to the field of fibre optic sensor technology.

Keiser, G. E. (1991) *Optical fiber communications* (McGraw-Hill) progresses systematically from descriptions of individual elements of a communication link to analysis of digital and analogue networks design.

Zhilin, V. G. (1990) *Optical fiber velocity and pressure transducers* (Hemisphere) covers lightguide optics, fabrication and design of sensitive elements.

Hasegawa, A. (1990) *Optical solitons in fibres* (Springer, 2nd edn.), starting at an elementary level, provides a clear exposition of the theoretical background and the most recent experimental results in this new and rapidly evolving field. Another book on the same subject is *Optical solitons* by F. K. Abdullaev, S. A. Darmanyan and K. Kabibullaev (Springer, 1991).

Taylor, J. R. (ed.) (1992) *Optical solitons* (CUP) describes the theoretical and experimental aspects of optical soliton generation, soliton properties and application to high-bit-rate communications.

Midwinter, J. E., Gambling, W. A. and Stewart, W. J. (1990) *Optical technology and wideband local networks* (CUP) offers insights into the problems of present day optical technology in optical fibre point-to-point systems.

Neumann, E. G. (1988) *Single-mode fibers: fundamentals* (Springer) covers physical explanations of waveguiding by single mode fibres and components of these systems.

Arditty, H. J., Dakin, J. P. and Kersten, R. (eds.) (1989) *Optical fibre sensors* (Springer) is the proceedings of the 6th International Conference held in Paris in 1989.

Dakin, J. P. (ed.) (1990) *The distributed fibre optic sensing handbook* (Springer).

Artech House publish books on fibre optics within their Optoelectronics Library Series, particular examples being:

Taylor, H. F. (ed.) (1984) *Advances in fibre optic communications*.

Cancellieri, G. and Ravaioli, U. (1984) *Measurement of optical fibres and devices*.

Dakin J. and Culshaw, B. (eds.) *Optical fibre sensors I. Principles and components* (1988) and *Optical fibre sensors II. Systems and applications* (1989).

Geckeler, S. (1987) *Optical fibre transmission systems*.

Marcel Dekker also publish books on fibre optics within their Series on Optical Engineering, thus:

Jeunhomme, L. B. (1989) *Single mode fibre optics: principles and applications* (vol. 23, 2nd edn.).

Murata, H. (1988) *Handbook of optical fibres and cables* (vol. 15).

DATA SOURCES

Wolf, H. F. (1981) *Handbook of fiber optics* (Granada) is a practical guide to communications systems based on optical fibres. Another useful source of data is by M. H. Weik (1981) *Fibre optics and lightwave communications standard dictionary* (Lewis).

Optoelectronics/fibre-optics applications manual by Hewlett and Packard (McGraw-Hill, 2nd edn., 1981) is another comprehensive data source in fibre optics, LED displays and associated systems.

Integrated optics bibliography by K. D. Mayne and J. P. Tomlinson (Academic, 1978) contains over 500 references.

Clarricoats, P. J. B. (ed.) *Progress in optical communications*, vol. I

(1980) and vol. II (1982) (IEE and Peregrinus) are reprints of the significant contributions on optical fibres appearing in the *IEE Electronics Letters* up to and including 1981. The same author has edited *Optical communications* (IEE Conference Publication 190, Peregrinus, 1980), which contains over 100 papers on fibres, fibre cables, devices, integrated optics equipment, techniques and systems.

Yeh, C. (1980) *Handbook of fibre optics* (Academic) provides state-of-the-art information covering fibre optics and semi-conductor light sources.

Karp, S., Gagliardi, R. M., Moran, S. E. and Stotts, L. B. (1988) *Optical Channels* (Plenum) is primarily conceived as a reference book for engineers working on optical communications systems. It deals with optical propagation in four domains, i.e. free-space, through atmospheric turbulence, scattering in clouds and through fibre optics.

Journals

Journals specializing in optical communications are:

Conference Publication Series (IEE)
Transactions in Communications (IEE)
NATO Advanced Study Institute Series (Plenum)
Journal of Lightwave Technology (IEEE/OSA)
Fibre and Integrated Optics (Crane Russak)
Fibre Optic Report (Advanced Technology)
Optics Communications (North-Holland)
Journal of Optical Computing (Wiley)
Microwave and Optical Technology Letters (Wiley)
Soviet Lightwave Communications (IOP)

Lasers

Since their invention in 1960 lasers have rapidly developed and are now used in a wide range of applications in industrial, medical and defence fields. As with optical communications, it is difficult to list all the publications.

Books and proceedings

LASER PHYSICS AND TECHNOLOGY

An introductory text by Brown, R. (1969) *Lasers* (Business Books) is a well-illustrated non-mathemetical presentation for the new reader.

Lengyell, B. A. (1971) *Lasers* (Wiley, 2nd edn.) is another basic book and Maitland, A. and Dunn, M. H. (1969) *Laser physics* (North-Holland) is a third. A modern book *Principles of lasers* by Svelto, O. (1989) (Plenum, 3rd edn.) explores the newest lasers and requires some knowledge of quantum mechanics.

Shimoda, K. (1991) *Introduction to laser physics* (Springer, 2nd edn.) covers the essential fundamentals and physical meaning of laser-related phenomena.

Haken, H. (1984) *Light. Vol. 2: Laser* (North-Holland) deals with laser physics in theoretical depth and gives practical applications.

Lasers: theory and applications by K. Thyagarajan and A. K. Ghatak (Plenum, 1981) is a basic text covering laser theory and selected applications; a novel feature is the inclusion of four Nobel lectures on the subject.

Hallard, T. (1977) *Exploring laser light*, (Optosonic, 3rd edn.) deals with experimental equipments and demonstrations, including beam-shaping technique, and contains a bibliography.

Young, M. (1986) *Optics and lasers* (Springer, 3rd edn.) includes fibres and optical waveguides.

Siegman, A. E. (1986) *Lasers* (Univ. Science) is a thorough review of this subject.

Hall, O. R. and Jackson, P. E. (eds.) (1989) *Physics and technology of laser resonances* (Hilger) gives an introduction to the physics of laser resonators and laser beams and to the design of laser resonators.

Das, P. K. (1991) *Lasers and optical engineering* (Springer).

Milonni, P. W. and Eberly, J. H. (1988) *Lasers* (Wiley) is a substantial treatment covering the principles underlying the operation of the laser with some applications; it is aimed for tutorial use and occasional reference.

Chester, A. N., Letokhov, V. S. and Martekucci, S. (1988) *Laser science and technology* (Plenum).

Specialized treatises are:

Roy, R. (ed.) (1991) *Laser noise* (SPIE, vol. 1376) is the proceedings of a conference held at Boston, USA, in 1990 and covers fundamental studies and noise reduction methods.

McDuff, G. G. (ed.) (1991) *Pulse power for lasers III* (SPIE, vol. 1411) is the proceedings of a conference held in Los Angeles in 1991 covering techniques, circuits, components and systems.

Weichel, H. (ed.) (1991) *Laser design* (SPIE, vol. MS 29) are selected papers covering historical aspects, optical materials for laser cavities, laser resonators and beam formation, resonator alignment and adaptive control, solid-state and liquid lasers.

Hindy, R. N. and Kohanzadeh, Y. (eds.) (1991) *Laser beam diagnostics* (SPIE, vol. 1414) is the proceedings of a conference held at Los Angeles in 1991 and covers instrumentation techniques, wavefront sensors and beam quality measurements, high energy lasers, analysis and experimental results.

Schnurr, A. D. (ed.) (1991) *Modelling and simulation of laser systems II* (SPIE, vol. 1415) is the proceedings of a conference held at Los Angeles, California, in 1991 and covers laser devices, beam control, acquisition, tracking, pointing, wavefront sensing and correction, non-linear optical systems and laser interactions.

Begley, D. L. and Seery, B. D. (eds.) (1991) *Free space laser communication technologies III* (SPIE, vol. 1417) is the proceedings of a conference held at Los Angeles, 1991, covering laser communication program overviews; system analysis performance and applications; laser beam control in pointing and tracking and laboratory demonstration systems.

Holmes, D. A. (ed.) (1990) *Optical resonators* (SPIE, vol. 1224) is the proceedings of a conference held in Los Angeles in 1990 and covers resonator physics, coupling devices, types of resonator and modelling.

Klein, C. A. (ed.) (1990) *Mirrors and windows for high power/high energy laser systems* (SPIE, vol. 1047) is the proceedings of a conference held in Los Angeles in 1990 and covers subsystems, measurements and characterization.

Reintjes, J. R. (ed.) (1989) *Laser wavefront control* (SPIE, vol. 1000) is the proceedings of a conference held at Boston in 1988 and examines corrective techniques, viz., stimulated Raman scattering, stimulated Brillouin scattering, phase conjugation and aberration correction, multi-wave mixing and adaptive optics.

Sanders, A. A. (ed.) (1988) *Laser beam radiometry* (SPIE, vol. 888) is the proceedings of a conference held in Los Angeles in 1988 and covers diagnostic instrumentation, laser power standards and laser beam measurements.

Sona, A. (ed.) (1989) *Beam diagnostics and beam handling systems* (SPIE, vol. 1024) is the proceedings of a conference held in Hamburg in 1988 and covers beam switching, transport and positioning equipment, beam shaping and focussing, and diagnostics.

Gosnell, T. R. (1991) *Ultrafast laser technology* (SPIE, vol. MS 44) gives a selection of papers reviewing the growth and state-of-the-art.

Seery, B. D. (ed.) (1989) *Optomechanical design of laser transmitters and receivers* (SPIE, vol. 1044) is the proceedings of a conference held in Los Angeles in 1989 covering, e.g., laser diode modules, transmitter/receiver technology, advanced technology diagnostics and materials.

Fauchet, P. M. and Guenther, K. H. (eds.) (1988) *Laser optics for intracavity and extracavity applications* (SPIE, vol. 895) is the proceedings of a conference held in Los Angeles in 1988 and covers design requirements, active elements and components.

LASER TYPES

Publications on specific laser types are:

Butler, K. J. (1980) *Semi-conductor injection lasers* (Wiley).
Thompson, G. H. B. (1980) *Physics of semi-conductor laser devices* (Wiley).
Kiessel, H. and Butler, K. J. (1977) *Semi-conductor lasers and heterodyne LEDs* (Academic).
Alferov, Z. I. (1992) *Soviet American workshop on the physics of semi-conductor lasers* (AIP Proceedings 240, IOP).
Fukada, M. (1991) *Reliability and degradation of semi-conductor lasers and LEDs* (Artech).
Buus, J. (1991) *Single frequency semi-conductor lasers* (SPIE, vol. TT 5) is a tutorial text giving a detailed description of properties, practical examples, experimental results and applications.
Ack, G. A. (ed.) (1989) *Semi-conductor lasers* (SPIE, vol. 1025) is the proceedings of a conference held in Hamburg, Germany, in 1988 covering trends in materials processing, laser diodes, quantum well lasers, phased-coupled arrays and semi-conductor laser amplifiers.
Fiqueron, L. (ed.) (1988) *High power laser diodes and applications* (SPIE, vol. 893) is the proceedings of a conference held in Los Angeles in 1988 and includes phased array semi-conductor laser technology and applications.
Fiqueron, L. (ed.) (1989) *Laser diode technology and applications* (SPIE, vol. 1043) is the proceeedings of a conference held in Los Angeles in 1989 and covers 47 papers on recent advances.
Ohtsu, M. (1992) *Highly coherent semi-conductor lasers* (Artech) is a thorough review of the subject.
Renner, D. (ed.) (1991) *Laser diode technology and applications III* (SPIE, vol. 1418) is the proceedings of a conference held in Los Angeles in 1991 and covers surface-emitting lasers, quantum well lasers, processes and materials for semi-conductor lasers, devices for optical amplifiers and high power semi-conductor lasers.
Yamamoto, Y. (ed.) (1991) *Semi-conductor lasers for coherent quantum optics* (Wiley) examines applications in communications, spectroscopy, optical amplifiers, and discusses quantum effects, squeezed state and spontaneous emission inhibition.

Cheo, P. K. *Handbook of solid state lasers* (Dekker, Opt. Eng. Series, vol. 18).

Brown, D. C. (1981) *High peak power Nd:glass laser systems* (Springer).

Koechrer, W. (1992) *Solid state laser engineering* (Springer, 3rd edn.) is completely revised and of interest to students of laser physics. Topics include diode-array pumping, tunable lasers, beam quality and system efficiency.

Badger, A. B., Esterowitz, L. and De Shazer, L. B. (eds.) (1986) *Tunable solid state lasers II* (Springer) is the proceedings of the OSA Topical Meeting in Oregon in 1986.

Byer, R. L., Gustafson, E. K., Trebino, R. (eds.) (1985) *Tunable solid state lasers for remote sensing* (Springer) is the proceedings of a NASA Conference in Stanford, USA, in 1984.

Powell, R. C. (ed.) *Solid state lasers* (SPIE, vol. MS 31) covers selected papers on fundamentals, transition metal ion lasers: lanthanide ion lasers and special laser systems, e.g. tunable, soliton and fibre lasers.

Dube, G. (ed.) *Solid state lasers* (SPIE, vol. I, 1990 and vol. II, 1991) are the proceedings of a conference held in Los Angeles in 1990 and 1991, respectively.

Ireland, C. L. M. (ed.) (1990) *High power solid state lasers and applications* (SPIE, vol. 1227) is the proceedings of a conference held at The Hague in 1990.

Dube, G. (ed.) (1989) *High power and solid state lasers II* (SPIE, vol. 1040) is the proceedings of a conference held in Los Angeles in 1989.

Weber, H. (ed.) (1989) *High power solid state lasers* (SPIE, vol. 1021) is the proceedings of a conference held in Hamburg, Germany, in 1988 and covers slabs and high power systems, glasses and crystals, resonators and transmission.

Sourer, J. M. (ed.) (1991) *High power lasers* (SPIE, vol. MS 43) is a selection of papers covering, e.g. high power Nd:glass lasers for fusion applications and Nd-doped laser materials, short pulse iodine laser amplifier and mode structure.

Kimball, R. and Wisoff, P. J. (eds.) (1989) *Metal vapour deep blue and ultraviolet lasers* (SPIE, vol. 1041) is the proceedings of a conference held in Los Angeles in 1989 and covers 35 papers on these topics.

Digonnet, M. J. F. (ed.) (1991) *Rare-earth-doped fibre laser sources and amplifiers* (SPIE, vol. MS 37) contains selected papers on amplification, fabrication and optical properties of different types of fibre laser.

France, P.W. (1991) *Optical fibre lasers and amplifiers* (Blackie) reviews developments over the past few years.

Duley, W. W. (1976) *CO_2 lasers, effects and applications* (Academic).

Witterman, W. J. (1987) *The CO_2 laser* (Springer).

Bradley, L. B. (1990) *Molecules and molecular lasers for electrical engineers* (Hemisphere) introduces the concept of molecules in the gas phase and how their quantum properties relate to the concept of molecular lasers.

Evans, J. D. and Locke, E. V. (eds.) (1989) *CO_2 lasers and applications* (SPIE, vol. 1042) is the proceedings of a conference held in Los Angeles in 1989.

Opower, H. (ed.) (1990) *CO_2 lasers and applications* (SPIE, vol. 1276) is the proceedings of a conference held in The Hague in 1990).

Quenzer, A. (ed.) (1989) *High power CO_2 laser systems and applications* (SPIE, vol. 1020) is the proceedings of a conference held in Hamburg, Germany, in 1988.

Loser, S. A. (1981) *Gas dynamic laser* (Springer).

Anderson, J. P. (1976) *Gas dynamic lasers: an introduction* (Academic).

Pike, E. R. (ed.) (1976) *High power gas lasers* (IOP).

Sauerbrey, R. A., Tillotson, J. H. and Chenavsky, P. P. (eds.) (1988) *Gas laser technology* (SPIE, vol. 894) is the proceedings of a conference held in Los Angeles in 1988 and covers excimer laser physics and technology, visible and infra-red gas laser technology; and applications.

Schuocker, D. (ed.) (1989) *Gas flow and chemical lasers* (SPIE, vol. 1031) is the proceedings of the 7th Biennial International Conference held in Vienna in 1988 and covers carbon dioxide lasers, carbon monoxide lasers, hydrogen fluoride and deuterium fluoride lasers, oxygen-iodine lasers, short wavelength chemical lasers, excimer lasers, discharge techniques, optics, propagation and interaction.

Domingo, C. and Orza, J. M. (eds.) (1991) *8th international conference on gas flow and chemical lasers* (SPIE, vol. 1397) covers current research developments in excimer, oxygen-iodine, vibrational chemical, carbon dioxide, carbon monoxide, nitrogen dioxide, short wavelength, and gas dynamic lasers, discharge and flow effects and laser-matter interactions.

Basov, N. G., Bashkin, A. S., Igashin, V. I., Oraevsaky, A. N. and Sheheglov, V. A. (1990) *Chemical lasers* (Springer).

Cheo, P. K. (ed.) *Handbook of molecular lasers* (Dekker, Opt. Eng. Series, vol. 14).

Bredaslow, G. (ed.) (1983) *The high power iodine laser* (Springer).

Jacob, S. J. (ed.) (1980) *Free electron lasers* (Addison-Wesley).

Martellucci, S. (1988) *Free electron lasers* (Plenum).

Petroff, Y. (ed.) (1989) *Free-electron lasers II* (SPIE, vol. 1133) is the proceedings of a conference held in Paris in 1989 covering new developments.

Prosnitz, D. (1990) *Free electron lasers and applications* (SPIE, vol. 1227) is the proceedings of a conference held in Los Angeles in 1990.

Marshall, T. C. (1985) *Free electron lasers* (McMillan) is concerned with the millimetre and sub-millimetre regions of the spectrum.

Luchini, P. and Motz, H. (1990) *Undulators and free electron lasers* (OUP) is an advanced theoretical analysis primarily covering the infra-red to ultraviolet and X-ray regions of the spectrum.

Elton, R. C. (1990) *X-ray lasers* (Academic) examines the fundamentals and principles of X-ray laser operation, principally using high temperature plasma as the working medium: a chapter is devoted to alternative approaches, viz. harmonic generation and frequency mixing. Free electron lasers and gamma-ray lasers are also discussed.

Tallento, G. (ed.) (1991) *X-ray lasers* (IOP Conference Series 116) is the proceedings of the 2nd International Conference at York in 1990.

Fill, E. E. (1992) *X-ray lasers 1992* (IOP) is the proceedings of the 3rd International Colloquium at Schlierise, Germany, in May 1991.

Jones, R. (ed.) (1988) *Short and ultra-short wavelength lasers* (SPIE, vol. 875) is the proceedings of a conference held in Los Angeles in 1988 and covers X-ray lasers, gamma-ray lasers, and short wavelength chemical lasers.

Richardson, M. C. (ed.) (1988) *X-rays from laser plasmas* (SPIE, vol. 831) is the proceedings of a conference held in San Diego in 1987 and covers spectroscopy, X-ray conversion, optics, instrumentation and applications.

Bischel, W. K. and Rahn, L. A. (eds.) (1988) *Pulsed single frequency lasers: technology and applications* (SPIE, vol. 912) is the proceedings of a conference held in Los Angeles in 1988 and covers polarization effects in laser pumped Nd:YAG lasers, tunable solid-state lasers, tunable dye lasers, UV and VUV radiation and Raman spectroscopy.

Schafer, F. P. (ed.) (1990) *Dye lasers* (Springer, 3rd edn.).

Stake, M. (1992) *Dye lasers: 25 years* (Springer).

Duarte, F. J. *High power dye lasers* (Springer Series in Optical Sciences, vol. 65).

Rhodes, C. K. (ed.) (1984) *Excimer lasers* (Springer, 2nd edn.).

Letardi, T. and Laude, L. D. (eds.) *Excimer lasers and applications II and III* (SPIE, vol. 1278, 1990 and vol. 1503, 1991) are the proceedings of conferences held at The Hague in 1990 and 1991 respectively, covering devices, techniques, systems and applications. The first conference was held in 1988 at Hamburg and covered in SPIE, vol. 1023 under the editor D. Basting.

Gaillaard, M. L. and Quenzer, A. (eds.) (1989) *High power lasers and laser-machining technology* (SPIE, vol. 1132) is the proceedings of a conference held in Paris in 1989 covering high power lasers, machining with excimer lasers, welding and cutting technologies, laser surface treatment and beam delivery.

Douglas, N. G. (1989) *Millimetre and sub-millimetre wavelength lasers. A handbook of CW measurements* (Springer).

Kaiser, W. (ed.) (1988), *Ultra-short laser pulses and applications* (Springer) is a series of papers on the generation, the nonlinearities, the interactions of ultra-short pulses, the instrumentation and spectroscopy.

Mollenauer, L. F., White, J. C. and Pollock, C. R. (1992) *Tunable lasers* (Springer, 2nd edn.) covers general principles, four-wave frequency mixing, stimulated Raman scattering, oscillators, excimer lasers, colour centre lasers, fibre Raman lasers, tunable high-pressure infra-red lasers and tunable paramagnetic-ion solid-state lasers.

Carruth, J. A. S. and McKenzie, A. L. (1986) *Medical lasers* (IOP) describes the physics and technology of surgical lasers, treatment and safety aspects.

Klim, J. J. and Tittel, F. K. (eds.) (1991) *Gas and metal vapor lasers and applications* (SPIE, vol. 1412) is the proceedings of a conference held at Los Angeles, 1991, covering types of metal vapour lasers, excimer lasers and applications.

Campbell, E. M. (ed.) (1990) *Femtosecond to nanosecond high-intensity lasers and applications* (SPIE, vol. 1229) is the proceedings of a conference held in Los Angeles in 1990 covering 23 papers on design, performance and applications.

THEORY

Barnes, F. S. (1972) *Laser theory* (IEEE) covers laser resonators, oscillators, amplifiers, mode locking and modulation treatments and includes a series of historic papers.

Maitland, A. and Dunn, M. H. (1969) *Laser physics* (North-Holland).

Sargent, M., Scully, M. O. and Lamb, E. (1974) *Laser physics* (McGraw-Hill).

Thyagarajan, K. *et al.* (1981) *Laser theory and applications* (Plenum), being a volume in Optical Physics and Engineering.

Bernheim, R. (1965) *Optical pumping: an introduction* (Addison-Wesley).

Willett, C. S. (1974) *Introduction to gas lasers: population inversion mechanisms* (Pergamon).

Weiss, C. O. and Vilaseca, R. (1991) *Dynamics of lasers* (VCH) is intended to familiarize those interested in non-linear dynamics in general, in laser physics or non-linear optics, with the dynamical properties of lasers. The emphasis is on intuitive concepts and the mathematics is not rigorous.

Eastham, D. (1986) *Atomic physics of lasers* (Taylor and Francis) is a monograph intended to be an introduction to lasers suitable for second or third year undergraduates.

Bandrauk, A. D. and Wallace, S. C. (eds.) (1992) *Coherence phenomena in atoms and molecules in laser fields* (Plenum) is the proceedings of a NATO Advanced Research Workshop (NATO, ARW 900857) held at McMaster University, Hamilton, Canada, in 1991.

Anan'ev, Y. A. (1992) *Laser resonators and the beam divergence problem* (Hilger) is a specialized analysis of the problems of high power/high quality/low beam divergence output beams. This is a very detailed and substantive treatment of high power laser resonators, their beam output profiles, and methods of wavefront correction.

LASER INTERACTIONS

Laser interaction with matter forms the basis of many techniques and technologies.

Ready, J. F. (1971) *Effects of high power laser radiation* (Academic) is a fundamental book, covering properties of lasers, measurement techniques, effects and damage to materials and applications.

Bertolotti, M. (ed.) (1982) *Physical processes in laser materials interactions*, is volume 84 in the NATO Advanced Study Institute series (Plenum), and covers characteristics of laser beams for machining, coupling of the laser beam to the material and the thermodynamics of structural changes. Sections on materials hardening and on annealing of semiconductor devices are included.

Rykalin, N., Gulov, A. and Kokara, A. (1978) *Laser machining and welding* is published by Pergamon.

Poate, J. M. and Mayer, J. W. (1982) *Laser annealing of semiconductors* (Academic) covers this subject in more detail, including processes of epitaxy, phase changes, solute trapping and diffusion processes.

Two further books are by Duley, W. W. (1982) *Laser processing and analysis* (Plenum) and Buss, M. (ed.) (1983) *Laser materials processing,* who also wrote *Laser processes of materials,* (1984) both published by North-Holland.

Appleton, B. R. and Celler, G. K. (eds.) (1982) *Laser and electron beam interactions with solids* (North-Holland) is the proceedings of the Materials Research Society Meeting, Boston, 1981.

Another report from this Society is by J. F. Gibbons, L. D. Hess and T. W. Sigmon (eds.) (1981) *Laser and electron beam solid interactions and materials processing* (North-Holland).

Laser Applications, Proceedings of the IEE, June 1982, reviews materials-processing aspects.

Mittleson, M. H. (1982) *Introduction to the theory of laser-atom interactions* (Plenum) is a fundamental treatment of atomic phenomena in the laser field.

Steinfeld, J. L. (1981) *Laser induced chemical processes* (Plenum) is a comprehensive picture of the chemical effects of high-power infra-red laser radiation, including a survey of the literature.

Laser Applications is a comprehensive review series published by Academic: vol. 1, 1971, vol. 2, 1974, vol. 3, 1977, vol. 4, 1980 and vol. 5, 1983, the latest volume dealing with laser photochemistry and processing of integrated circuits and microelectronic materials.

Lasers in chemistry by M. A. West (ed.) is the proceedings of a Conference at the Royal Institution, London, 1977 (North-Holland, 1977).

More recent publications are:

Boyd, I. W. (ed.) (1992) *Laser surface processing and characterisation* (North-Holland) covers the proceedings of a conference by this name held at Strasbourg, France, in May 1991.

Prohkorov, A. M., Konov, V. I., Ursu, I. and Mihailescu (1990) *Laser heating of metals* (Hilger) presents the basic physical and chemical processes in this interaction.

Niku-Lari, A. (1989) *High power lasers* (Pergamon) covers the application of lasers to materials processing.

Von Allmen, M. (1988) *Laser beam interactions with materials* (Springer) covers absorption and heating by laser energy, melting and solidification, evaporation and plasma formation, laser supported combustion waves and laser supported detonation waves.

Basov, N. G. *et al.* (1986) *Heating and compression of thermonuclear targets by laser beam* (CUP) is an advanced treatise concerned with controlling a thermonuclear fusion reaction by using a high power laser.

Belforte, D. and Levitt, M. (eds.) (1992) *The Industrial Laser Handbook 1992–1993 edition* (Springer) is a review of laser material processing techniques including beam delivery and measuring equipment and technology trends.

Wood, R. M. (1986) *Laser damage in optical materials* (Hilger) is a comprehensive review of laser induced damage, reflecting the author's personal experience. Some theory is invoked but the treatment is mainly phenomological and morphological, with pictorial record.

Mulser, P. (1991) *High power laser-matter interaction* (Springer).

Houldcroft, P. T. (1991) *Lasers in materials processing* (Elsevier).

Ream, S. L. and Fujioka, T. (eds.) (1991) *ICALEO '90 laser materials processing* (SPIE, vol. 1601) is the proceedings of a conference held in Boston, USA, in 1990 covering CO_2 laser beam diagnostics and cutting, Nd:YAG materials processing, etc.

Konov, V. I., Luk'yanchuk, B. S. and Boyd, I. W. (eds.) (1990) *First international school on laser surface microprocessing* (SPIE, vol. 1352) is the proceedings of a conference held at Tashkent, 1989, and covers chemical etching, vapour deposition, phase processes, photophysical phenomena and diagnostics.

Laude, L. D. (ed.) (1990) *Laser-assisted processing* (SPIE, vol. 1279) is the proceedings of a conference held in The Hague in 1990 and covers fundamental processes, ceramics and polymers, ablation and plasmas, vapour deposition and microprocessing.

Gaillard, M. L. and Quenzer, A. (eds.) (1989) *High power laser-machining technology* (SPIE, vol. 1132) is the proceedings of a conference held in Paris in 1989 and covers, for example, machining with excimer laser, welding and cutting technologies, laser surface treatment, diagnostics and beam delivery.

Piwczyk, B. P. (1991) *Excimer laser materials processing and beam delivery systems* (SPIE, vol. 1377) is the proceedings of a conference held at Boston in 1990 and covers devices and instrumentation.

Bennett, H. E. *et al.* (eds.) *Laser-induced damage in optical materials* (SPIE, vol. 1438, 1990 and SPIE, vol. 1441, 1991) are the proceedings of conferences held at Boulder, Colorado, in 1989 and 1990 respectively.

Campbell, E. M. and Balding, H. (1988) *High intensity laser-matter interactions* (SPIE, vol. 913) is the proceedings of a conference held in Los Angeles in 1988 and covers the physics and applications of intense laser-matter interactions, laser-plasma X-ray sources, ultra-fast phenomena and diagnostics.

Laude, L. D. and Rauscher, G. (eds.) (1989) *Laser assisted processing* (SPIE, vol. 1022) is the proceedings of a conference held in Hamburg, Germany, in 1988, and covers metallurgical applications and chemical reactions, high precision laser machining, ablation and polymers, synthesis and oxidation.

There are numerous publications on laser-plasma interactions and only a selection can be given here:

Hughes, T. P. (1975) *Plasmas and laser light* (Hilger) is a substantial textbook.

Laser interaction and related plasma phenomena, 5 vols to 1981 (Plenum) is the proceedings of a workshop on this subject.

Ballan, R. and Adam, J. C. (eds.) (1982) *Laser-plasma interaction* (North-Holland) reports the proceedings of the Les Houches Summer School and is devoted to the study of the non-linear phenomena and their role in laser-matter interaction.

Hora, H. (ed.) (1983) *Laser and Particle Beams* is a CUP Journal, which covers the generation of high-intensity laser beams and their interaction with materials, including plasmas, with application to nuclear fusion.

Hora, H. and Miley, G. H. *Laser Interaction and Related Plasma Phenomena* (Plenum) vol. 8, 1988, and vol. 9, 1991.

Radziemski, L. J. and Cremers, D. A. *Laser induced plasmas: physical, chemical and biological applications* (Dekker, Opt. Eng. Series, vol. 21).

The Scottish Universities Summer School run short courses on laser-plasma interactions which are published through the Institute of Physics, vol. 1, 1980, vol. 2, 1982, vol. 3, 1986 and vol. 4, 1989.

REMOTE SENSING

The use of a laser for remote sensing is reviewed in the section on 'Atmospheric optics' but the following books are reported here:

Durst, F., Melling, A. and Whitelaw, J. H. (1976) *Principles and practice of laser-doppler anemometry* (Academic). This book serves as a basis for research workers in fluid mechanics, engineers in industry and as a student text.

Drain, L. E. (1980) *The laser doppler technique* (Wiley) discusses the principles of velocity measurements using this technique and the instrumentation which is used.

Thompson, H. D. and Stevenson, W. H. (1979) *Laser velocimetry and particle sizing* (McGraw-Hill) covers the development of laser velocimetry.

Bachman, C. G. (1979) *Laser radar systems and techniques* (Artech) is a comprehensive review of LADAR, including the range equation, propagation, target echo and detection, information processing and system applications.

LASER APPLICATIONS

A selection of publications is given below, taken from this vast and rapidly developing field:

Ohshiro, T. and Calderhead, R. G. (1988) *Low level laser therapy: a practical introduction* (Wiley) reviews the present state-of-art.

Ohshiro, T. (ed.) (1991) *Low reactive level laser therapy: practical application* describes clinical application in a wide range of complaints and conditions (Wiley).

Muller, G. J. and Bierlien, H. P. (eds.) (1990) *Second German symposium on laser angioplasty* (SPIE, vol. 1462), held in Berlin in 1990, includes, e.g. interventional angiocardiology, surgical angioplasty, and excimer laser angioplasty.

Jacques, S. L. (1990) *Laser tissue interaction* (SPIE, vol. 1202) is the proceedings of a conference held in Los Angeles in 1990 covering photothermal and photo-acoustic mechanisms, photochemical effects and tissue interaction with various laser types.

Konov, V. L. (1990) *First international conference on lasers and medicine* (SPIE, vol. 1353) held in Tashkent in 1990, covers interaction with tissue photodynamic therapy, medical laser systems and diagnostics.

Gomer, C. J. (1990) *Future directions and applications in photodynamic therapy* (SPIE, vol. IS 6) covers clinical status, chemical mechanisms and system spectrometry.

Bradley-Moore, C. (1977) *Biological applications of lasers*, vol. III (Academic) includes laser isotope separation and multiphoton infra-red laser photochemistry.

Wolbarsht, M. *Laser applications in medicine and biology* (Plenum vol. 4, 1989 and vol. 5, 1991).

Chester, A. N., Martellucci, S. and Scheggi, A. M. (1988) *Laser systems for photobiology and photomedicine* (Plenum).

Akhmanov, S. A. and Poroshina, M. Yu. (eds.) (1991) *Laser applications in life sciences* (SPIE, vol. 1403) is the proceedings of a conference held in Moscow in 1990 and covers applications to biomedical optics, in particular Raman scattering spectroscopy of biomedical objects.

O'Brien, S. J. *et al.* (1991) *Lasers in orthopaedic, dental and veterinary medicine* (SPIE, vol. 1424) is the proceedings of a conference held in Los Angeles in 1991.

Abela, G. S. (ed.) (1991) *Diagnostic and therapeutic cardiovascular interventions* (SPIE, vol. 1425) is the proceedings of a conference held in Los Angeles in 1991 covering fluorescence spectroscopy and excimer laser clinical experience, etc.

Dougherty, T. J. (ed.) (1991) *Optical methods for tumor treatment and early diagnosis: mechanisms and techniques* (SPIE, vol. 1426) is the proceedings of a conference held in Los Angeles in 1991 covering early detection (clinical and pre-clinical).

Tan, O. T., White, R. A. and White, J. V. (eds.) (1991) *Lasers in dermatology and tissue welding* (SPIE, vol. 1422) is the proceedings of a conference held in Los Angeles in 1991 and covers use of the carbon dioxide laser and medical laser systems.

Pietrafitta, J. J., Steiner, R. W. and Watson, G. M. (1991) *Lasers used in urology, laparoscopy and general surgery* (SPIE, vol. 1421) is the proceedings of a conference held in Los Angeles in 1991. As well as specializing in these areas, additional subjects are stone fragmentation and laser endoscopy using fibre optics.

Jacques, S. L. (ed.) (1991) *Laser-tissue interaction II* (SPIE, vol. 1427) is the proceedings of a conference held in Los Angeles in 1991 covering, e.g. photothermal and photochemical mechanisms, pulsed infra-red laser ablation, and photo-acousticss.

Puliafito, C. A. (ed.) (1991) *Ophthalmic technologies* (SPIE, vol. 1423) is the proceedings of a conference held in Los Angeles in 1991 and covers ophthalmic measurement techniques, new surgical procedures and technologies and laser photo-ablation techniques.

Kim, Y. (1991) *Medical imaging V: image capture formatting and display* (SPIE, vol. 1444) is the proceedings of a conference held in Los Angeles in 1991. An International Symposium on Biomedical Optics Europe 93 was announced by the Biomedical Optics Society in conjunction with the European Optical Society and SPIE, to take place in Budapest in September 1993.

Boggan, J. E., Cerullo, L. J. and Smith, L. C. (1991) *Three dimensional bio-imaging systems and lasers in neurosciences* (SPIE, vol. 1428) is the proceedings of a conference held in Los Angeles in 1991 covering lasers in neurosciences and techniques for bio-imaging.

Chance, B. (ed.) (1991) *Time resolved spectroscopy and imaging of tissue* (SPIE, vol. 1431) is the proceedings of a conference held in Los Angeles in 1991 covering time resolved studies of photon migration, theory and simulations and frequency-domain studies.

Antos, R. L. and Krisiloff, A. J. (eds.) (1991) *CAN-AM Eastern '90* (SPIE, vol. 1398) is the proceedings of a conference held at Rochester, N.Y., in 1990 which contains a section on applications of optics to biology and medicine.

Antonosov, P. (ed.) (1991) *Lasers: physics and applications* (World Scientific) is the proceedings of the 6th International School on Quantum Electronics held at Varna, Bulgaria, in 1990.

Svanberg, S. and Mecha, H. (eds.) (1988) *Laser technology in chemistry* (Springer, 1988).

Andrews, D. L. (1990) *Lasers in chemistry* (Springer, 2nd edn.).

Evans, D. K. (ed.) *Laser applications in physical chemistry* (Marcel Dekker, Opt. Eng. Series, vol. 20).

Luxton, J. T. and Parker, D. E. (1992) *Industrial lasers and their applications* (Prentice-Hall, 2nd edn.).

Wiedmann, J. (1984) *Laser doppler anemometry* (Springer).

Turner, J. T. (ed.) (1990) *Laser anemometry* (Springer) is the proceedings of the 3rd International Conference held at Swansea, UK, in 1989.

Iinuma, K. *et al.* (1987) *Laser diagnostics and modelling of combustion* (Springer).

Yen, W. M. and Levenson, M. D. (eds.) (1987) *Lasers, spectroscopy and new ideas: a tribute to Arthur L. Schawlow* (Springer).

Jain, K. (1990) *Excimer laser lithography* (SPIE, vol. 1122) is the first book in this field intended for researchers and others who wish to obtain state-of-the-art familiarity.

Lenk, J. D. (1992) *Lenk's laser handbook* (McGraw-Hill) provides practical advice on trouble-shooting for laser-based consumer electronic equipment.

Tam, S. C., Silva, D. E. and Kuok, M. H. (1991) *Optical systems in adverse environments* (SPIE, vol. 1399) is the proceedings of a conference held in Singapore in 1990, covering environmental effects, lasers and laser techniques, field instrumentation and metrology.

Long, M. B. (ed.) (1989) *ICALEO '89 Optical methods in flow and particle diagnostics* (SPIE, vol. 1404) is the proceedings of a conference held at Orlando, Florida, in 1989 and covers laser doppler velocimetry, particle imaging velocimetry, field measurements and volumetric imaging.

Dibble, R. W., Fourquette, D. and Ketterle, W. (1990) *ICALEO '90 Optical methods in flow and particle diagnostics* covers recent developments in laser diagnostics.

Billon, J. P. and Fabre, E. (eds.) (1991) *Industrial and scientific uses of high-power lasers* (SPIE, vol. 1502) is the proceedings of a conference held at The Hague in 1991 and presents state-of-the-art research covering YAG and CO_2 lasers.

Pol, V. (ed.) (1991) *Optical/laser microlithography III* (SPIE, vol. 1463) is the proceedings of a conference held at San Jose, California, in 1990 and by the same editor *Optical/laser microlithography IV* at San Jose in 1991.

Forrest, G. T. and Levitt, M. R. (eds.) (1991) *The laser market place in 1991* (SPIE, vol. 1520) is the proceedings of a conference held in Los Angeles in 1991 covering new trends in laser technology and applications to scientific, industrial, medical and defence markets.

Belforte, D. A. and Levitt, M. R. (eds.) (1991) *The laser market place for industrial lasers* (SPIE, vol. 1517) is the proceedings of a conference held in Chicago in 1990.

Becherer, R. J. (ed.) (1991) *Laser radar VI* (SPIE, vol. 1416) is the proceedings of a conference held in Los Angeles in 1991 covering laser sources, advanced components and subsystem technology. The same author edited *Laser radar III* (SPIE, vol. 999, 1989).

Baldis, H. A. (ed.) (1991) *Short-pulse high-intensity lasers and applications* (SPIE, vol. 1413) is the proceedings of a conference held in Los Angeles in 1991 and consists of five papers on design and performance and twelve papers on applications.

Antonetti, A. (ed.) (1990) *Applications of ultra-short laser pulses in science and technology* (SPIE, vol. 1268) is the proceedings of a conference held at The Hague in 1990 and covers the generation of high-intensity femtosecond pulsed lasers, etc.

Werner, C. and Bilbro, J. W. (eds.) (1989) *Coherent laser radar: technology and applications* (SPIE, vol. 1181) is the proceedings of a conference held at Munich in 1989, covering general applications of coherent systems, system technology, ranging and imaging.

Soares, O. D. D., Almeida, S. P. and Bernardo, L. M. (eds.) (1989) *Laser technologies in industry* (SPIE, vol. 952) is the proceedings of a conference held in Porto, Portugal, in 1988, and covers techniques and applications.

Pirri, A. N. and Piwczyk, B. P. (eds.) (1988) *Excimer beam applications* (SPIE, vol. 998) is the proceedings of a conference held in Boston in 1988 and covers excimer laser beam systems, physical mechanisms of excimer beam interactions and excimer laser materials processing.

Another-SPIE sponsored event was *Optical tools for advanced manufacturing* which is the proceedings of a conference held in Boston in November, 1992, and covered optical robotics, laser machining and gauging, optical inspection and quality control, vision and videometrics, etc.

Frankowski, G., Abramson, N. and Fuzessy, Z. (eds.) (1991) *Applications of metrological laser methods in machines and systems* (Akademic-Verlag) is a detailed survey at a practical level.

Steen, W. M. (ed.) (1989) *Lasers in manufacturing* (IFS and Springer) is the proceedings of the 6th International Conference held at Birmingham, UK, in May 1989.

HOLOGRAPHY AND COHERENCE

The advent of the laser as a coherent source heralded the practical realization of optical holography, and this technique has found use on a wide range of applications.

Strokes, G. W. (1969) *An introduction to coherent optics and holography* (Academic, 2nd edn.) and Troup, G. J. (1967) *Optical coherence theory* (Methuen) are introductory books.

Harvey, A. F. (1970) *Coherent light* (Interscience) is a comprehensive survey for electrical engineers, and Perina, J. (1972) *Coherence of light* (Van Nostrand) is a theoretical treatment reissued as a 2nd edn. (Reidel, 1985).

The following publications are selected from the vast number on holography:

Butters, J. N. (1971) *Holography and its technology* (Peregrinus).

Collier, R. J., Burckhardt, C. B. and Lin, L. H. (1971) *Optical holography* (Academic).

Lehmann, M. (1970) *Holography – technique and practice* (Focal).

Caulfield, H. J. (1979) *Handbook of optical holography* (Academic).

Wenyen, M. (1978) *Understanding holography* (David and Charles).

Collier, R. J. (1977) *Optical holography* (Academic).

Smith, H. M. (1975) *Principles of holography* (Wiley).

Saxby, G. (1980) *Holograms* (Focal).

Abramson, N. (1982) *The making and evaluation of holograms* (Academic).

Soroko, L. M. (1980) *Holography and coherent optics* (Plenum).

Schumann, W. (1979) *Holographic interferometry* (Springer).

Vest, C. M. (1979) *Holographic interferometry* (Wiley).

MacAdam, D. L. (1980) *Interferometry by holography* (Springer).

Ostrovsky, Y. I., Butusov, M. M. and Ostrovskaya, G. V. (1980) *Interferometry by holography* (Springer).

Marom, F. *et al.* (eds.) *Holography and optical data processing* (Jerusalem, 1976; Pergamon, 1977).

Catley, W. T. (1974) *Optical information processing and holography* (Wiley).

Bally, G. von (ed.) (1979) *Holography in medicine and biology* (Workshop Munster, 1979) (Springer).

Smith, H. M. (ed.) (1977) *Holographic recording materials* (Springer).

Yaroslavskii, L. P. and Merzlyakov, N. S. (1981) *Methods of digital holography* (Plenum).

Optical and acoustical holography, Proceedings of the NATO Advanced Study Institute (Milan, 1971) (Plenum, 1972).

Chuguy, J. V. and Kolesova, N. T. (1976) *Bibliography on holography* is a translation from the Russian (Scientific Information Services).

The following books on holography and coherence have been selected from those published since the second edition of this book was produced:

Harihon, P. (1984) *Optical holography* (CUP) covers principles, techniques and applications.

Syms, R. R. A. (1990) *Practical volume holography* (Clarendon) is a dedicated treatise on theory and practice of this little documented field.

Jones, R. and Wykes, C. (1989) *Holographic and speckle interferometry* (CUP) covers the basic optical principles, techniques, experimental design and applications.

Kasper, J. E. and Feller, S. A. (1987) *The complete book of holograms* (Wiley) shows how they work and how to make them.

Robillard, J. and John, H. (1990) *Industrial applications of holography* (OUP) emanates from a conference and covers various technological applications.

Ostrovsky, Yu. I., Shchepinov, V. P. and Yakovlev, V. V. (1991) *Holographic interferometry in experimental mechanics* (Springer).

Schumann, W., Zurcher, J. P. and Cuche, D. (1985) *Holography in deformation analysis* (Springer).

Greguss, P. and Jeong, T. H. (eds.) (1991) *Holography* (SPIE) commemorates the 90th birthdate of Dennis Gabor.

Stone, T. W. and Thompson, B. J. (eds.) (1991) *Holographic and diffractive lenses and mirrors* (SPIE, vol. MS 34) is a thorough review of this subject.

Vikram, C. S. (1992) *Particle field holography* (CUP) is a first book to describe the entire subject of this field which is concerned with the study of very small objects in a dynamic medium.

Saxby, G. (1991) *Manual of practical holography* (Focal) demonstrates the principles and practice, with illustrations.

NON-LINEAR OPTICS

The laser relies upon, and has substantially aided research into, the non-linear optical behaviour of materials. Publications in this area are listed as follows:

Baldwin, G. C. (1968) *An introduction to non-linear optics* (Plenum).

Akhananov, S. A. and Khoklov, R. V. (1972) *Problems of non-linear optics* (Gordon and Breach).

Zernike, F. and Midwinter, J. E. (1973) *Applied non-linear optics* (Wiley).

Shen, Y. R. (ed.) (1977) *Non-linear infra-red generation* (Springer).

Fled, M. S. and Letchkov, V. S. (1980) *Coherent non-linear optics* (Springer).

Reinfjer, J. F. (1983) *Non-linear optical parametric processes in liquids and gases* (Academic).

Haken, H. (1984) *Non-linear optics*, vol. 3 of *Light* (North-Holland).

Mills, D. L. (1991) *Non-linear optics* (Springer) is an introduction to this field covering many topics including wave propagation in optical fibres and chaos.

Schubert, M. and Wilhelmi, B. (1986) *Non-linear optics and quantum electronics* (Wiley). A review, based upon a course of lectures for advanced students.

Laud, B. B. (1991) *Lasers and non-linear optics* (Wiley, 2nd edn.) is an advanced text.

Butcher, P. and Cotter, D. (1990) *The elements of non-linear optics* (CUP) gives an account of the most important principles of non-linear optics.

Dmitriev, V. G., Gurzadyan, G. G. and Nikogosyan, D. N. (1991) *Handbook of non-linear optical crystals* (Springer) is a reference source containing a complete description of the properties and applications of all non-linear crystals reported in the literature up to the beginning of 1990.

Keller, O. (ed.) (1990) *Non-linear optics in solids* (Springer) is a broadbased treatment of the theoretical and experimental aspects of the non-linear interaction of light with condensed matter.

Lyons, M. H. (ed.) (1989) *Materials for non-linear and electro-optics* (IOP) is the proceedings of an international conference held in Cambridge in 1989.

Olver, P. J. and Sattinger, D. H. (1990) *Solitons in physics, mathematics and non-linear optics* (Springer) is an overview of several research directions including optics.

Prasad, P. N. and Williams, D. J. (1991) *Introduction to non-linear effects in molecules and polymers* (Wiley) surveys the field of non-linear optics in organic materials at a basic level.

Sarker, S. (ed.) (1986) *Non-linear phenomena and chaos* (Malvern Physics Series, 1986) considers fluctuations, oscillations and instabilities in liquid crystal devices, cryogenic fluids, isothermal systems, resonators, etc., and analytical methods of predicting chaos.

Flytzanis, C. and Oudar, J. L. (eds.) (1988) *Non-linear optics: materials and devices* (Springer) is the proceedings of the International School of Materials, Science and Technology held in Erice, Sicily, in 1985.

Gibbs, H. M. *et al.* (eds.) (1986) *Optical bistability III* (Springer) is the proceedings of the Topical Meeting held at Tuscon, Arizona, in 1985.

Kobayashi, T. (ed.) (1989) *Non-linear optics of organics and semi-conductors* (Springer, 1989) is the proceedings of an international symposium held in Tokyo in 1988.

Vilaseca, R. and Corbalan, R. (eds.) (1991), *Non-linear dynamics and quantum phenomena in optical systems* (Springer).

Gibbs, H. M., Khitrova, G. and Peyghambarian, N. (eds.) (1990) *Non-linear photonics* (Springer).

Brandt, H. E. (ed.) (1991) *Non-linear optics* (SPIE, vol. MS 32) is a comprehensive and definitive review collated from selected papers.

Fisher, R. A. and Reintjes, J. F. (1991) *Non-linear optics* (SPIE, vol. 1409) is the proceedings of a conference held at Los Angeles in 1991, covering quantum confined structures and frequency conversion, phase conjugation and photorefractive effects, and non-linear optical effects and materials.

Peyghambarian, N. (ed.) (1990) *Non-linear optical materials and devices for photonic switching* (SPIE, vol. 1216) is the proceedings of a conference held in Los Angeles in 1990 and covers, e.g. non-linear waveguide devices, electro-optic non-linearities and applications, semi-conductor lasers and arrays, organic non-linear materials and devices, etc.

Ostrowsky, D. B. and Reinisch, R. (1992) *Guided wave non-linear optics* (Kluwer)

D. B. Ostrowsky and R. Reinisch were editors for the NATO Advanced Science Institute Summer School on *Guided wave non-linear optics* held in Corsica, France, in August, 1991 and the proceedings were published by Plenum in 1992.

Miyata, S. (ed.) (1992) *Nonlinear optics* (North Holland) is the proceedings of the 5th Toyota Conference on Nonlinear Optical Materials held in Aichi-Ken, Japan, in October 1991.

Newall, A. C. and Moloney, J. V. (1992) *Nonlinear optics* (Addison-Wesley) is the first monograph in the Advanced Topics in Inter-disciplinary Mathematical Sciences series, reflecting its advanced mathematical treatment.

Sakai Jun-Ichi (1992) *Phase conjugate optics* (McGraw-Hill) addresses the basic concepts and theoretical foundations of phase conjugation and four-wave mixing, the materials which produce phase conjugate light and applications.

Fisher, R. A. (1983) *Optical phase conjugation* (Academic) is an advanced vectorial treatment of this subject.

Zel'dovich, B. Y. (1985) *Principles of phase conjugation* (Springer).

Walther, H., Koroteev, N. and Scully, M. (1993) *Frontiers in nonlinear optics* (IOP) gives an overview of the field by experts in memory of Sergei Akhmanov.

GENERAL REVIEWS

Applications of lasers in general are covered in *The third conference on the laser,* published by the New York Academy of Sciences, vol. 267, 1976, updating earlier conferences in 1965 and 1969. Topics cover laser systems, lasers and nuclear energy, laser biology, laser medicine, laser communications, lasers and information handling.

Monte, R. *Laser applications,* vol. I, 1971 and vol. II, 1974, published by Academic, contains selected papers on several major applications of lasers, including tracking, scanning, monitoring, communications, metrology and the laser gyroscope.

Harry, H. E. (1974) *Industrial lasers and their applications* (McGraw-Hill) is a general interest, non-mathematical, review of industrial and medical applications.

Jerrard, H. G. (ed.) (1983) *Electro-optics/laser international 1982 UK* (Butterworths) is the proceedings of an international conference, Brighton, 1982. covering a wide spectrum of topics including optical fibre and optical information processing, holography, laser safety and optical system design.

Wherrett, B. S. (1980) *Laser advances and applications* (Wiley) is the proceedings of the 4th National Quantum Electronics Conference, 1979, covering a wide field of subjects.

Progress in Soviet laser research is described by N. G. Basov (ed.) in *Lasers and their applications in physical research* (Plenum, 1979).

Laser handbook (Elsevier) is a substantive authority on new research initiatives and developments. Vol. 4 (1985), edited by M. L. Stitch and M. Bass, covers free electron lasers, colour centre lasers, ring laser gyroscopes, non-linear optical phase conjugation and optical bistability. Vol. 5 (1986) with the same editors, covers coherent and vacuum ultra-violet sources, tunable paramagnetic-ion lasers, superhigh resolution spectroscopy, laser chemistry and laser diagnostics of magnetically confined thermonuclear plasmas. Vol. 6 (1991) edited by W. B. Colson, C. Pellegrini and A. Renieri is concerned with free electron lasers.

Tam, A. S., Gale, J. L. and Stwalley, W. C. (1989) *Advances in laser science III* (AIP Conf. Proc. 172) is the proceedings of the 3rd International Laser Science Conference in Atlantic City in 1987.

Belforte, D. and Levitt, M. (1989) *1989 Industrial laser handbook* (SPIE, vol. 1122) provides an overview of new technologies relevant to laser materials processing, presents useful data for industrial laser processing and provides a comprehensive product and service directory for the industry.

Weber, M. V. (ed.) (1987) *CRC handbook of laser science and technology. Vol. I lasers and masers* (CRC) is the first volume in a series in laser science and technology.

Meyers, R. A. (ed.) (1991) *Encyclopedia of lasers and optical technology* (Academic) provides concise reviews over a wide field.

SAFETY

Sliney, D. and Wolbarsht, M. (1982) *Safety with lasers and other optical sources* (Plenum) is a comprehensive handbook on optical safety.

Grandolfo, M., Rindi, A. and Sliney, D. H. (1991) *Light, lasers and synchroton radiation. A health risk assessment* (Plenum).

Johnson, A. M. (1991) *Eyesafe lasers: components systems and applications* (SPIE, vol. 1419) is the proceedings of a conference at Los Angeles in 1991 covering eyesafe laser components, systems and applications.

Muller, G. J. and Sliney, D. H. (eds.) (1989) *Dosimetry of laser radiation in medicine and biology* (SPIE, vol. IS 5) addresses dosimetry and tissue interaction with laser radiation, laser delivery systems, photodynamic therapy and international safety regulations.

Galoff, P. K. and Sliney D. H. (eds.) (1990) *Laser safety, eyesafe laser systems and laser eye protection* (SPIE, vol. 1207) is the proceedings of a conference held in Los Angeles in 1990 covering eye damage mechanisms, retinal and corneal effects, eyesafe laser systems and laser safety.

Berlien, H. P. *et.al.* (eds.) (1989) *Advances in laser medicine II, safety and laser tissue interaction* (SPIE, vol. 1143) is the proceedings of a conference held in Berlin in 1988 and addresses the safe use of lasers, training and international regulations.

BSEN 60825 is the British Standard on laser safety, following the European Standard EN 60825.

Journals

The following is a selection of journals specializing in laser science and associated subjects, but many others in the general list given in Chapter 6 also cover these fields and should be consulted:

Optics and Laser Technology (Butterworths)
Optics and Lasers in Engineering (Applied Science)
Current Laser Abstracts (Journal of Inst. Las. Doc.)

Journal of Soviet Laser Research (Plenum)
Laser Focus with Fibre Optic Technology (Advanced Technology)
Laser Report (Advanced Technology)
Lasers and Unconventional Optics (European Abstracts)
Electro-Optics incorporating *Laser Review* (Milton)
Laser und Optoelektronik (AT-Fachverlag)
The AIAA Journal
Opto and Laser Europe (IOP)
Optics and Laser Technology (Butterworths and Heinemann)
Laser and Particle Beams (CUP)
Laser Therapy (Wiley)
Lasers in Surgery and Medicine (Wiley)
Physica D. Non-linear Phenomena (North-Holland)
Non-linear Optics (Gordon and Breach)
Journal of Non-linear Science (Springer)
SPIE Proceedings (SPIE)

Atmospheric optics and remote sensing

Books and proceedings

ATMOSPHERIC OPTICS

The study of the interaction of optical radiation with the atmosphere is of interest in diverse applications including meteorology and remote sensing.

McCartney, E. J. (1976) *Optics of the atmosphere: scattering by molecules and particles* and *Absorption and emission by atmosphere gases* (1983) by the same author are published in the Wiley series of Pure and Applied Optics.

Optical properties of the atmosphere is an AGARD Conference (1976).

Liou, K. N. (1983) *An introduction to atmospheric radiation* is a monograph in the Academic International Geophysics series.

Iqbal, M. (1983) *An introduction to solar radiation* (Academic) discusses quantitative methods for assessing the amount of solar radiation reaching the earth. A general reader is by A. and M. Meinel *Sunsets twilights and evening skies* (CUP, 1983), which gives well-illustrated explanations of these phenomena together with an extended treatment of the atmospheric and optical effects of volcanic eruptions.

Marchuk, G. I. *et al. The Monte Carlo methods in atmospheric physics* (Springer Series in Optical Sciences 12) was published in 1980.

Leland, R. P. (1989) *Stochastic models for laser propagation in atmospheric turbulence* (Springer) is based on lecture notes.
Ulrich, P. B. and Wilson, R. E. (eds.) (1991) *Propagation of high-energy laser beams through the earth's atmosphere II* (SPIE, vol. 1408) is the proceedings of a conference held at Los Angeles in 1991 covering effects of winds, cloud, turbulence: use of adaptive optics for minimizing effects, and small instabilities.
Weichel, H. (1990) *Laser beam propagation in the atmosphere* (SPIE, vol. TT 3) is a tutorial text showing how an initially well-defined laser beam is changed as it propagates through the atmosphere by absorption, scattering, haze, fog and rain, by turbulence and by thermal blooming.
Fisher, R., A. and Wilson, L. E. (1989) *Non-linear optical beam manipulation and high energy beam propagation through the atmosphere* (SPIE, vol. 1060) is the proceedings of a conference held in Los Angeles in 1989, including optical phase conjugation, Raman and Brillouin scattering.
Bissonnette, L. R. and Miller, W. B. (1991) *Propagation engineering: fourth in series* (SPIE, vol. 1487) is the proceedings of a conference held at Orlando, Florida, in 1991 and addresses the characterization of optical turbulence, its mitigation, characterization of aerosol effects and characterizing the propagation environment.

REMOTE SENSING

The expanding subject of remote sensing is covered in many books and journals:

Manual of remote sensing (American Society of Photogrammetry, 2nd edn., 1983) is published in two volumes, 1 *Theory instrumentation and techniques*, and 2 *Interpretation and applications*.
Slater, P. N. (1980) *Remote sensing* (Addison-Wesley) describes the optical principles underlying the Landsat satellite programme.
Kahle, A. B. *et al.* (eds.) (1981) *Sessions on remote sensing* (Pergamon) is the proceedings of the Topical Meeting of the COSPAR Scientific Commission Budapest (1980) and *Optical techniques for remote probing of the atmosphere* is another topical meeting run and published by the Optical Society of America (1983).
Houghton, J. T., Taylor, F. W. and Rogers, C. D. (1984) *Remote sounding of the atmosphere* (CUP) covers the visible, infra-red and microwave imaging aspects, complementing a book by Houghton, J. T. (1977) *The physics of atmospheres* (CUP).
Another book is by Camagni, P. and Sandroni, S. (eds.) (1984) *Optical remote sensing of air pollution* (North-Holland).

Publications which refer specifically to laser sensing of the atmosphere and propagation are:

Killinger, D. K. and Mooradian, A. (1983) *Optical and laser remote sensing* (Springer).

Measures, R. M. (1984) *Laser remote sensing fundamentals and applications* (Wiley).

Hinkley, E. (ed.) (1976) *Laser monitoring of the atmosphere*, in the Springer series, Topics in Applied Physics, vol. 14.

Strohbehn, J. W. (1978) *Laser beam propagation in the atmosphere* (Springer).

Zuev, V. E. (1982) *Laser beams in the atmosphere* (Plenum).

Ruquist, R. D. *Meteorological effects on high energy laser propagation* and Reilly, J. *et al. Pulse laser propagation through atmospheric dusts at 10.6μm*, are both published by the AIAA (1978).

Measures, R. M. (ed.) (1988) *Laser remote chemical analysis* (Wiley) covers the fundamentals of remote sensing by chemical analysis, laser propagation through the atmosphere, environmental sensing based upon infra-red absorption of laser radiation, use of LIDAR for pollution mapping, measurement of trace constituents in the atmosphere, oceanic and terrestrial surveillance and fibre optic remote sensing.

Zeuv, V. E. and Naats, I. E. (1983) *Inverse problems of LIDAR scanning of the atmosphere* (Springer).

Santoleri, J. J. (ed.) (1991) *Environmental sensing and combustion diagnostics* (SPIE, vol. 1434) is the proceedings of a conference held at Los Angeles in January 1991.

Sokoloski, M. M. (1989) *Laser applications in meteorology and Earth and atmospheric remote sensing* (SPIE, vol. 1062) is the proceedings of a conference held in Los Angeles in 1989 and covers advances in laser technology for remote sensing, meteorological applications, surface applications, trace species and velocity fields.

Beer, R. (1992) *Remote sensing by Fourier transform spectrometry* (Wiley) covers the development of instrumentation and applications to planetary and earth-atmospheric research.

Spectroscopy

Books and proceedings

GENERAL

Clark, G. T. (1960) *The encyclopedia of spectroscopy* (Reinhold) is a comprehensive introduction to the field of spectroscopy.

Condon, E. U. and Shortley, G. H. (1935) *The theory of atomic spectra* (CUP) is an early book which examines the classical methods for analysing the properties of atoms.

Slater, J. C. (1960) *Theory of atomic structure* (McGraw-Hill), Candler, C. (1964) *Atomic theory and the vector model* (Hilger, 2nd edn.) and Shore, B. W. (1968) *Principles of atomic spectra* (Wiley) deal with the modern theory of atomic structure and spectra.

Kuhn, H. G. (1969) *Atomic spectra* (Longman), Sobelman, I. I. (1972) *Introduction to the theory of atomic spectra* (Pergamon), Dixon, R. N. (1965) *Spectroscopy and structure* (Methuen) and King, G. W. (1964) *Spectrosopy and molecular structure* (Holt-Rinehart) all cover the general structural aspects of spectroscopy.

Klosa, E. and Wilhelmi, B. (eds.) (1989) *Ultrafast phenomena in spectroscopy* (Springer) is the proceedings of the 6th International Symposium held at Neubrandenburg in 1990.

Harris, C. B. *et al.* (eds.) *Ultrafast phenomena VII* is the proceedings of the 7th International Conference held at Monterey, Ca., in 1990 (Springer).

Svanburg, S. (1991) *Atomic and molecular spectroscopy* (Springer) is a textbook review of modern techniques at a fundamental and applied level.

Andrews, D. L. (ed.) (1990) *Perspectives in modern chemical spectroscopy* (Springer) with contributions by numerous experts.

Hollas, I. M. (1991) *Modern spectroscopy* (Wiley, 2nd edn.) is a textbook covering basic principles and techniques.

Brown, D. W., Floyd, A. J. and Sainsbury, M. (1988) *Organic spectroscopy* (Wiley) is described as a user-friendly undergraduate textbook covering four types, including ultra-violet/visible and infra-red spectroscopy.

Struve, W. S. (1989) *Fundamentals of molecular spectroscopy* (Wiley) provides a concise introduction to the spectroscopy of atoms and molecules.

Recent books in the Wiley Advances in Spectroscopy Series (editors: R. J. H. Clark and R. E. Hester) are vol. 17 *Spectroscopy of matrix isolated species* (1989), vol. 18 *Time resolved spectroscopy* (1989) and vol. 19 *Spectroscopy of advanced materials* (1991).

John Wiley produce a series of open learning texts in analytical chemistry (ACOL) covering basic concepts, classical methods, instrumental techniques and applications. Of particular relevance are:

Denny, R. and Sinclair, R. (1987) *Visible and ultra-violet spectroscopy*.

Rendell, D. (1987) *Fluorescence and phosphorescence*.

George, B. and McIntyre, P. (1987) *Infra-red spectroscopy*.

Metcalfe, E. (1987) *Atomic absorption and emission spectroscopy*.

Yersin, H. and Vogler, A. (eds.) (1989) *Photochemistry and photo-physics of co-ordination compounds* (Springer) is the proceedings of the 7th International Symposium on this subject held at Elmau in 1987.

Di Bartolo, B. (1987) *Spectroscopy of solid-state laser-type materials* (Plenum).

Moore, G. L. (1989) *Introduction to inductively coupled plasma spectroscopy* (Elsevier).

Sneddon, J. (ed.) (1990) *Sample introduction in atomic spectroscopy* (Elsevier).

Schiff, H. I. (ed.) (1991) *Measurement of atmospheric gases* (SPIE, vol. 1433) is the proceedings of a conference held at Los Angeles in 1991 and covers optical absorption spectroscopy, laser-induced fluorescence and tunable diode-laser spectrometry.

Saperstein, D. D. (ed.), *Applied spectroscopy in materials science* (SPIE, vol. 1437) is the proceedings of a conference held at Los Angeles in 1991 and covers carbon thin films, infra-red fibre optics, lasers in mass spectrometry and new techniques and applications.

Books dealing with spectroscopic techniques and instrumentation are:

Candler, C. (1949) *Practical spectroscopy* (Hilger), a basic introduction based upon the Hilger Spectrometer.

Blackburn, J. A. (ed.) (1970) *Spectra analysis: methods and techniques* (Dekker) and Bousquet, P. (1971) *Spectroscopy and its instrumentation* (Hilger).

Kortum, G. (1969) *Reflectance spectroscopy* (Springer) deals with principles, methods and applications.

Davis, S. P. (1970) *Diffraction grating spectrographs* (Holt-Rinehart).

Some of the more recent books in this field are the following:

Thorne, A. P. (1988) *Spectrophysics* (Chapman & Hall, 2nd edn.) describes the methods of experimental spectroscopy and their use in the study of physical phenomena at an elementary level. The main addition to the second edition is a chapter on laser spectroscopy. Intended for undergraduates or for postgraduates beginning research in this field.

Mich, I. J. and Thulstrup, E. W. (1986) *Spectroscopy with polarised light* (VCH) is an introduction to the application of orientational dependence of optical properties of molecules to the solution of problems in chemistry, biology and polymer science. Requires knowledge of quantum mechanics, optics and spectroscopy.

Busch, K. (1990) *Multi-element detection systems for spectrochemical analysis* (Wiley) presents a unified treatment of multi-channel detection systems in the ultra-violet and visible regions of the spectrum for spectrochemical analysis.

Thompson, M. and Walsh, J.N. (1989) *Handbook of inductively coupled plasma spectrometry* (Blackie) is an evaluation of this technique and application to geological, geochemical, archaeological and metallurgical analysis.

LASER SPECTROSCOPY

Of the many publications on laser spectroscopy the following are selected:

Smith, R. A. (1976) *Very high resolution spectroscopy* (Academic), covering the visible, infra-red and ultra-violet regions.

Shirmoda, K. (1976) *High resolution laser spectroscopy*, in the Springer Topics in Applied Physics Series, covering absorption studies and fluorescence.

Atkinson, G. H. (ed.) (1983) *Time resolved vibrational spectroscopy* (Academic).

Cummins, H. Z. and Pike, E. R. *Photon correlation spectroscopy and velocimetry* (Plenum, 1974, 1977) which contains the lectures presented to the NATO Advanced Study Institute on this subject, Capri, 1973 and 1976.

Levenson, M. D. and Kano, S. S. (1988) *Introduction to non-linear laser spectroscopy* in the Quantum Electronics Principles and Applications Series (Academic, rev. edn.) describes the theory and practice at a level suitable for graduate students, requiring a general knowledge of quantum mechanics, addressing the dynamics of the interaction and emphasizing high-resolution techniques with copious references.

Letokhov, V. S. and Chebotayev, V. P. (1977) *Non-linear laser spectroscopy* (Springer).

Klinger, D. S. (1983) *Ultra sensitive laser spectroscopy* is another volume in the series Quantum Electronics – Principles and Applications (Academic) and examines very weak spectral transitions.

Jacobs, S. F., Sargent, M. and Scully, M. O. (eds) (1974) *Laser application to optics and spectroscopy* (Addison-Wesley) is based upon lectures of a summer school and covers semiconductors, diode lasers and dye lasers, principally, and their application to spectroscopy.

Corney, A. *Atomic and laser spectroscopy* (OUP, 1977, 1980).

Arecchi, F. T. (1984) *Advances in laser spectroscopy*, proceedings of a NATO Advanced Study Institute, volume 95, held in Tuscany, 1983 (Plenum).

Laser Spectroscopy is the title of a series of conferences held at Megeve, France, 1975, Wyoming, USA, 1977, Rottach-Egern, Germany, 1979, and Alberta, Canada, 1981, published by Springer, which review new advances in laser technology and spectroscopy.

Walther, H. (ed.) (1976) *Laser spectroscopy of atoms and molecules* (Springer).

Mittleman, M. H. (1982) *Introduction to the theory of laser-atom interactions* (Plenum) covers the scattering and radiation by atoms in a laser field from a basis of understanding elementary quantum mechanics.

Wherrett, B. S. (ed.) (1980) *Laser advances and applications* (Wiley) is the proceedings of 4th National Quantum Electronics Conference and covers laser spectroscopy as a major theme.

Alfano, R. R. (1982) *Biological events probed by ultrafast laser spectroscopy* (Academic) deals with application of nanosecond and picosecond laser techniques to systems of biological interest and covers photosynthesis, haemoproteins and DNA.

Pratesi, R. and Sacchi, C. A. (1980) *Lasers in photomedicine and photobiology* is a conference proceedings on these subjects held in Florence, 1979, and published by Springer in the Optical Science Series.

Kliger, D. S. (ed.) (1983) *Ultra-sensitive laser spectroscopy* (Academic) is a study of techniques for detecting very weak spectral transitions.

Stenholm, S. (1984) *Foundations of laser spectroscopy* (Wiley) is a textbook designed to teach theoretical ideas needed to do calculations in laser spectroscopy.

Yen, W. M. and Seizer, P. M. (eds.) (1986) *Laser spectroscopy of solids* (Springer, 2nd edn.).

Dermtröder, W. (1988) *Laser spectroscopy* (Springer, 3rd edn.) is a key book aimed at physicists and chemists who want to study laser spectroscopy in depth and requires prior knowledge of atomic and molecular physics, electrodynamics and quantum optics. Also by the same author *Laserspektroskopie, Grundlagen und Techniken 2 Aufl* (Springer, 1991).

Hirota, E. (1985) *High resolution spectroscopy of transient molecules* (Springer) covers the application of laser spectroscopy (principally) to study chemical reactions.

Kaplyanskii, A. A. and Macfarlane, R. M. (ed.) (1987) *Spectroscopy of solids containing rare earth ions* (North-Holland) is an advanced treatise comprising research papers in key areas.

Letokhov, V. S. (1986) *Laser analytical spectrochemistry* (Hilger) is an overview of research at USSR Academy of Sciences. The same author has written *Laser picosecond spectroscopy and photo-chemistry of biomolecules* (Hilger, 1987) which is a critical review of applications to biophysics and photobiology and, thirdly, *Laser spectroscopy of highly vibrationally excited molecules* (Hilger, 1989).

Moenke-Blankenburg, L. (1989) *Laser micro-analysis* (Wiley) focuses on the use of the laser as a tool for chemical analysis covering the source properties, the interaction phenomena and the measurements involved. Applications to optical emission spectrometry, atomic absorption spectrometry, atomic fluorescence spectrometry, mass spectrometry and plasma excitation spectroscopy.

Radhakrishna, B. C. and Tan, B. C. (eds.) (1991) *Laser spectroscopy and non-linear optics of solids* (Springer) is the proceedings of an international workshop held at Kuala Lumpur, Malaysia, in 1991.

Persson, W. and Svanberg, S. (eds.) (1987) *Laser spectroscopy VIII* (Springer) is the proceedings of the 8th International Conference held at Are, Sweden, in 1987.

Bunkin, F. V., Karlov, N. V. and Lukjancuk, B. S. (1988) *Laser induced chemistry* (Springer).

Yen, W. M. (1989) *Laser spectroscopy of solids II* (Springer).

Gacetz, B. A. (1983) *Advances in laser spectroscopy* is volume 2 of a Wiley series concerned with progress in this subject, and aimed at spectroscopists and chemists.

The following are classified according to technique and may include the use of lasers:

ABSORPTION SPECTROSCOPY

Lothian, G. E. (1969) *Absorption spectrophotometry* (Hilger, 3rd edn.).

Slavin, W. (1968) *Atomic absorption spectroscopy* (Interscience).

Edbon, L. (1983) *Introduction to atomic absorption spectroscopy* (Hayden), a self-teaching approach.

Blass, W. E. and Halsey, G. W. (1981) *Deconvolution of absorption spectra* (Academic) examines the technique of resolution recovery in recorded absorption spectra by use of signal processing techniques.

FLUORESCENCE SPECTROSCOPY

Guilbault, G. G. (1973) *Practical fluorescence, theory, methods and techniques* (Dekker).

Pesce, A. J. (ed.) (1971) *Fluorescence spectroscopy: an introduction for biology and medicine* (Dekker).

Wehry, E. L. (1981) *Modern fluorescence spectroscopy*, vols 3 and 4 (Plenum).

Lakowicz, J. R. (1983) *Principles of fluorescence spectroscopy* (Plenum) describes the principles of fluorescence spectroscopy, polarization lifetimes, quenching, energy transfer and excited state reactions, with set problems at the end of each chapter to aid understanding.

Hemmila, I. A. (1991) *Applications of fluorescence in immuno-assays* (Wiley) describes the technique and clinical applications.

Ichinose, N. *et al.* (1991) *Fluorometric analysis in biomedical chemistry* (Wiley) address basic principles of fluorescence measurements in biochemical and biomedical applications.

RAMAN SPECTROSCOPY AND SCATTERING

Colthup, N. B., Daly, L. H. and Wiberley, S. E. (1964) *Introduction to infra-red and Raman spectroscopy* (Academic).

Advances in Raman spectroscopy, vol. I, is the proceedings of the 3rd International Conference on Raman Spectroscopy, University of Reims, September 1972.

Anderson, A. (1971) *The Raman effect*, 2 vols (Dekker).

Koningstein, J. A. (1972) *Introduction to the theory of the Raman effect* (Reidel).

Szymanski, H. A. *Raman spectroscopy: theory and practice* (Plenum, vol. I, 1967, vol. II, 1970).

Tobin, M. C. (1971) *Laser Raman spectroscopy* (Interscience) covers techniques and instrumentation.

Clark, R. J. H. and Hester, R. E. (eds.) *Advances in infra-red and Raman spectroscopy*, vol. 11 is a volume in the Wiley Series on Advances in Infra-Red and Raman Spectroscopy which is intended to examine critically the fundamental and applied aspects in all branches of the subject, integrating theory and practice at an advanced level. This volume deals with bio-molecular spectroscopy.

Weber, A. (ed.) (1979) *Raman spectroscopy of gases and liquids* (Springer) deals with high-resolution rotational Raman spectra of gases, Raman cross-sections in gases and liquids, resonance and anti-Stokes Raman spectroscopy and intermolecular forces revealed by Raman scattering.

Chang, R. K. (1982) *Surface enhanced Raman scattering* (Plenum).

Barov, N. G. (ed.) (1982) *Stimulated Raman scattering* (Plenum) presents five papers on recent research covering the generation and spectral intensity distribution of stimulated Raman scattering in condensed media and angular distribution of scattered light.

Parker, F. S. (1983) *Applications of infra-red Raman and resonance Raman spectroscopy in biochemistry* (Plenum) is a reference work for advanced students and research workers in the field of spectroscopy.

Gardiner, D. J. and Graves, P. R. (eds.) (1992) *Practical Raman spectroscopy* contains contributions by various experts on Fourier transform Raman spectroscopy (Elsevier).

Bower, D. I. and Maddams, W. F. (1992) *The vibrational spectroscopy of polymers* (CUP) describes the theory and practice of infra-red and Raman spectroscopy to the study of the physical and chemical characteristics of polymers. It is written as an introduction to more advanced treatises.

Durig, J. R. and Sullivan, J. F. (1990) *XII International conference on Raman spectroscopy* (Wiley) is the proceedings of the conference held at the University of South Carolina in 1990.

The *Journal of Raman Spectroscopy* reviews current progress in the field.

ULTRA-VIOLET SPECTROSCOPY

Jaffe, H. H. and Orchin, M. (1962) *Theory and application of ultra-violet spectroscopy* (Wiley).

McIlrath, T. J. and Freeman, R. R. (eds.) (1982) *Laser techniques for extreme ultra-violet spectroscopy* (AIP) contains 57 papers on the examination of atomic and molecular properties through the study of energy levels greater than 6eV by both direct and parametric up-conversion laser techniques.

INFRA-RED AND MILLIMETRIC SPECTROSCOPY

Chantry, G. W. (1971) *Submillimetre spectroscopy: a guide to the theoretical and experimental physics of the far infra-red* (Academic).

Miller, R. G. J and Stace, B. C. (eds.) (1972) *Laboratory methods in infra-red spectroscopy* (Heyden, 2nd edn.).

Möller, K. D. and Rothschild, W. G. (1971) *Far-infra-red spectroscopy* (Wiley/Interscience).

Stewart, J. E. (1970) *Infra-red spectroscopy: experimental methods and techniques* (Dekker).

Houghton, J. T. and Smith, S. D. (1966) *Infra-red physics* (OUP) is an introductory text discussing the fundamentals of the techniques and physics of infra-red spectroscopy.

Alpert, N. L., Keiser, W. E. and Szymanski, H. A. (1970) *Theory and practice of infra-red spectroscopy* (Plenum, 2nd edn.) is an introductory text for undergraduates dealing with instrument design and structural analysis.

Schultz, G. and Moss, T. S. (eds.) (1977) *High resolution infra-red and submillimetre spectroscopy* (Pergamon) gives a selection of papers presented at a workshop held in Bonn, 1977.

Möller, K. D. and Rothschild, W. G. (1971) *Far-infra-red spectroscopy* (Wiley).

Bellamy, L. J. (1975) *The infra-red line spectra of complex molecules* (Chapman and Hall, 3rd edn.) covers vibrational spectra of large molecules.

Gaydon, A. G. (1976) *Identification of molecular spectra* (Chapman and Hall, 2nd edn.).

Wollrab, J. E. (1967) *Rotational spectra and molecular structure* (Academic).

Chantry, G. W. (1979) *Modern aspects of microwave spectroscopy* (Academic) covers microwave spectrometers, double-resonance techniques, scanning spectroscopy and interferometric spectrometry.

Rebane, K. K. (1970) *Impurity spectra of solids* (Plenum).

Sherwood, P. M. A. (1972) *Vibrational spectroscopy of solids* (CUP).

Miller, T. A. and Bondybey, V. E. (1983) *Molecular ions: spectroscopy, structure and chemistry* (North-Holland) deals with the physics and chemistry of molecular ions by photon techniques.

Woodward, L. A. (1972) *Introduction to the theory of molecular vibrations and vibrational spectroscopy* (OUP).

Straughan, B. P. (1976) *Spectroscopy* vol. 2 (Chapman and Hall) covers molecular, microwave, infra-red, far infra-red and Raman spectroscopy, force constants, group theory and thermodynamic functions.

Ingram, D. J. E. (1967) *Spectroscopy at radio and microwave frequencies* (Butterworths, 2nd edn.).

Bratos, S. and Pick, R. M. (eds.) (1980) *Vibrational spectroscopy of molecular liquids and solids*, NATO Advanced Study Institute series, vol. 56 (Plenum).

Davydov, A. A. (1990) *Infra-red spectroscopy of absorbed species on the surface of transition metal oxides* (Wiley) is concerned mainly with investigations of oxide catalysts.

Clarke, R. J. H. and Hester, R. E. (1988) *Spectroscopy of surfaces* (Wiley) is the fourth volume of this series and provides an up-to-date review of surface spectroscopic techniques.

Schmid, E. D., Siebert, F. and Schneider, F. W. (1988) *The spectroscopy of biological molecules* (Wiley) is the proceedings of the second European conference on this subject.

FOURIER TRANSFORM SPECTROSCOPY

Bell, R. J. (1972) *Introductory Fourier transform spectroscopy* (Academic).

Mertz, L. (1965) *Transformations in optics* (Wiley) is concerned with Fourier and Fresnel transformations.

LUMINESCENCE SPECTROSCOPY

Lumb, M. D. (ed.) (1978) *Luminescence spectroscopy* (OUP) gives a theoretical and experimental introduction to the subject.

Ozawa, L. (1990) *Cathodoluminescence* (VCH) covers the fundamentals of cathodoluminescence in crystals, the preparation and characterization of luminescent materials and their application.

Kricka, L. and Stanley, P. (eds.) (1991) *Sixth international symposium on bioluminescence and chemiluminescence* (Wiley) contains the proceedings of this Symposium held at Cambridge in September 1990.

OPTOGALVANIC SPECTROSCOPY

Stewart, R. S. and Lawler, J. E. (ed.) (1991) *Optogalvanic spectroscopy*, 1990 (IOP Conference Series 113) reviews recent progress, new developments and applications in this field.

OPTOACOUSTICS

Optoacoustic spectroscopy and detection by Yoh-Han Pao (Academic, 1977) deals with the spectroscopic examination of gases and solids.

Berg, N. J. and Lee, J. N. (eds.) (1983) *Acousto-optic signal processing* (Dekker) covers theory of acoustic optic interactions, basic devices and systems.

Saporiel, J. (1979) *Acousto-optics* (Wiley) discusses propagation of light in crystals and the theory of acousto-optical interactions and devices.

Gottlieb, M. *et al.* (1983) *Electro-optic and acousto-scanning and deflection* (Dekker).

Heiss, P. and Pelzl, J. (eds.) (1987) *Photoacoustic and photothermal phenomena* (Springer) is the proceedings of the 5th International Topical Meeting at Heidelberg in 1987.

Tsai, C. S. (ed.) (1990) *Guided wave acousto-optics: devices and applications* (Springer).

Zharov, V. P. and Letokhov, V. S. (1986) *Laser optoacoustic spectroscopy* (Springer).

Scruby, C. B. and Drain, L. E. (1990) *Laser ultrasound* (IOP) is a comprehensive description of the generation and application of laser ultrasonics.

Willis, H. A., Van der Maas, J. H. and Miller, R. G. J. (1987) *Laboratory methods in vibrational spectroscopy* (Wiley) is a laboratory manual for the practical infra-red spectroscopist.

Xu, J. and Stroud, R. (1991) *Acousto-optic devices* (Wiley) gives a comprehensive explanation of the principles of acousto-optics, devices and signal processing.

Korpal, A. (ed.) (1990) *Acousto-optics* (SPIE, vol. M 516) consists of selected papers covering theory, techniques, devices and materials. Also published in the Marcel Dekker Optical Engineering Series, vol. 16.

The *Journal of Photoacoustics* is published by Dekker.

UNCLASSIFIED

Buertolucci, M. D. and Harris, D. C. *Symmetry and spectroscopy* (OUP, 1978/1980).

Arecchi, F. J. and Bonifacio, R. (eds.) (1978) *Coherence in spectroscopy and modern physics,* NATO Advanced Study Institute series, vol. 37 (Plenum).

Intermolecular spectroscopy and dynamical properties of dense systems, proceedings of the International School of Physics 'Enrico Fermi', Course LXXV (1978), van Kranendouk, J. (ed.) (North-Holland, 1980) contains a review of the theoretical and experimental aspects of collision-induced spectroscopy and light scattering, and application to rotational and vibrational properties of molecules in dense systems.

Mitra, S. (1989) *Fundamentals of optical and X-ray mineralogy* (Wiley) explains how to predict opticamineralogical properties from first principles and introduces modern analytical techniques.

DATA

Source data are found in:

Clarke, A. (1972) *The Sadtler standard spectra – a guide to the literature on spectral data* (NRLSI, 2nd edn.).

Massachusetts Institute of Technology wavelengths tables (MIT, 1969) and Denny, R. C. (1982) *Dictionary of spectroscopy* (Wiley-Interscience, 2nd edn.).

Beck, R., English, W. and Guis, K. (1980) *Table of laser lines in gases and vapours* is published by Springer (3rd revision).

Morok, H. and Torok, S. (1971) *Technical dictionary of spectroscopy and spectral analysis* (multi-lingual) (Lewis).

Journals

Journals in spectroscopy are:

Journal of Quantitative Spectroscopy and Radiative Transfer (Pergamon)
Journal of Physics B (IOP)
Physica C (Elsevier)
Journal of Molecular Spectroscopy (Academic)
Spectrochimica Acta Part A Molecular Spectroscopy and Part B Applied Spectroscopy (Society of Applied Engineering)
Progress in Atomic Spectroscopy (Plenum)
Journal of Raman Spectroscopy (Wiley)
Journal of Fluorescence (Plenum)
Canadian Journal of Applied Spectroscopy (Polyscience)
Optical Spectra (Optical Pub. Co.)
Spectroscopy Letters (Dekker)
Spex Speaker (Spex)
Journal of Applied Spectroscopy, translated from *Zhurnal Prikladnov Spectroskopii* (Consultants)
Optics and Spectroscopy (OSA)
Journal of the Spectroscopical Society of Japan (Bunko Kenkya)

CHAPTER SEVENTEEN

Physics of materials

F. L. PRATT

This chapter and Chapter 18 on semiconductor physics both look at information sources available for the condensed matter physicist; note that additional coverage of condensed matter properties will also be found in many of the other chapters in this book, e.g. electricity and magnetism, experimental heat, chemical physics and optoelectronics, quantum optics and spectroscopy.

Many different properties of a material will be of interest to the physicist, ranging from electronic, optical and magnetic to structural, mechanical and thermodynamic. An important trend in recent years has been the development of ever more sophisticated means to tailor these physical properties, e.g. by growing multilayer structures or by molecular engineering. As a branch of physics, condensed matter is very broad in scope, with a correspondingly large and expanding body of literature; note, for example, that it is a condensed matter journal *Physical Review B* that supplies the greatest number of contributions to the INSPEC Physics Abstract Service (see Appendix A). It is not only the quantity of literature being produced that poses a problem for the information seeking physicist; the interdisciplinary nature of many research fields and the blurring of boundaries between fields means that relevant information may, for example, appear under materials science, chemistry or engineering classifications. It is thus particularly important that in searching out information in this field, the physicist makes good use of all the assistance available in the form of abstracting services and online databases and becomes familiar with the journals relevant to the field of interest.

The next section gives a general survey of the literature of condensed matter physics. This is followed by more detailed discussion of the information sources available for several key areas in the physics of materials; superconducting materials, organic materials and magnetic materials. Since it would not be possible to cover comprehensively the broad range of subjects that makes up condensed matter physics in a limited space, these topics have been selected as representative fields where there has been much recent activity.

General sources

Review journals and review series

Review articles provide a valuable starting point for surveying the literature in a field. Although condensed matter review articles sometimes appear in the normal physics journals, they are mainly to be found in a number of specialist review journals:

Advances in Physics (AIP)
Contemporary Physics (Taylor and Francis)
Physics Reports (North-Holland)
Reports on Progress in Physics (IOP)
Reviews of Modern Physics (AIP)

Besides these journals, which cover the full range of physics, there are also review journals and review series devoted specifically to condensed matter topics:

Comments on Condensed Matter Physics (Gordon and Breach)
Critical Reviews in Solid State and Materials Science (CRC)
Solid State Physics: Advances in Research and Applications
    H. Ehrenreich and D.Turnbull (eds.) (Academic)

Handbooks

The most comprehensive set of numerical data can be found in the Landolt-Börnstein series of handbooks: *Numerical data and functional relationships in science and technology*, Group 3, *Crystal and solid state physics*, 28 volumes (Springer).

Conference proceedings

Published conference proceedings give a useful overview of the work being carried out in the subject area. The information provided by the short conference papers may be used to seek out more detailed reports that appear in the primary literature. Many scientific publishers produce conference proceedings but some publishers produce them as extensive series, for example:

Springer Proceedings in Physics (Springer)
Springer Series in Solid State Sciences (Springer)
NATO Advanced Research Workshops (Kluwer)
NATO Advanced Study Institutes: Series B Physics (Plenum)

Primary research journals

The primary research journals generally provide the most detailed and important source of literature. The main journals dedicated to publication of basic research in condensed matter physics are:

Journal of Physics: Condensed Matter (IOP)
Journal of the Physics and Chemistry of Solids (Pergamon)
Physica B,C (Elsevier)
Physica Status Solidi (b) (Academie Verlag)
Physical Review B (APS)
Solid State Communications (Pergamon)
Soviet Physics: Solid State (AIP, translated from Russian)
Zeitschrift für Physik B: Condensed Matter (Springer)

General physics journals which publish a significant number of papers on basic research in condensed matter physics are:

Europhysics Letters (EPS)
JETP Letters (AIP, translated from Russian)
Journal of the Physical Society of Japan (Phys. Soc. Jpn.)
Journal de Physique (French Phys. Soc.)
Philosophical Magazine A,B (Taylor and Francis)
Philosophical Magazine Letters (Taylor and Francis)
Physical Review Letters (APS)

For papers oriented more towards applied physics, the following journals can be consulted:

Applied Physics A: Solids and Surfaces (Springer)
Applied Physics Letters (Springer)
Physica Status Solidi (a) (Academie Verlag)
Japanese Journal of Applied Physics (Phys. Soc. Jpn.)
Journal of Physics D: Applied Physics (IOP)

For any given topic in condensed matter physics a number of other journals will generally be a source of relevant papers, these journals may be specialized physics journals or journals in fields that overlap with condensed matter physics such as chemistry, materials science and engineering. There are also now a number of inter-disciplinary journals which are being produced in response to the multi-disciplinary nature of many research areas in condensed matter physics, for example:

Advanced Materials (VCH)
Synthetic Metals (Elsevier)
Journal of Macromolecular Science (Dekker)

Superconducting materials

Since the discovery of high temperature oxide superconductors in 1986, research in all aspects of superconductivity has been intensive and a correspondingly large body of literature has been generated. A number of physics journals are now entirely devoted to super-conductivity and many solid state journals now have separate sections grouping together papers on superconductivity. The journals devoted to superconductivity are:

Physica C (Elsevier)
Journal of Superconductivity (Plenum)
Superconductor Science and Technology (IOP)

Further discussion of information sources for some of the more novel classes of superconductor is given in the following sections.

Oxide superconductors

An introduction to the oxide superconductors at graduate level is given in J. W. Lynn (ed.) *High temperature superconductivity* (Springer, 1990). A useful series of volumes containing collected review articles is edited by D. M. Ginsberg; the most recent in this series is *Physical properties of high temperature superconductors III* (World Scientific, 1992).

A large number conferences and symposia have been held on oxide superconductors. The series of conferences *Materials and Mechanisms of High Temperature Superconductivity* gives perhaps the most comprehensive coverage of the field; these are published in the journal *Physica C* (Elsevier), the first conference was held in Interlaken in 1988 (volumes 153–155), followed by Stanford in 1989 (volumes 162–164) and Kanazawa in 1991 (volumes 185–189). A special abstract journal for the oxide superconductors is provided by *Key Abstacts: High Temperature Superconductors* (INSPEC).

Molecular superconductors

Molecular superconductors currently include organic charge transfer salts and alkali metal doped Fullerenes. A good introduction to the field of organic superconductivity is given in *Organic superconductors* by T. Ishiguro and K. Yamaji (Springer, 1990). Recent conference proceedings on organic superconductors are:

Low dimensional conductors and superconductors D. Jérome and L. G. Caron (eds.) NATO-ASI Series (Plenum, 1987)
The physics and chemistry of organic superconductors, G. Saito and S. Kagoshima (eds.) (Springer, 1990)
Organic superconductivity, V. Z. Kresin and W. A. Little (eds.) (Plenum, 1991)

Many papers on organic superconductors are presented at the International Conference on Synthetic Metals which is held every two years. The proceedings are published in the journal *Synthetic Metals* (Elsevier), the more recent conferences and volume numbers were Santa Fe, 1988, 27–29; Tübingen, 1990, 41–43 and Gothenburg, 1992, 55–57.

Fullerenes are a recent addition to the class of molecular superconductors. As in any relatively new subject, most information is to be found in the primary journals and in special sessions in conferences on related topics. *Physics and chemistry of Fullerenes*, P. W. Stephens (ed.) (World Scientific, 1992) is a guide to the early literature of the subject including reprints of many important papers. Conference proceedings currently available are:

Fullerenes: status and perspectives, C. Taliani, G. Ruani and R. Zamboni (eds.) (World Scientific, 1992)
Clusters and Fullerenes, V. Kumar, T. P. Martin and E. Tosati (eds.) (World Scientific, 1992)

Heavy fermion superconductors

A number of actinide systems are found to be superconductors with very unusual properties. These heavy fermion systems and their superconductivity have attracted considerable interest in recent years. A survey of the field is given by *Theoretical and experimental aspects of valence fluctuations and heavy fermions*, L. C. Gupta and S. K. Malik (Plenum, 1987). Recent conferences in this field have been published in *Journal of Magnetism and Magnetic Materials* (North-Holland), the 1986 Grenoble conference *Anomalous rare earths and actinides: valence fluctuation and heavy fermions* appears as volumes 63–64, and the 1988 Frankfurt conference *Crystal field effects and heavy fermion physics* appears as volumes 75–76. Useful reference data is found in the six-volume series *Handbook on the physics and chemistry of the actinides*, A. J. Freeman, G. H. Lander and C. Keller (eds.) (North-Holland, 1984–1991).

Organic materials

Organic systems provide a potentially vast array of materials with a wide range of physical properties. Recently, organic metals, super-conductors and ferromagnets have been developed to stand along-side the more traditional organic materials, i.e. polymeric insulators and organic semiconductors. The chemical physics of organic materials is covered in Chapter 5. Here we concentrate on the electronic and optical properties of these materials.

An introduction to the field may be gained from the proceedings of a NATO Advanced Study Institute covering a range of organic systems, which is published in *Lower dimensional systems and molecular electronics*, R. M. Metzger, P. Day and G. Papavassiliou (eds.) (Plenum, 1990). For non-linear optics, see *Introduction to nonlinear optical effects in molecules and polymers*, P. N. Prasad and D. J. Williams (Wiley, 1991). Specialist journals relevant to the area are:

Synthetic Metals (Elsevier)
Journal of Molecular Electronics (Wiley)
Molecular Crystals and Liquid Crystals Science and Technology (Gordon and Breach)

Conjugated polymers

An overview of the properties of conjugated polymers is provided by the proceedings of a NATO Advanced Research Workshop *Conjugated polymeric materials: opportunities in electronics, opto-electronics and molecular electronics*, J. L. Bredas and R. R. Chance (eds.) (Kluwer, 1989). More general background reference material is provided by *Comprehensive polymer science*, G. Allen and G. C. Bevington (eds.) (Pergamon 1989), which covers all aspects of polymer science; volume 1 (*Polymer characterisation*) and volume 2 (*Polymer properties*) are particularly relevant for polymer physics. *Handbook on conducting polymers*, T. A. Skotheim (Dekker, 1986) provides a good reference.

A series of winter schools held in Kirchberg provides a useful set of monographs:

Electronic properties of polymers and related compounds, H. Kuz-many, M. Mehring and S. Roth (eds.) (Springer Series in Solid State Science, vol. 63, 1985).
Electronic properties of conjugated polymers II H. Kuzmany, M. Mehring and S. Roth (eds.) (Springer Series in Solid State Science, vol. 76, 1987).

Electronic properties of conjugated polymers III, H. Kuzmany, M. Mehring and S. Roth (eds.) (Springer Series in Solid State Science, vol. 91, 1989).

Molecular crystals

Useful introductions to molecular crystal properties are given by:

Molecular crystals, J. D. Wright (CUP, 1987).
Structure and properties of molecular crystals, M. Pierrot (Elsevier, 1990).
Magnetism and optics of molecular crystals, J. W. Rohleder and R. W. Munn (Wiley, 1992).

The journal *Molecular Crystals and Liquid Crystals Science and Technology* (Gordon and Breach) is split into four sections with molecular crystals covered in Section A. The section above on 'Molecular superconductors' also provides many relevant references.

Liquid crystals

Good introductory texts for the field of liquid crystals are given by *Liquid crystals*, S. Chandrasekhar (CUP, 1992) and *Liquid crystals: nature's delicate phase of matter*, P. J. Collings (Hilger, 1990). Recent developments are covered by *Physics of liquid crystalline materials*, I.-C. Khoo and F. Simoi (eds.) (Gordon and Breach, 1991). Liquid crystalline polymers are dealt with in *Liquid crystallinity in polymers: principles and fundamental properties*, A. Ciferri (VCH, 1991), and *Liquid crystalline polymers*, A. M. Donald and A. H. Windle (CUP, 1992). Section A of *Molecular crystals and liquid crystals science and technology* (Gordon and Breach) covers liquid crystals.

Langmuir-Blodgett films

For an introduction to the field, consult *Langmuir Blodgett films* G. G. Roberts (ed.) (Plenum, 1990). Articles on Langmuir Blodgett films appear in *Thin solid films* (Elsevier) and in Section C of *Molecular crystals and liquid crystals science and technology* (Gordon and Breach).

Magnetic materials

Magnetic materials research continues to be an active area, driven by technological applications and newly developed techniques for producing thin films and multilayers.

Recent introductions to magnetic materials are given in *Solid state magnetism*, J. Crangle (Edward Arnold, 1991), and *Introduction to magnetism and magnetic materials*, D. Jiles (Chapman and Hall, 1991).

Much information is available in various series of handbooks, e.g. *Handbook of magnetic materials*, E. P. Wohlfarth, K. H. J. Buschow (eds.), 6 volumes (North-Holland, 1988–1991); *Handbook on the physics and chemistry of the actinides*, A. J. Freeman, G. H. Lander and C. Keller (eds.), 6 volumes (North-Holland, 1984–1991); *Handbook on the physics and chemistry of rare earths*, K. A. Gschneider Jr. and L. Eyring (eds.), 15 volumes (North-Holland, 1978–1991), and *Concise encyclopaedia of magnetic and superconducting materials*, J. Evetts (ed.) (Pergamon, 1992).

The main journal covering the physics of magnetic materials is *Journal of Magnetism and Magnetic Materials* (North-Holland).

The International Conference on Magnetism is held every three years and covers the full range of magnetic materials and phenomena; recent conferences were held in San Francisco in 1985 (*Journal of Magnetism and Magnetic Materials*, volumes 54–57, North-Holland), Paris in 1988 (*Journal de Physique Colloque*, volume C8-1988, French Phys. Soc.) and Edinburgh in 1991 (*Journal of Magnetism and Magnetic Materials*, volumes 104–107, North-Holland).

Organic magnetism

Magnetism in organic materials is a rapidly developing field. Recent progress is covered in the proceedings of the 1992 Tokyo symposium *Chemistry and physics of molecular-based magnetic materials* published in *Molecular crystals and liquid crystals science and technology* (Gordon and Breach).

Magnetic films and multilayers

Magnetic film and multilayer research has recently developed into a very active field. A review is given by *Interface magnetism* (Surface Science Reports, vol. 12(2), T. Shinjo (North-Holland 1991). Useful symposium proceedings covering this area are *Magnetic films, multilayers and superlattices* (*Journal of Magnetism and Magnetic Materials*, volume 93, North-Holland, 1990) and the proceedings of a NATO Advanced Study Institute *Science and technology of nanostructured magnetic materials*, G. C. Hadjipanayis and G. A. Prinz (NATO-ASI: B 259, Plenum, 1991).

CHAPTER EIGHTEEN

Semiconductor physics

R. J. NICHOLAS

Introduction

The field covered by the term 'semiconductor physics' is rather vague for two reasons: firstly, the study of the physics of semi-conducting materials is merely a part of the wider field of condensed matter physics in general, and many of the properties, techniques and problems of semiconductor physics are repeated for all types of solid. Wide band-gap materials, such as diamond for example, often make an appearance at semiconductor conferences, although they would more usually be thought of as insulators. On the other hand, heavily doped 'degenerate' semiconductors and semimetals frequently need to be treated as metals. One result of this is that there is no well-defined section of the *Physics Abstracts Classification Scheme*, which can be said to refer solely to semiconductors in a comprehensive way, and one must be prepared to look at most of the sections on condensed-matter physics.

The second cause of difficulty is that there is a diffuse boundary into electronics, engineering and materials science. This is further complicated by the huge commercial and military significance of semiconductor technology, which means that certain areas begin to be affected by the constraints of confidentiality. A few years ago it would have been easy to say that the physicist explained how the transistor worked, but that its production and use would be the province of the materials scientist and the electronic engineer. Such a distinction is no longer possible, since the vast range of different devices now being designed and produced require the integration of several disciplines, and in many cases this is probing whole new fields in condensed matter physics.

The consequences of these remarks will be seen in the following pages, where the emphasis has been placed upon particular subject areas, without worrying too much about the intrusion of a few materials not normally thought of as semiconductors, and spreading out into materials science and engineering where they are relevant.

The selection of literature given below was done with two criteria in mind. It is hoped that the selection given will cover most areas of interest, and will do this in a readable and informative way. The other important factor was to choose works which would give a good starting point for a much more extensive search through the literature. Consequently, the age of a work has been an important factor, and so one or two excellent, if a little elderly, texts have been omitted through reasons of space.

Handbooks and review series

The outstanding reference work in which to find numerical data on the properties of semiconducting compounds is still the excellent work in the Landolt-Börnstein series: *Numerical data and functional relationships in science and technology, New series, Group III: Crystal and solid state physics*, vol. 17: *Semiconductors*, O. Madelung *et al.* (eds.) (Springer) subdivided into the following volumes:

17a: *Physics of group IV elements and III-V compounds* (1982)
17b: *Physics of II-VI and I-VII compounds, semimagnetic semiconductors* (1982)
17c: *Technology of Si, Ge and SiC* (1984)
17d: *Technology of III-V, II-VI and non-tetrahedrally bonded compounds* (1984).
17e: *Physics of non-tetrahedrally bonded elements and binary compounds I* (1983)
17f: *Physics of non-tetrahedrally bonded binary compounds II* (1983)
17g: *Physics of non-tetrahedrally bonded binary compounds III* (1984)
17h: *Physics of ternary compounds* (1984)
17i: *Special systems and topics* (1984)

For those materials which have been treated, the EMIS (Electronic Materials Information Service) of INSPEC (IEE) has published a very useful series of reviews of data, references and discussion. To date, these cover: *Gallium arsenide* (2nd. ed, 1990), *Mercury cadmium telluride* (1987), *Silicon* (1988) and *Amorphous silicon* (2nd. ed, 1990) and *Indium phosphide* (1991).

Another very useful reference work, containing a large number of separate articles by experts in their various fields is the *Handbook on semiconductors*, edited by T. S. Moss (North-Holland, 1980–1982), in four volumes covering:

1 *Band theory and transport properties* (ed. W. Paul, 1982)
2 *Optical properties of solids* (ed. M. Balkanski, 1980)
3 *Materials, properties and preparation* (ed. S. P. Keller, 1980)
4 *Device physics* (ed. C. Hilsum, 1982)

These have recently been revised, and a second edition will be published in 1993.

The continuing series *Semiconductors and semimetals*, edited by R. K. Willardson and A. C. Beer (Academic, beginning in 1966) and now E. Weber, plus additional guest editors for each volume, contains some excellent review articles covering the whole range of semiconductor physics which are often the best place to find a review of a particular field; with the early volumes having some, by now, classic accounts of materials, methods and techniques. Recent volumes keep up the very high standard, as in:

Vol. 25 *Diluted magnetic semiconductors* (eds. J. K. Furdyna and J. K. Kossut, 1988)
Vol. 26 *III-V Compound semiconductors and semiconductor properties of superionic materials* (eds. R. K. Willardson and A. C. Beer, 1988)
Vol. 27 *Highly conducting quasi-one-dimensional organic crystals* (ed. E. M. Conwell, 1988)
Vol. 28 *Measurement of high speed signals in solid state devices* (ed. R. B. Marcus, 1990)
Vol. 29 *Very high speed integrated circuits: gallium arsenide LSI* (ed. T. Ikoma, 1990)
Vol. 30 *Very high speed integrated circuits: heterostructures* (ed. T. Ikoma, 1990)
Vol. 31 *Indium phosphide: crystal growth and characterisation* (eds. R. K. Willardson and A. C. Beer, 1990)
Vol. 32 *Strained layer superlattices: physics* (ed. T. P. Pearsall, 1990)
Vol. 33 *Strained layer superlattices: materials science and technology* (ed. T. P. Pearsall, 1990)
Vol. 34 *Hydrogen in semiconductors* (eds. J. I. Pankove and N. M. Johnson, 1991)
Vol. 35 *Nanostructured systems* (ed. M. Reed, 1992)
Vol. 36 *Spectroscopy of semiconductors* (eds. D. G. Seiler and C. Littler, 1992)
Vol. 37 *The mechanical properties of semiconductors* (eds. K. T. Faber and K. J. Malloy, 1992)

Vol. 38   *Imperfections in III-V materials*. (Illus.) (eds. E. R. Weber, R. Eicke and R. K. Willardson, 1992)
Vol. 39   *Minority carriers in III-V semiconductors*. (Illus.) (eds. R. K. Willardson *et al.*, 1993).

Lists of all the past articles will be found in the latest volumes. Two other useful, although less specific, series are the Springer books published as *Topics in applied physics and solid-state sciences*. The majority of the Springer books referred to in the following pages have been published as volumes in one or the other of these two series. Another good series which frequently contains useful articles which are relevant, is *Solid state physics* edited by F. Seitz *et al.* (Academic, beginning in 1955); again earlier articles are listed in the later volumes.

Another excellent series often featuring semiconductor topics is *Modern problems in condensed matter sciences*, general editors V. M. Agranovich and A. A. Maradudin (North-Holland), which includes recent volumes on *Nonradiative_recombination in semi-conductors*, eds. V. N. Abakumov, V. I. Perel and I. N. Yassievich, vol. 33 (1991), *Mesoscopic phenomena in solids*, eds. B. L. Altshuler, P. A. Lee and R. A. Webb, vol. 30 (1991), and *Landau level spectroscopy*, eds. G. Landwehr and E. I. Rashba, vol. 27, pts I and II (1991).

Journals

Following the remarks in the introduction, it is not surprising to find that semiconductor work is spread over a wide range of journals. The 'purer' type of work usually appears in general solid state journals, such as:

Physical Review B15 (the issue appearing on 15th. of each month) and
Physical Review Letters (AIP)
Journal of Physics C. – Solid State Physics (IOP)
Physica Status Solidi a &b (Akademie)
Solid State Communications (Pergamon)
Journal of the Physical Society of Japan (Phys. Soc. Jpn)
Journal of the Physics and Chemistry of Solids (Pergamon)
Soviet Physics – Solid State (in translation by AIP).

More specialized journals include:

Semiconductor Science and Technology (IOP)
Superlattices and Microstructures (Academic)
Journal of Crystal Growth (North-Holland)
Soviet Physics – Semiconductors (in translation by AIP)

Journal of Luminescence (North-Holland)
Surface Science (North-Holland)
Journal of Non-crystalline Solids (North-Holland)
Philosophical Magazine A and *B* (Taylor and Francis), which has a
   strong interest in disordered systems.

The more applied work tends to appear in:

Journal of Applied Physics and Applied Physics Letters (AIP)
Solid State Electronics (Pergamon)
Electronics Letters (IEE)
IEEE Transactions on Quantum Electronics and Electron Devices
   (IEEE)
Japanese Journal of Applied Physics (Phys. Soc. Jpn)
Journal of Vacuum Science and Technology A and B (AIP).

Conference proceedings

A good way to gauge the areas of interest and current level of
progress in a particular subject is to examine the proceedings of
recent conferences devoted to a particular area or topic. The most
comprehensive semiconductor meeting is the biennial 'International
Conference on the Physics of Semiconductors'. The last three meetings to be held have published their proceedings as:

19th International Conference on the Physics of Semiconductors,
   Warsaw, 1988, Zawadzki, W. (ed.) (Institute of Physics, Polish
   Academy of Sciences, Warsaw, 1989).
The Physics of Semiconductors, Thessaloniki (1990), Anastassakis,
   E. M. and Joannopoulos, J. D. (eds) (World Scientific, 1990).
Physics of Semiconductors, Beijing 1992, to be published.

Other regular conference series are given below, together with the
details of the published proceedings of the most recent event.

*Physics of Narrow Gap Semiconductors, Southampton, 1992, Semi-
   conductor Science and Technology*, vol. 8 (1993).
Gallium Arsenide and Related Compounds, 1992, Karuizawa, Japan
   (I.O.P, Bristol, UK, 1993).
Semi-Insulating III-V Materials, Mexico, 1992, ed. C. J. Miner, W.
   Ford and E. R. Weber (IOP, Bristol, U.K., 1993).
Amorphous Silicon Materials and Solar Cells, Colorado, 1991, ed. B.
   Stafford (AIP Conf. Proc. 234, AIP, New York, 1991).
Wide Gap II-VI Semiconductors, ed. T. K. Moustakas, J. I. Pankove
   and Y. Hamakawa (Symposium Proc. Ser. vol 242, M.R.S., U.S.A.,
   1992).
*Electronic Properties of Two-Dimensional Systems, Nara, Japan,
   1991*, ed. M. Saitoh, Surf. Sci. 263 (1992).

Modulated Semiconductor Structures, Nara, Japan, 1991, eds. Hiyamizu, S. and H. Nakashima, Surf. Sci. 267 (1992).

17th International Conference on Hot Carriers in Semiconductors, eds. C. Hamaguchi and M. Inoue, *Semiconductor Science and Technology*, 3B vol. 7 (1992).

European Solid State Device Research Conference, 1990, Nottingham, eds. W. Eccleston and P. J. Rosser (IOP, Bristol, UK, 1990).

Applications of High Magnetic Fields in Semiconductor Physics 1992, Chiba, Japan, N. Miura. (ed.) *Physica b*(1993).

Defect Recognition in Semiconductors before and after Processing, Wilmslow, UK, 1991, eds. M. R. Brozel and D. J. Stirland (*Semiconductor Science and Technology*, 1A vol. 7, 1992).

Insulating Films on Semiconductors, Liverpool, 1991, eds. W. Eccleston and M. J. Uren (IOP, Bristol, UK, 1991).

Microscopy of Semiconducting Materials 1991, Oxford, ed. A. G. Cullis (IOP Conference Series, 117, 1991).

Metal Organic Vapour Phase Epitaxy, 1992, Stringfellow, G. B. and Colemann, J. J. (eds.) *J. Cryst. Growth*, vol. 124(1–4) (1992).

Molecular Beam Epitaxy, 1991, Tu, C. W. and Harris, J. S. (eds.) *J. Cryst. Growth*, vol. 111(1–4) (1991).

General texts

Much of the basic physics necessary to describe semiconductors is part of the general understanding of solid state physics. This would cover such important areas as band structure, elementary excitations, coupling between electrons, phonons and photons, impurities and localized states and group theory. Such material may be found in the classic graduate student texts such as: *Quantum theory of solids* by C. Kittel (Wiley, 1963), *Principles of the theory of solids* by J. M. Ziman (CUP, 2nd edn.), *Quantum theory of the solid state* by J. Callaway (Academic, 2nd edn., 1991) or in the excellent *Introduction to solid state theory* by O. Madelung (Springer, 1978) (Springer Series in Solid-state Sciences, 2).

Of the books devoted entirely to semiconductors some good examples are:

Smith, R. A. *Semiconductors* (CUP, 2nd edn., 1978), which is a well-updated classic text.

Seeger, K. *Semiconductor physics: an introduction* (Springer, 5th edn., 1991) (Springer Series in Solid-state Sciences, 40).

Ridley, B. K. *Quantum processes in semiconductors* (OUP, 1982).

It should be pointed out, however, that for partly historical reasons all three of these texts probably overemphasize the importance of transport phenomena. The treatment of *Semiconductor statistics*, by J. S. Blakemore (Pergamon, 1962), is very comprehensive and to date has no rivals.

There is now a large number of textbooks devoted to the operation of semiconductor devices; of these, *Solid state electronic devices* by B. G. Streetman (Prentice-Hall, 2nd edn., 1980) provides a simple introduction, and includes an extensive bibliography. The most comprehensive and useful book in this area, particularly since the publication of its second edition, is the *Physics of semiconductor devices* by S. M. Sze (Wiley, 2nd edn., 1981). An excellent book treating the new applications which have developed from the growth of semiconductor heterostructures is *Quantum semiconductor structures – fundamentals and applications* by C. Weisbuch and B. Vinter (Academic, 1991). This also contains an excellent recent bibliography of the field.

An interesting attempt to summarize the field of semiconductor physics research, and to try to encourage common themes with areas such as superconductivity and surface physics can be seen in the proceedings of the NATO forum, *Highlights in condensed matter physics in the eighties and future prospects*, L. Esaki (ed.) (NATO-ASI vol. B285, Plenum, 1992).

Specialized texts

This section comprises a compilation of more specialized works, which cover specific topics. These are frequently in the form of collections of articles by experts, such as the proceedings of symposia or summer schools. The different areas of interest have been grouped under the following headings:

Transport and hot-electron phenomena
Optical properties
Impurities
Surfaces
Band structure
Disordered systems
Semiconductor heterostructures
Devices
Growth
Particular materials.

Transport and hot-electron phenomena

Nag, B. R. (1980) *Electron transport in compound semiconductors* (Springer) (Springer Series in Solid-state Sciences, 11). A good general text.

Tauc, J. and Henish, H. (1962)*Photo and thermoelectric effects in semiconductors* (Pergamon).

Conwell, E. M. (1967) *High field transport in semiconductors* (Academic) (Solid State Physics Series, supplement 9).

Hobson, G. S. (1974) *The Gunn effect* (Clarendon).

Ferry, D. K. and Jacoboni, C. (eds.) (1992) *Quantum transport in semiconductors* (New York, Plenum). The lecture notes from a summer school updated and expanded to cover most aspects of non-equilibrium transport.

Devreese, J. T. (ed.) (1972) *Polarons in ionic crystals and polar semiconductors* (North-Holland). The proceedings of a NATO Advanced Study Institute on Frohlich polarons and the electron-phonon interaction in polar semiconductors.

Optical properties

More general texts include:

Pankove, J. I. (1975) *Optical processes in semiconductors* (Dover).

Greenaway, D. L. and Harbeke, G. (1968) *Optical properties and band structure of semiconductors* (Pergamon).

Abeles, F. (ed.) (1972) *Optical properties of solids* (North-Holland).

Seraphin, B. O. (ed.) (1976) *Optical properties of solids: new developments* (North-Holland).

Specific works which are very useful include:

Bube, R. H. (1992) *Photoelectronic properties of semiconductors* (CUP).

Cho, K. (ed.) (1979) *Excitons* (Springer).

Cardona, M. and Guntherodt, G. (eds.) *Light scattering in solids*, in 4 vols (Springer, 1975–1983) (Topics in Applied Physics, 8, 50, 51, 54). This covers almost all aspects of Raman and Brillouin scattering, with much of the discussion directly related to semiconductors.

Cardona, M. and Ley, L. *Photoemission in solids I and II* (Springer, 1978, 1979) (Topics in Applied Physics, 26 and 27).

D'Andrea, A. *et al.* (1992) *Optics of excitons in confined systems* (IOP) is the proceedings of a meeting devoted to a wide range of materials and structures.

Moss, T. S. *et al.* (1973) *Semiconductor optoelectronics* (Butterworths).

Landsberg, P.T. (1992) *Recombination in semiconductors* (CUP).

Devreese, J. T. and Peeters, F. (ed.) (1984) *Polarons and excitons in polar semiconductors and ionic crystals* (Plenum). Another NATO summer school covering Frohlich polarons, excitons and electron-hole condensation.

Impurities

Milnes, A. G. (1973) *Deep impurities in semiconductors* (Wiley).
Jaros, M. (1982) *Deep levels in semiconductors* (Hilger).
Stoneham, A. M. (1975) *Theory of defects in solids* (Clarendon).
Herman, H. (ed.) (1980) *Treatise on materials science and technology, vol. 18 – Ion implantation*, edited by J. K. Hirbonen (Academic).
Jantsch, W. and Stradling, R. A. (eds.) (1991) D(X) centres and other metastable defects in semiconductors, *Semiconductor Science and Technology*, 10B, vol. 6 the proceedings of a school devoted entirely to this topic.

Surfaces

Prutton, M. (1983) *Surface physics* (Clarendon, 2nd edn.) (Oxford Physics Series, 11). A short and simple introduction, with a good bibliography.
Blakely, J. M. (1973) *Introduction to the properties of crystal surfaces* (Pergamon) (International Series on Materials Science and Technology, 12).
Garcia-Moliner, F. and Flores, F. (1979) *Introduction to the theory of solid surfaces* (CUP).

Band structure

Long, D. (1968) *Energy bands in semiconductors* (Wiley). A simple and clear introduction.
Smith, R. A. (1969) *Wave mechanics of crystalline solids* (Chapman and Hall, 2nd edn.). A graduate-level textbook.
Phillips, J. C. (1973) *Bonds and bands in semiconductors* (Academic).
Haydock, R., Inglesfield, J. E. and Pendry, J. B. (1991) *Bonding and structure of solids* (CUP) not really a semiconductor book, but an up to date summary of this field.
Callaway, J. (1964) *Energy band theory* (Academic) (Pure and Applied Physics Series, 16).
Cornwell, J. F. (1969) *Group theory and electronic energy bands in solids* (North-Holland).
Jones, H. (1975) *The theory of Brillouin zones and electronic states in crystals* (North-Holland, 2nd edn.).
Cohen, M. L. and Chelikowsky, J. R. (1988) *Electronic structure and optical properties of semiconductors* (Springer) (Series in Solid State Sciences, vol 75). A very comprehensive survey of the band structures of a wide variety of semiconductors.

Heterostructures and superlattices

This has been the most rapidly growing area in recent years, with a huge growth in both academic and commercial interest. General introductory articles and texts are:

Jaros, M. (1990) *Physics and applications of semiconductor microstructures* (Clarendon). A good basic introduction.

Weisbuch, C. and Vinter, B. (1991) *Quantum semiconductor structures* (Academic).

Klipstein, P. K. and Stradling, R. A. (eds.) (1990) *Growth and characterisation of semiconductors* (IOP). A wide ranging introduction to both growth and assessment techniques.

Electronic properties of two-dimensional systems (Ando, T. *et al.*, 1982). A somewhat elderly but still outstanding review article which concentrates mainly on the properties of silicon inversion layers.

Beeby, J. L. (ed.) (1991) *Condensed systems of low dimensionality* (Plenum) (NATO-ASI, vol. B253) has a very comprehensive selection of work covering a wide range of systems and properties, including transport, optics, strained layers and layered materials.

Specialist works tend to age quite rapidly in this area, so it is usually a good idea to consult the recent lists of NATO-ASI proceedings, and the more recent issues of the review series discussed above. Some good recent works include:

Bastard, G. (1988) *Wave mechanics applied to semiconductor heterostructures* (Les Editions de Physique, Les Ulis, France). An excellent book which gives many useful quantum mechanical treatments of important structures.

Long, A. R. and Davies, J. H. (1991) *Physics of nanostructures* (IOP). The proceedings of a NATO-ASI on this important area.

Prange, R. E. and Girvin, S. M. (1987) *The quantum Hall effect* (Springer).

Disordered systems

Mott, N. F. and Davis, E. A. (1979) *Electronic processes in non-crystalline materials* (Clarendon, 2nd edn.). An excellent and comprehensive review covering both theory and experiment.

Mott, N. F. (1993) *Conduction in non-crystalline materials* (Clarendon, 2nd edn.).

Brodsky, M. H. (ed.) (1986) *Amorphous semiconductors* (Springer, 2nd edn.) (Topics in Applied Physics, 36).

Kamimura, H. and Aoki, H. (1989) *The physics of interacting electrons in disordered systems* (Clarendon).

Nagaoka, Y. and Fukuyama, H. (eds.) (1982) *Anderson localization* (Springer) (Proceedings of the 4th Taniguchi International Symposium, 1981) gives a good overview at a rather advanced level.

Friedman, L. R. and Tunstall, D. P. (eds.) (1978) *The metal-non-metal transition in disordered systems* (Proceedings of the 19th Scottish Universities Summer School in Physics) (SUSSP).

Street, R. A. (1991) *Hydrogenated amorphous silicon* (CUP) is a good general introduction to both the basics and the applications of this important material.

Devices

Some good introductory texts are:

van der Ziel, A. (1976) *Solid state physical electronics* (Prentice-Hall, 3rd edn.).

Milnes, A. G. (1979) *Semiconductor devices and integrated electronics* (Reinhold).

Yang, E. S. *Fundamentals of semiconductor devices* (McGraw-Hill, 1978); *Microelectronic devices* (McGraw-Hill, 1988).

More specialized works are:

Casey (Jr.), H. C. and Panish, M. B. (1978) *Heterostructure lasers, parts A and B* (Academic).

Kressel, H. (ed.) (1990) *Semiconductor devices for optical communication* (Springer, 2nd (updated) edn.).

Barbe, D. F. (ed.) (1980) *Charge-coupled devices* (Springer) (Topics in Applied Physics, 38).

Orton, J.W. and Blood, P. (1990) *The electrical characterization of semiconductors: measurement of minority carrier properties* (Academic).

Herman, M. A. (ed.) (1980) *Semiconductor optoelectronics* (Wiley). Lectures given at 2nd International School on Semiconductor Optoelectronics, Wladyslawowo, 1978.

Nicollian, E. H. and Brews, J. R. (1982) *MOS (metal oxide semiconductor) physics and technology* (Wiley). A very comprehensive textbook and handbook.

Growth

Astles, M. G. (1990) *Liquid-phase epitaxial growth of III-V compound semiconductor materials and their device applications* (IOP).

Chang, L. L. and Ploog, K. (eds.) (1985) *Molecular beam epitaxy* (NATO-ASI) (Nijhoff).

Dryburgh, P. M., Cockayne, B. and Barraclough, K. E. (1987) *Advanced crystal growth* (Prentice-Hall). Proceedings of a summer school covering many different aspects of both bulk and epitaxial growth techniques.

Stringfellow, G.B. (1989) *Organometallic vapor phase epitaxy* (Academic).

Suntola, T. and Simpson, M. (1988) *Atomic layer epitaxy* (Blackie).

Particular materials

III-V materials form the specific subject matter for the first four volumes of the series *Semiconductors and semimetals*, edited by R. K. Willardson and A. C. Beer (Academic), and make very frequent appearances in subsequent volumes. For numerical data on almost any material the reader is referred to the Landolt-Börnstein volumes on semiconductors and the EMIS volumes on specific materials, as mentioned in the 'Handbooks' section above.

Some works devoted to specific families or classes of semiconductors are

Pearsall, T. P. (ed.) (1982) *GaInAsP alloy semiconductors* (Wiley).

Aven, M. and Prener, J. S. (eds.) (1967) *Physics and chemistry of II-VI materials* (North-Holland).

Jain, M. (ed.) (1992) *II-VI semiconductor compounds* (World Scientific).

Shay, J. L. and Wernick, J. H. (1975) *Ternary chalcopyrile semiconductors: growth, electronic properties and applications* (Pergamon).

Gerlach, E. and Grosse, P. (eds.) (1979) *Proceedings of the international conference on the physics of selenium and tellurium* (Königstein, Springer).

Ravich, Yu. I., Efimova, B. A. and Smirnov, I. A. (1970) *Semiconducting lead chalcogenides* (Plenum).

Nimitz, G., Schlicht, B. and Dornhaus, R. (1983) *Narrow gap semiconductors* (Springer). In two parts, the first on IV-VI materials, the second on HgCdTe (Springer Tracts in Modern Physics, 98).

Reference

Ando, T., Fowler, A. B. and Stern, F. (1982) *Reviews of Modern Physics*, **54** (2), 437–672.

CHAPTER NINETEEN

Grey literature

J. P. CHILLAG

Other chapters of this book use a subject approach to various types of literature in a number of specialized fields. Here, however, it is intended to cover just one type of literature relevant to all the subject areas discussed. This material is referred to as 'grey literature'.

A significant and increasing amount of information will be found in material described by this term. Grey literature has been with us for a long time, though perhaps not in such volume as now. 'Non-conventional' and 'unpublished literature' are just two other terms by which this material is known.

A widely accepted definition of 'grey literature' is that it is literature which is not readily available through normal bookselling channels, and it is therefore difficult to identify and obtain (Wood, 1990a). Examples of grey literature include report literature, technical notes and specifications, conference proceedings and preprints, translations, official publications, supplementary publications and data, trade literature, etc.

It should be recognized, however, that not all material in all of these categories is 'grey' as defined above: conference proceedings may be published as books or in journals, many official publications may be commercially available, and so are, for example, cover-to-cover translated journals.

Poor availability is only one characteristic of grey literature. Others include poor bibliographic information and control, non-professional layout and format, and low print runs.

As the definition of grey literature is vague, it is not easy to assess the output. In the USA, it can be gauged by the fact that the National Technical Information Service (NTIS), the National Aeronautics and Space Administration (NASA) and the US Department of

Energy (DOE) between them announce annually some 70 000 unique grey documents, mostly reports, conference papers and proceedings. A more than fair share of this output will be of interest to those working in various fields of physics.

Some specialized bibliographic guides to certain categories of grey literature exist. These are usually produced by central clearing houses linking announcement, with (or without) abstracts, to document delivery services. Most major printed guides of this type are offered with parallel online search opportunities, and in a number of instances CD-ROMs, of which later. However, much of the grey literature produced is seriously lacking bibliographic control.

Workers trying to use the physics literature will be glad to know that grey literature in their fields of interest is probably more accessible than that in many other disciplines. It is hoped that the details given below, describing grey literature by category of material, will be helpful.

Reports

The beginnings of report literature are generally thought of as coinciding with the end of the Second World War, soon after which the Office of Technical Services (OTS) and the US Atomic Energy Commission (USAEC) – predecessors of the present NTIS and DOE, respectively – started to release technical reports on a large scale. However, one should go back much further. In Britain, forerunners of the Aeronautical Research Council (defunct since 1980) started issuing their *Reports and Memoranda* in 1909, the US National Bureau of Standards (NBS) (now the National Institute for Standards and Technology – NIST) began their *Technologic Papers* series in 1910. The National Advisory Committee for Aeronautics (NACA) – now NASA – was a relative latecomer to this field in 1915.

However, as much in physics research changes constantly and rapidly, one should turn to the present day and to the more current sources and retrieval tools for reports.

Reports can be defined as documents which contain results of, or progress made with, research and/or development work, investigations and surveys. Typically, the report is issued by the funding or performing body, it is usually not commercially published and its contents escape any formal refereeing. It may be only available in microform and it is usually identified by unique numeric or alphanumeric codes (Wood, 1990b and Auger, 1989). Users of report literature should be aware of the importance of quoting *all* report numbers (alpha-numerics) at their disposal together with any other bibliographic information. Full references given by the authors of today will be of immense help for access by tomorrow's readers. The alpha-numeric identification code usually provides a unique identi-fication: NASA-CR = National Aeronautics and Space Administra-tion – Contractor Report, JINR-R = Joint Institute for Nuclear Research – Report. A valuable aid for identification is the *Report Series Codes Dictionary* (Aronson, 1986). Similar, more current information is provided in 'authority lists' produced, for example, by the International Nuclear Information System *INIS Authority List for Corporate Entries and Report Prefixes,* (INIS-6, Rev 23), by NTIS, and by the System for Information on Grey Literature in Europe (SIGLE).

Bibliographic control of report literature is better in the USA than elsewhere. There, the four main agencies concerned with the collec-tion, abstracting and announcement, and supply of reports are NTIS, NASA, DOE and the Defense Technical Information Centre (DTIC). NTIS, having progressed through a number of organiza-tional name changes since the early days of OTS, produces its fort-nightly *Government Reports Announcements and Index (GRA & I)*; collecting information and documents from 4700 organizations listed in *Corporate Author Authority List* (CAAL) (Kane, 1987), NTIS is involved in all subject fields. However, in addition to material accessed from research departments and establishments, etc., well known in the physics fraternity, its high coverage in the physics field stems from the fact that *GRA & I* also includes a very high proportion of the reports originating in NASA and the DOE. *AD Reports*, the reports released by DTIC for more public use, are another significant part of listings in *GRA & I*. Each issue of *Government Reports Announcements and Index* contains multi-access indexes. Annual indexes and various cumulative indexes, including a 15–year *NTIS Title Index* (keyword, author, report number) on microfiche (1964–1978) (plus biennial supplements), make 'manual' searching easy. The NTIS database can also be accessed online from 1964 and on CD-ROM from 1980 onwards. Most documents listed can be obtained from NTIS singly or within subject profiles as *Selected Research in Microfiche (SRIM)*. The microfiches and enlargements are also available from a number of centres in the USA, the British Library Document Supply Centre (BLDSC) in the UK and more selectively in other countries. Other NTIS products include weekly *Abstracts Newsletters* (in 28 subject groups, including physics), monthly *NTIS Technical Notes* and *Published Searches (NTIS-PS)* of the NTIS Bibliographic Data Base. One such is *Handbooks on Physics* (73 citations).

Twice a month NASA publishes *Scientific and Technical Aerospace Reports (STAR)* with a coverage implied by its title. Reports listed in *STAR* are usually shown as 'availability NTIS', a computer tag sometimes misapplied. The tag should certainly be valid for entries prefixed by an asterisk, e.g. *N92–26265, and many others indicating microfiche with a # symbol. Again, the BLDSC and other centres are likely supply points and the BLDSC has microfiche and hard copy holdings of NASA and NACA reports right back to 1915. NACA and early NASA reports are listed in *Index of NACA Technical Publications 1915–1949* and updates. *International Aerospace Abstracts (IAA)* is described later.

The first abstracting journal regularly listing reports on a significant scale was *Nuclear Science Abstracts (NSA)*, started in 1947. Changing its title in 1976 to *ERDA Research Abstracts*, and again in 1977 to *Energy Research Abstracts*, *ERA* covers not only grey literature (reports, conferences, etc.) but also the published literature on all energy topics (nuclear, fossil, solar, wind, etc.). Coverage to 1976 was only in the field of nuclear energy. It has indexes in every issue as well as annual indexes, and the Cumulative Report Number Index to volumes 1–18 and later cumulations are amongst the most-thumbed indexes in many physics laboratories and libraries. In the early 1980s, the numbering system used has undergone a change from the well known alpha-numerics to a DE number, i.e. BNL-52317 has now also the number DE 92014911. The documents, microfiche and hard copy bear both numbers, and retrieval is made easier if users quote both references.

At one time the then USAEC maintained a very large number of depository libraries mainly in the USA, Canada and the UK. The number of such centres has been greatly reduced, their character has been changed and their geographical spread improved. The centres are listed on the inside covers of *Energy Research Abstracts* and include organizations such as INIS and the BLDSC. Useful Department of Energy tools to identify report series include *Energy Information Database – Report Number Codes* (Wallace). Public online access to the DOE Energy Data Base (EDB) is more restricted than similar access to the NTIS and NASA databases.

Other useful listings of American reports include those of government agencies such as *Publications of the National Bureau of Standards* (NBS) starting with a 1901–1947 cumulation. The latest multi-annual cumulation is published by NBS' successor: the Institute of Standards and Technology (NIST) as *NIST Special Publication NIST SP 790* = PB136507 and covers the years 1977–87. Annual updates have been issued as *NBS/NIST Special Publication 305–Suppl-nn*. Others include the Environmental Protection Agency's *EPA Cumulative Bibliography 1970–1976* and the quarterly *EPA Publications Bibliography*; the Electric Power Research Institute's *EPRI Guide*, a cumulative index updated quarterly; and *Selected RAND Abstracts* published by the independent Rand Corporation.

In the UK, the British Library Document Supply Centre (BLDSC), known in earlier years as BLLD (British Library Lending Division) and before that the National Lending Library for Science and Technology (NLLST), has always paid particular attention to grey literature and has built up over the years a report collection now numbering over three million documents. To improve bibliographic control of British grey literature and to help to promote its use, *British Research and Development Reports (BRDR)* was launched in 1969, later to become *British Reports,Translations and Theses (BRTT)*. The reports listed form part of the UK and Republic of Ireland input into the System for Information on Grey Literature in Europe (SIGLE), to be described below. *BRTT* and SIGLE subject categories and those in the major US 'tools' are based on COSATI (Committee on Scientific and Technical Information) classifications, but over the years significant changes have been made. As can be seen in *SIGLE Manual Part 2: Subject Category List* changes in *BRTT* and SIGLE are most noticeable in subject groups: energy, physics and environment (EAGLE, 1991). Each monthly part of *BRTT* has a keyterm index. Author, report number and keyterm indexes are cumulated quarterly and annually on 48x comfiche.

Unrestricted documents listed in *R & D Abstracts*, which ceased in 1981 – duplicating to a large extent BLDSC holdings – have been transferred to, and are available from, the British Library.

There are also other lists of reports in particular subject fields. The *UKAEA List of Publications Available to the Public*, for 34 years published monthly and cumulated annually, ceased publication with the 1989 volume, as has Rutherford Appleton Laboratory's *Catalogue of Rocket and Satellite Data* and various listings of research associations, e.g. the Electrical Research Association (ERA), the British Hydrodynamics Research Association (BHRA), etc. Research association reports are often restricted – at least initially – to their membership.

In Germany, the Fachinformationszentrum Energie, Physik, Mathematik (FIZ), together with Hannover University Library and Technical Information Bureau (UB & TIB) prepares the input for *Forschungsberichte aus der Technik und Naturwissenschaften – Reports in the Fields of Science and Technology*. Research centres and agencies produce their own listings such as the 1965–1983 cumulation of *BMFT Forschungsberichte*, distributed for the Bundesministerium für Forschung und Technologie by FIZ. The

BMFT series ceased publication in 1987. Various Forschungs-vereinigungen (research associations) publish lists of reports they produce or disseminate.

From records created by the Centre Nationale de la Recherche Scientifique – Institut de l'Information Scientifique et Technique (CNRS-INIST), the databases PASCAL and SIGLE list French reports and other grey literature originating in France. CEDOCAR – the Centre de Documentation de l'Armament – produces similar 'bulletins signalétiques' on documents which are more defence oriented.

FIZ together with UB & TIB, CNRS-INIST and DG XIII of the European Communities (see below) are responsible for SIGLE input from Germany, France and the European Communities, respectively.

International bodies producing and disseminating report literature include INIS at the International Atomic Energy Agency (IAEA). It produces twice a month *INIS Atomindex* which, apart from published literature in the nuclear field, concentrates heavily on nuclear energy reports and conference material worldwide. The published literature listed should be requested through the users' own library; the reports in most instances are available on microfiche from INIS, the BLDSC and a number of other centres. The *INIS Authority List* is a useful aid for the identification of report series.

Through its DG XIII: the Directorate General Telecommunications, Information Industries and Innovation, the European Communities also commission, produce and disseminate a large number of reports. These are listed in the monthly *Euro-Abstracts, Scientific and Technical Publications: Section 1–Euratom and EEC* and *Section 2–Steel, Coal and Social Research*, respectively. The publication has annual indexes, but a more comprehensive index to the EUR reports is the *Catalogue EUR Documents, 1968–1982* (EUR 7500 and supplement), revived from 1989 onwards as the annual *Research Publications*. EUR reports also appear in the EC Publication Office's more general multi-annual *Publications 1985–1991*, the Commission's SCAD and CATEL databases, and on CD-ROMs.

Many other series of European Communities documents also fall into the grey literature category. Some will be listed in various newsletters and announcements which the European Communities produce, but those of a serial nature are included in *European Communities publications: a guide to British Library resources*.

Before the break-up of the former USSR and the other political changes in Central and East Europe, large numbers of reports emanated from the Joint Institute for Nuclear Research (JINR) and the national laboratories and institutions of the member countries of the now defunct Council for Mutual Economic Cooperation (COMECON). At the end of 1992 some of these centres still exist under their 'old' name, others function under new banners and many more disappeared altogether. Research, albeit on a reduced scale, is continuing and reports are produced and listed in institution pamphlets and brochures. Researchers in the English-speaking countries may find it more convenient to use abstracting tools such as *INIS Atomindex*. Documents from these sources are likely to be available from the BLDSC, from INIS or NTIS.

Reports from the European Organization for Nuclear Research (CERN) are announced in the annual *CERN Publications*, and also in the other earlier-mentioned indexes.

The Advisory Group for Aeronautical Research and Development, better known by the acronym AGARD, and its parent body, the North Atlantic Treaty Organization (NATO), are responsible for many reports relevant to research in physics. NATO Science Committee publications, better known as the *NATO ASI Series* are shown not only in publisher catalogues but also on the NATO-PCO DATABASE ON CD-ROM. AGARD reports, conference proceedings and preprints in particular, appear not only in *AGARD Index of Publications 1952–1970* and its periodic updates, but are also cited regularly in *Scientific and Technical Aerospace Reports (STAR)*. *STAR* is also the main announcement medium for documents of the European Space Agency (ESA).

Indexes containing large numbers of citations to reports and other grey literature in specialized fields are mentioned in other chapters of this book. Examples of these include *High Energy Physics Index* and *Computer Index of Neutron Data (CINDA)*. Commercially produced abstracting journals such as *Physics Abstracts* and *Engineering Index (EI)* are including an increasing proportion of report and other grey literature entries. *EI* report citations are sometimes included in *NTIS Published Searches*, some issues of which may be indicated entirely to *EI* citations.

The major 'tools' specializing in report abstracting are, however, the most likely sources for identifying and locating reports, and searching of these databases can usually be facilitated by online and/or CD-ROM access.

Dissertations and theses

Anyone who has tried to identify and locate dissertations and doctoral theses will know how apt the term grey literature is for this category of material. Boyer (1972) observed that since the doctoral dissertation embodies the results of extended research and is an original contribution to knowledge, it is surprising that the various guides to theses are overlooked by so many researchers as an information source. Announcement aids, online and now also CD-ROM access facilities to theses have improved since then, but the comments are still essentially true.

In discussing theses and dissertations and their availability, it is convenient to divide them into American, British and others. The majority of American dissertations have for many years been abstracted in *Dissertations Abstracts International (DAI)*. Relevant to readers here is *DAI – Section B – The Sciences & Engineering*, and its annual cumulations and indexes. There is a retrospective index to volumes 1–29 of *DAI* (1938 to June 1969), published in nine separate parts (part 1: mathematics and physics, part 9 is the author index). Produced by the same publishers, University Microfilms International (UMI), is the retrospective searching aid *Comprehensive Dissertations Index 1861–1972* as well as *American Doctoral Dissertations* and *Masters Abstracts*.

Many, but not all institutions whose dissertations are announced in *DAI*, etc. make their dissertations available through UMI. Some – for example, the University of Chicago and Harvard University – have to be approached directly. In the UK, many of the American dissertations are available through the BLDSC, which acquired from 1970 the total output of current dissertations in the UMI programme. However, rapidly escalating costs and improved delivery speeds from the USA led in 1978 to the cancellation of this subscription. Most requests for US dissertations are now satisfied by the BLDSC either from its stock, by on-demand purchase or through interlibrary loans.

The BLDSC is the main source for most of the post-1970 British doctoral theses. Before then, the only access for these used to be the department or library of the university where the thesis was presented. In 1970, in an attempt to promote the use of this material and relieve the universities of the document supply load, they were invited to send their doctoral theses to the BLDSC for microfilming and listing in *British Reports, Translations and Theses (BRTT)* (Smith, 1993). From a nucleus of two universities in 1970, the scheme grew to the current intake from over 70 universities. In 1992, following the restructuring of tertiary educational institutions in the UK, former polytechnics acquired university status, and the BLDSC receives doctoral theses from the 'new' universities, too. Theses announced in *BRTT* are available from BLDSC, which now holds some 100 000 British doctoral theses. Replacing earlier sectional subject catalogues of British doctoral theses available from the BLDSC, Information Publications International (IPI) published

BRITS Index 1971–1987, and thus far two later annual supplements. *Index to Theses (with Abstracts)*, which replaces the earlier *Index to Theses Accepted for Higher Degrees by the Universities of Great Britain and Ireland and the Council for National Academic Awards*, lists not only titles, but provides also brief abstracts to the theses. The theses listed are usually only available from the BLDSC, or in some cases from the university concerned. There is also a *Retrospective Index to Theses of Great Britain and the Republic of Ireland 1716–1950*; volume 4 of this work covers the physical sciences.

In France, theses have been recorded since 1884 in the *Catalogue des Thèses et Ecrits Academiques* and in supplement *D* of *Bibliographie de la France*; also, more recently, on the *TeleThèses* database, the SIGLE database and CD-ROM, as well as the CD THÈSES CD-ROM. Availability of scientific theses in France is from CNRS-INIST.

Since 1991, German theses are listed in the *Deutsche Nationalbibliographie Reihe H*. Until the reunification of Germany, such theses were included in the *Jahresverzeichnis der Hochschulschriften der DDR, der BRD und West Berlin*, bringing together material from both the former Federal Republic and the Democratic Republic of Germany. Theses in physics and science subjects originating in the Federal Republic appear from 1983 in *Forschungsberichte aus der Technik und Naturwissenschaften*. They are now included, together with similar material from the UK and France, in the SIGLE database.

Since 1990, Canadian theses are included in *DAI* (see preceding paragraphs). In 1984–1989, the theses were listed in *Canadian Theses* published by the National Library of Canada (NLC), which also provided access to the theses; but already long before that, Canadian theses were brought together under a special programme of the National Research Council of Canada (NRCC). The theses were available on microfiche with details listed in *Canadian Theses on Microfiche – Catalogue*.

In the former Soviet Union, dissertations were only rarely cited as such. The few which were, usually appeared only in summary form as *VINITI Depository Papers*, described later. For such documents the Library of Congress, the BLDSC or UB & TIB may prove to be the best or only source.

In addition to the sources described above, most universities also provide lists and information about their own theses and dissertations. Listings of particular subjects prepared by interested bodies may also be available. To list these in the space available is, however, not possible.

Conferences – proceedings, papers, preprints . . .

It is a long and complicated journey from the first thought of holding a conference, through organizing its venue and programme, inviting papers, chasing authors for their contributions – the preprint to be – to the climax of the conference itself and the publication of the proceedings afterwards (Singleton, 1976).

Some publications announcing conferences to be held are mentioned here only to show that many such tools exist. They include *World Meetings, Meetings on Atomic Energy, Forthcoming International Scientific and Technical Conferences, International Congress Calendar, Energy Meetings*. It must be remembered, however, that many a plan to hold a conference and publish its proceedings will change drastically before the appointed day.

The purpose of this chapter is to help in the identification, location and supply of documents – such as may exist – of a given conference. It is notoriously difficult to identify conference proceedings because they can be referred to in different ways which rarely conform to those used in standard bibliographic tools (Chillag, 1980 and Oseman, 1989). In Britain, the BLDSC aims to acquire conference proceedings as comprehensively as possible. Every effort is made to track down, acquire and enter into a database all proceedings, irrespective of subject, language and place of publication, whether they appear as a regular series, as one-off publications, separate papers or parts or supplements to original journals or monographs. The collection to date amounts to over 300 000 titles and the annual input is about 16 000. Around 25 per cent of this annual input can be classed as grey literature, i.e. the proceedings are not commercially published, and in many cases are preprints distributed only in limited number. Cole (1978) poses the question 'Conference publications: serials or monographs?'. They can be many other things, too, and most of them are grey of varying shades. The conferences collected at the BLDSC are listed in *Index of Conference Proceedings (ICP)* published monthly and cumulated annually. There is also an 18–year cumulation for the years 1964–1981 on microfiche. The database is also accessible online and on CD-ROM. A recent scan of this database revealed details of over 3500 conferences which had the word physics included in their title. Of course, a much larger number of conferences in the collection will deal with topics in the field of physics. All conferences listed on the database are held by the BLDSC.

There are various patterns in the publication, such as it is, of conference material. Many conference proceedings are fully published through commercial publishing houses or professional, learned societies or institutions. It may still be difficult to find out of their existence, so inclusion in *ICP* will assist with this task. Some other publications will list published proceedings only, will cover conference material of just their own institution, will be union catalogues of conferences held by certain libraries, groups of libraries or countries and cover considerably fewer conferences than appear in *ICP*. On the other hand, more detailed information may be given in such directories and guides, some of which offer a supply service enabling subscribers to acquire proceedings listed in their guides. *InterDok Directory of Published Proceedings – Series 1: SEMT – Science/Engineering/Medicine/Technology* is one such publication. The Institute of Electrical and Electronics Engineers' annual *Index to IEEE Publications* is another. Other useful listings include the annually cumulating *Bibliographic Guide to Conference Publications,* a one-off *Union List of Conference Proceedings in Libraries of the Federal Republic of Germany and Berlin (West),* and the *Samkatalog over nyanskaffat Konferenstryck.*

Of course, not all conferences will finish up with tidy, published proceedings. Material from many will appear only as *papers* of various institutions, or as *preprints,* or make their phantom appearance in these or in *poster sessions* at conferences.

To tackle problems like these, the then Technical Information Center (now the Office of Science and Technology Information – OSTI) of DOE has set up a conference register, described by Pflueger (1980). The Energy Data Base (EDB) allocates a CONF number to bring together in its files all the individual papers of a conference, even when they might be received at different times. The CONF numbering system allows continuing expansion for individual papers received only at a later date. For example, CONF 9212135 indicates the 135th conference held in December 1992 entered in the records. The CONF numbers identify the very large number of conferences listed in *Index to Conference Titles: Selected Conferences Cited in the Energy Data Base 1977–1982* (DOE report *DOE/TIC-4045*) and its updates. These conferences are also listed in *Energy Research Abstracts* and elsewhere. This illustrates that the dividing line between different categories of grey literature is itself often grey: whether one considers the CONF as reports or conferences is a matter for individual choice.

A similar but less systematic approach has been adopted for conference papers by the Commissariat à l'Energie Atomique for their CEA-CONF series.

On contract to NASA, the American Institute of Aeronautics and Astronautics (AIAA) produces twice a month *International Aerospace Abstracts (IAA)* in parallel with *STAR*. Whilst *STAR* concentrates on reports, most entries in *IAA* relate to conferences and conference papers in the field of aerospace. In the past, most papers listed in *IAA* were available on microfiche. This has now changed: only *AIAA Papers* are now reproduced for wider distribution.

Papers series by institutes such as IEEE and ASME (American Society of Mechanical Engineers), relate to papers for presentation at conferences, and may be the only available access. The annual *ASME Index to Publications* is just one of a number of similar listings. *Conference Papers Index (CPI)* will list individual papers in many fields, including physics and astronomy. Information on the conference itself precedes the data in the individual papers. *(Journal of) Conference Papers in Applied Physical Sciences (CONPAPS)* is another publication of a similar type.

There are a number of subject registers of preprints, for example the weekly *Preprints in Particles and Fields*. This publication, produced by the Stanford University Linear Accelerator Center, also includes an *anti-preprint* list, giving the open literature references of those preprints (about 80 per cent) that are eventually published. Alas, some preprints are likely to become more 'grey'. The International Centre for Theoretical Physics (ICTP), Trieste, has announced in mid-1992 that it is 'temporarily' suspending distribution of its *ICTP Preprints*.

Papers and preprints can be very elusive. What about the author who submits an abstract of a paper he then never completes? What about papers which are written but in the end are not accepted for, or included in, the published proceedings? How can one gain access to a brief note pinned to a notice board at what one refers to as a *poster session* of a conference. These are all situations where grey literature suddenly becomes 'black'. But even then there may be hope. For example, the BLDSC holds almost twice as many papers from a high-energy physics conference as its organizers will admit to exist!

Translations

Only about one half of the world's literature is published in English, and many scientists in the English-speaking world will on occasion face the problem of language barriers (Wood, 1986). It should be remembered that very large numbers of articles, reports and books have been translated, and that locating a translation is a very much cheaper task than to commission a translation to be done. Not all translations fall into the category of grey literature. For example, there are many commercially translated and published books on the market. There have been hundreds of – mainly Russian – journals which were regularly translated in their entirety or selectively.

Journals in translation will provide details of these (Chillag, 1991).
In 1991, when the 5th edition of this publication was published, over
a quarter of the 1350 titles listed were current. The former USSR
and East Europe provided the source of a large proportion of these
translations. At the time of writing it is still too early to speculate on
how many of these titles – originals or translations – will survive in
their present form, or at all, owing to the many changes which have
occurred in that part of the world. Most of the translations
mentioned above are part and parcel of open literature.

However, there is a vast number of *ad hoc* translations of
individual journal articles, reports, etc., produced, often for internal
purposes, by government agencies, universities, industrial concerns
and laboratories. Translations of this kind are certainly not conventionally published. Translations, or information about them, are
collected by a number of centres described below.

In the UK, the BLDSC has collected over the years well over half a
million such translations, maintaining an internal index for their
retrieval. Current input of translations from British sources is also
included in *British Reports, Translations and Theses.* The former
Commonwealth Index to Scientific and Technical Translations
maintained at Aslib – duplicating to a very large extent the BLDSC
indexes – ceased in 1984 and has since been incorporated into other
indexes (Chillag, 1988).

In the USA, too, access to translations, and information and
indexes to them, have undergone numerous changes over the years.
The role played until the mid-1980s by the Office of Technical
Services, publisher of *Technical Translations* and the translation
series *TT-Yr-number*, was absorbed, partly at least, into the well
established activities of the National Translations Center (NTC) at
the University of Chicago John Crerar Library. Thus, in 1964, NTC
became the focal point for scientific-technical translations in the
USA. Its *Consolidated Index of Translations into English (CITE)*
cumulated pre-1966 translations information pooled from a
number of USA, British and Commonwealth sources. A further 17-
year cumulation *CITE 2* for the years 1967–1983 was also
published. From 1967 until 1987, NTC also published *Translations
Register-Index (TRI)*, discontinued and incorporated into *WTI*, of
which later.

Close co-operation between OTS, NTIS, NTC and BLDSC
facilitated access to combined resources of well over half a million
ad hoc translations on both sides of the Atlantic for well over
quarter of a century. Developments of the last few years, including
the role of the International Translations Centre, are described
below. To complete a description of the translations 'scene' in the
USA, one must mention NTIS, DOE, NASA and DTIC, also known

as the CENDI Group (Commerce, Energy, NASA and Defense Information Group), who commission, announce and disseminate vast numbers of translations from their various centres. One must also single out the nearly 100 000 translations of *JPRS*, the Joint Research Publications Service, announced monthly in *Transdex*. The very large *JPRS* microfiche collection includes a number of series of particular interest in the fields of physics. One such is *JPRS Report: Science and Technology/Central Eurasia: Physics and Mathematics* (JPRS-UPM).

And so to the International Translations Centre (ITC), Delft. ITC was founded in 1961, in those days as the European Translations Centre (ETC), under the auspices of the Organisation for Economic Cooperation and Development (OECD).

In 1978, three European translations announcements journals (*World Index of Scientific Translations, Transatom Bulletin* and *Bulletin des Traductions*) combined into *World Transindex*. In 1987, also absorbing NTC's *Translations Register-Index*, the title changed to *World Translations Index (WTI)*. *WTI* provides not only a printed monthly announcement tool for scientific-technical translations from many, including inter-Western source languages, but is also the major online database for translations (Risseeuw, 1990).

In 1988, the University of Chicago decided to close down NTC operations. The vast NTC translations collection was transferred to the Library of Congress. The bibliographic record creation switched to ITC and its *WTI*, with the last backlog of 1989–91 NTC records now also included in *WTI*. Copies of most of the translations are available from BLDSC, LC, ITC or other centres elsewhere.

EI PAGE ONE database provides an interactive link with *WTI* translations. It provides a 'marker' from ITC records to over 30 000 physical sciences, technology and engineering related non-English-language articles.

Supplementary publications and data

Supplementary publications are yet another category of grey literature. While there may be a large audience for descriptions of the general methods of a research project or its conclusions, only a small number of scientists is likely to be interested in the finer details (BLDSC, 1989). Also, for a variety of reasons, publishers of scientific journals, etc., may decide not to print lengthy articles in their entirety, and need to find ways to make these unpublished parts accessible. In yet other situations there may be difficulties in including important ephemeral material into a report series. All such material fits into the supplementary publications category. Some publishers produce such material as a microfiche supplement to their main publications.

In the UK, in consultation with editors of learned journals, the BLDSC operates a scheme for the storage and supply of detailed data which supplements articles in these journals. The collection now numbers over 16 000 such supplementary items, the largest single group of which is crystallographic data.

Not as centralized as in the UK, some professional institutes in the USA and elsewhere maintain similar repositories for supplementary material.

Although it may cover a slightly different type of *data*, this may be the right place to mention *Physik Daten – Physics Data* of FIZ, Karlsruhe. The series consists of data compilations in selected fields of physics. Details of these, also of grey literature and other material from FIZ can be found in *Verzeichnis der Veröffentlichungen und weiterer Dienstleistungen*.

Until recent times many former Soviet journals included abstracts, or abridged versions only, of certain articles, the additional/full text appearing only in what was and still is known as *VINITI Depository Papers*. Many thousands of these, including dissertations, have been distributed in extremely small quantities by VINITI (The acronym VINITI previously expanded into 'All-Union Institute for Scientific and Technical Information'. For the time being at least VINITI is retaining its acronym). The *Depository Papers* were and still are listed monthly in *Deponirovannye Nauchnye Raboty Estestvennye: Tochnye Nauki Tekhnika*. In the West, only BLDSC, UB & TIB and Library of Congress are known to have appreciable numbers of these documents, mainly on microfiche.

It was considered appropriate to include another group of documents in this section, rather than under the heading of reports: the US Nuclear Regulatory Commission (NUREG) produces reports which are listed in the NUREG Regulatory and Technical Reports (NUREG-0304). However, NUREG is also responsible for the release of hundreds of thousands of more ephemeral documents relating to the operation and operational needs of nuclear power stations. These documents, the *NUREG DOCKET* series, are very numerous, and present great difficulties in bibliographic and physical control and retrieval. Although a number of centres hold collections of the *DOCKET*s, it is advisable to try to negotiate access to them with the organizations which use *DOCKET*s extensively for their own internal purposes, such as NUREG. In the UK, an approach to AEA Technology, Risley may be more appropriate.

System for Information on Grey Literature in Europe (SIGLE)

The preceding paragraphs have already highlighted many of the problems associated with grey literature. These problems of acquisition, bibliographic control and access to grey literature led the Commission of the European Communities to assist with the setting up of SIGLE in 1981 (Hasemann, 1991). The main aim of SIGLE is document supply, and in general to improve the bibliographic coverage and accessibility of grey literature, by combining the resources offered by important national information/document supply centres in a number of European Community countries. The main participants in the scheme are the BLDSC (UK), the FIZ Karlsruhe in Germany and the CNRS-INIST in France. The BLDSC also inputs some grey literature from the Republic of Ireland. Input is also provided by Belgium, Italy, the Netherlands, the Commission itself and, from 1992, Spain. By the end of 1992, annual input had grown to around 40 000 and SIGLE had some 300 000 documents on the database accessible through three online database hosts: STN International, BLAISE and SUNIST. The database was also issued on CD-ROM during 1992. Every document listed in SIGLE should be available from at least the national centre which provided the input (Wood, 1993).

CD-ROMs and . . . on disc

Online access and information retrieval through an already vast and still expanding number of databases has been with us for a long time now. The major database hosts (BRS, DATASTAR, DIALOG, JICST-JOIS, STN International, TELESYSTEMES, etc.) and the databases they offer will have been described, where relevant, in other chapters of this book.

In addition to database access online, the last few years also saw the rise of a new medium: CD-ROM (Compact Disc – Read Only Memory). CD-ROM enables researchers to access databases off-line where online access is not practical for technical or economical reasons. They also offer a very fast alternative to 'manual' searching, having powerful and flexible search capabilities.

CD-ROM provision is a real growth industry, and the listing below can be no more than just an indication of the many 'titles' available in the field of physics. A comprehensive listing of CD-ROMs can be found in the *CD-ROM Directory* and – one should have guessed it – in the CD-ROM DIRECTORY ON DISC (Finlay, 1992). Some titles are quite general, others deal with very specific topics; few contain just grey literature, coverage of many more include 'published' literature (books, journals) in addition to reports, conference material and other grey literature. Table 19.1 lists some CD-ROMs of interest to physicists available at the end of 1992.

Conclusion

Wood (1984) and many others before him explained how difficult it is to quantify grey literature output. Whatever its real size, one can be fairly confident that grey literature is increasing as the world's output of literature increases, and, because of the economics of publishing, the proportion of 'grey' to 'white' goes up all the time. The vast amount of knowledge and information contained in the still very much underused grey literature costs infinitely more to generate than to disseminate. Having got the information, it surely makes sound economic sense to exploit it.

Means and working tools to provide access to the 'grey literature' have been described above. One should note, however, that many of the problems of grey literature would be easier to overcome if authors and producers of such material realized its potential value and became more concerned about its fate. Among the steps that could be taken are:

1 To produce documents to better physical and bibliographic standards.
2 To be less restrictive as to what is released, and announce openly only material which is for unlimited distribution (this chapter concerns itself only with open, unlimited, unrestricted grey literature).
3 To announce documents through local publicity and by sending copies to appropriate secondary services and national grey-literature centres.
4 To have larger print runs to meet the demand that more publicity will generate.
5 To send copies to national depositories, repositories, copyright libraries and specialist collections, which can provide biblio-graphic and physical access to the documents concerned and lessen demand on the originating source, particularly when copies from the originator are no longer available.

Table 19.1 Some CD-ROM titles in physics and related subjects currently available (as at end-1992)

Title	Updating frequency	Access from	Publisher[1]
AQUALINE ABSTRACTS	Quarterly	1960–	CSA
ARCTIC AND ANTARCTIC REGIONS	Semi-annual	1985–	NISC
BOSTON SPA CONFERENCE PROCEEDINGS	Semi-annual	1963–	BLDSC
BOSTON SPA SERIALS ON CD-ROM	Semi-annual	Total	BLDSC
CD THÈSES	Annual	1972–	LASER
CENBASE/MATERIALS	Quarterly	1988–	JWS
COMPUTING ARCHIVE FROM ACM	Annual	1982–	ACM
CTI PLUS	Quarterly	1982–	BS
DARPA Discs (various series)	One-off	1989–	NTIS
DEFENSE LIBRARY ON DISC	Quarterly	1991–	NTIS
DIALOG ONDISC AEROSPACE DATABASE	Quarterly	1986–	DIALOG
DIALOG ONDISC COMPENDEX PLUS	Quarterly	1987–	DIALOG
DIALOG ONDISC EI EE DISC	Quarterly	1980–	DIALOG
DIALOG ONDISC EI ENERGY & ENVIRONMENT	Quarterly	1980–	DIALOG
DIALOG ONDISC METADEX	Quarterly	1985–	DIALOG
DIALOG ONDISC NTIS	Archival	1980–84	DIALOG
DIALOG ONDISC NTIS	Quarterly	1985–	DIALOG
DIALOG BLUE SHEETS ON DISC	Quarterly		DIALOG
DIALOG DISCOVERY PREVIEW	Demo.disc		DIALOG
DISSERTATION ABSTRACTS ON DISC	Archival	1861–1984	UMI

Dissertation Abstracts On Disc	Semi-annual	1985–	UMI
Earth Sciences on SilverPlatter	Quarterly	1990–	SP
Enviro EnergyLine Abstracts Plus	Quarterly	1970–	BS
ESPACE Access	Quarterly	1978–	EPO
GeoArchive	Semi-annual	1974–	NISC
IC Discrete Parameter Database	Two-monthly	1988–	KIBV
IEEE IEE Publications OnDisc	Monthly	1988–	UMI
INIS Database on SilverPlatter	Archival	1976–88	SP
INIS Database on SilverPlatter	Quarterly	1989–	SP
INSPEC OnDisc: Physics OnDisc	Quarterly	1989–	UMI
Electronics & Computing	Quarterly	1989–	UMI
Japan Technology	Archival	1985–88	DIALOG
Japan Technology	Quarterly	1989–	DIALOG
MathSci Disc	Semi-annual	1981–(1940-)	SP
NATO PCO-ASI	Annual	1983–	SPRINGER
NIST CrystalData Identification File	Annual	1987–	ICDD
Powder Diffraction File	Annual	1987–	ICDD
Registry of Mass Spectral Data	One-off	–	JWS
Science Citation Index	Annual	1980–1989	ISI
Science Citation Index	Quarterly	1988–	ISI
Selected Water Research Abstracts	Quarterly	1967–	SA
SIGLE on SilverPlatter	Semi-annual	1983–	SP
Voyagers to the Outer Planets	by planet	–	LASP
Water Resources Abstracts on Disc	Quarterly	1980–	SP
World Research Database	Semi-annual	1991–	LOC

Note [1] For publishers' addresses see Appendix C

References

Aronson, E. J. (ed.) (1986) *Report series codes dictionary* 3rd ed. Gale Research.

Auger, C. P. (1989) *Information sources in grey literature* 2nd ed. London: Bowker-Saur.

BLDSC (1989) *Supplementary publications scheme.* Boston Spa: British Library Document Supply Centre.

Boyer, C. V. (1972) *An analysis of the doctoral dissertation as an information source* (ERIC report, ED 065 172). Texas U.P.

Chillag, J. P. (1980) 120000 conference proceedings from stock: the conference collection and database at the British Library Lending Division. In: *Conference literature: its role in the distribution of information*, proceedings of the Workshop on Conference Literature in Science and Technology (Albuquerque, 1980), ed. G. J. Zamora, pp. 148–157. Learned Information.

Chillag, J. P. (1988) Translations. In: *British librarianship and information work 1981–1985*, vol. 2, ed. D. Bromley and A. Allott, pp. 99–100. London: Library Association.

Chillag, J. P. (ed.) (1991) *Journals in translation* 5th ed. Boston Spa: British Library Document Supply Centre and International Translations Centre.

Cole, J. E. (1978) Conference publications: serials or monographs? *Library Resources and Technical Services*, **8**(2), 168–173.

EAGLE (European Association for Grey Literature Exploitation) (1991) *SIGLE Manual, part 2: subject category list* 3rd ed. Technical Committee of EAGLE.

Finlay, M. (ed.) (1993) *The CD-ROM directory* 9th ed. TFPL Publishing.

Hasemann, C. (1991) SIGLE: access to grey literature in Europe. In: *Interlending and document supply*, proceedings of the Second International Conference (London, 1990), ed. A. Gallico, pp. 71–73. IFLA Office for International Lending).

Kane, A. (1987) *Corporate author authority list* 2nd ed. Gale Research.

Oseman, R. (1989) *Conferences and their literature.* London: Library Association.

Pflueger, M. L. (1980) Conference papers as reports. In: *Conference literature: its role in the distribution of information*, proceedings of the Workshop of Conference Literature in Science and Technology (Albuquerque, 1980), ed. G. J. Zamora, pp. 83–96. Learned Information.

Risseeuw, M. (1990) The International Translations Centre (ITC): an international network facilitating access to scientific and technical translations. *Alexandria*, **2**(1), 51–66.

Singleton, A. (1976) Physics conferences in the United Kingdom. *Aslib Proceedings*, **28**(5), 204–219.

Smith, A. W. (1993) Grey literature. In: *British librarianship and information work 1986–1990*, vol. 2, eds. D. Bromley and A. Allott, pp. 111–122. London: Library Association.

Wallace, G. G. (ed.) *Energy Information Data Base: report number codes.* (DOE Report DOE/TIC-85 Rev. 13 (or later)). DOE Office of Science and Technology.

Wood, D. N. (1984) Availability and bibliographic control of grey literature with specific reference to the UK. In: *Access to published information*, proceedings of the Access to Published Information Conference (Torquay, 1983), pp. 45–51. London: Library Association.

Wood, D. N. and Smith, B. (1986) Overcoming the language barrier at the British Library Document Supply Centre. In: *International Translations Centre 1961–1986.* International Translations Centre.

Wood, D. N. (1990a) Management of grey literature. In: *Management of recorded information: converging disciplines*, comp. C. J. Durance, pp. 61–68. London: Bowker-Saur.

Wood, D. N. and Chillag, J. P. (1990b) Acquisitions: grey literature. In: *Academic library management*, ed. M. B. Line, pp. 84–89. London: Library Association.

Wood, D. N. and Smith, A. W. (1993) SIGLE; a model for international cooperation? (a paper presented at the sixth ECCSID Conference, Canterbury, 1991). *Interlending and Document Supply*, **21**(1), 18–22.

CHAPTER TWENTY

Patent literature*

STEPHEN VAN DULKEN

Introduction

The whole field of technological achievement is recorded in the patent specifications of the world. They provide detailed and up-to-date information on work not previously made public, and less than 10 per cent of this ever becomes available elsewhere. They are moderately priced, and many are readily available for consultation and photocopying at principal public libraries. Yet their very existence is often ignored, especially by academic institutions.

One reason for this is that the specification, being primarily a legal document, is less attractive to read than an illustrated article in a journal, where, for commercial reasons, preference must be given to subjects of wide public interest.

There exist over 32 million patent specifications in many languages. These are being added to at the rate of more than a million a year, so some expertise is needed to find those of interest.

In this chapter, the principal features of the specification are explained so that the reader can readily identify those of interest and make a quick assessment of their relevance. Specifications, like books in a library, are classified by subject and the classification systems used are described as well as the modern searching aids.

Patent legislation is only mentioned where it is needed for an understanding of the literature, but a short bibliography of reference books is given at the end of the chapter.

---

* based on the second edition chapter written by B. M. Rimmer

The patents system

A patent is a bargain between the State and the inventor, who, in return for disclosing his or her invention, is granted a monopoly in its use for a specific period, generally a maximum of 20 years. The system encourages research and development and ensures a period of protection in which to recover research expenditure and to exploit the invention, either by using it or by licensing it to others. It provides a worldwide pool of information without which much time and effort would be expended solving problems already encountered by others. A firm spent many thousands of pounds developing a process for the manufacture of float glass, and when applying for a patent in 1952 ran into problems because there was already an American patent dated 1902 for substantially the same process. Where consumer items are concerned, the public would be faced with a bewildering profusion of variations on the same technical theme, and there would be free-for-all competition to make the cheapest copy of any new item on the market.

Inventions may take a long time to become commercialized. The idea may be ahead of its time and either its potential is not realized or it may need to wait for other technologies to catch up. Biro developed his ballpoint pen in 1938, but its full exploitation did not begin until a satisfactory ink was available in 1955. Not many inventors make a fortune nowadays. One who did was the late Percy Shaw, who invented the self-cleaning 'cat's eyes', which are so much part of our lives. Most inventions are made by teams of research workers who are employed for that purpose. Recent legislation in European countries has emphasized employees' rights with respect to inventions. In the United States, only the inventor may apply for a patent, though he may assign his rights at any time.

Three conditions of patentability are still the essential requirements for obtaining a patent in the principal countries of the world:

1 Novelty: the invention must be new and not have become public knowledge anywhere in the world. This is difficult to prove, but in practical terms it means that the inventor must make an application for a patent before speaking about his invention in public, or publishing details of it, or even displaying it at an exhibition without special permission. Note that (except in the United States) the important date is the date of the original filing of the patent specification (something easy to prove), and not the date of invention (something more difficult to prove).
2 The invention must be useful and capable of being manufactured or used industrially.

3 The invention must involve an appreciable inventive step; it may not be a trivial improvement which would be obvious to others working in the field.

Procedure for obtaining a patent

Patenting procedures have become increasingly standardized in patent offices internationally, but differences can still occur. The description below holds true for Britain and many other patent systems.

A description of the invention, called the specification, together with the drawings is filed at the Patent Office. The date on which this receipt is recorded is the priority date or the official date of birth of the invention. The Patent Office will then make a search to ascertain whether any information relevant to the invention has already been published in other patent specifications, books, periodicals or even trade literature. A report is sent to the applicant, who can then withdraw the application before it is published, so keeping it confidential. About 18 months after the priority date, the specification is published together with a list of references turned up in the search. The applicant then has several months in which to decide whether to request examination for patentability. Clearly, if amongst the list of references there is a description of a similar invention of which the would-be inventor was unaware, it would be wiser to abandon the application at this stage.

If, however, it is decided to continue and the invention is judged to meet the required conditions, the patent is granted and the specification is published a second time, incorporating any amendments which may have been needed. Fees are payable at each stage and also periodically to keep the patent in force for the full term of 20 years from the date of filing.

A patent for the invention must be applied for in a similar way for each country where protection is sought. If published, they are called equivalents of each other and together they form a patent family. They can be valuable if, for example, a German invention appears in an English-language equivalent. These applications must be filed within 12 months of the priority date. Failure to do so will normally mean that the invention can be manufactured in, or exported to, that country.

The United States is the only significant country to retain the practice of publishing only at grant.

Utility models and industrial designs

Utility models or petty patents are granted for the protection of implements, tools or devices capable of improving efficiency of working. Their duration is much shorter than that of patents but there are fewer formalities. This form of protection is available in about half a dozen countries, the principal ones being Japan and Germany, where they are known as Gebrauchsmuster.

Industrial designs may be registered in nearly all countries for the protection of the external appearance of an object capable of industrial manufacture. In some countries, three-dimensional objects are called models, not to be confused with utility models, and designs are essentially two-dimensional such as lace and wallpaper. Design registration in Britain is now for 25 years.

The specification or an abstract of utility models is nearly always published and illustrations of industrial designs are included in most official gazettes (except in Britain).

The patent specification

The content of the specification is based on information provided by the inventor, but its main purpose is to define the scope of the invention in legal terms, so it is drawn up by an expert in this field, a patent agent. Patent specifications are becoming much more acceptable as a source of information to the non-expert although the carefully chosen wording can be difficult to understand.

When reading a patent specification it is important to remember that by law the invention must be completely described in sufficient detail for it to be 'performed' by others working in the field. A patent can be revoked if 'insufficient disclosure' is proved. There is no limit to the length of a patent specification, so economy of style is not a requirement. Nevertheless, it is assumed that the reader will be aware of basic principles or have ready access to a textbook.

Figure 20.1 shows the front page of EP 0 468 415, a European Patent Convention (EPC) specification. This has become, since it began in 1978, the common route for foreign companies seeking protection in Britain, British national patents having become much less important. More details are given later.

The suffix A after the publication number in the top right corner signifies that it is being published for the first time, 18 months after its priority date; a second publication, a granted patent, carries the suffix B. The bibliographic details, such as dates, and names and addresses of inventor and applicant, are indicated against an international code (INID) below. These codes assist when scanning specifications in foreign languages. The priority data at (31), (32) and (33) indicate that a patent application with respect to this invention was originally made in Britain. At (51) is the International Patent Classification and, in British or American patents, (52) would give the domestic or national classification. The significance of these as an aid to searching patents is explained later.

Europäisches Patentamt

European Patent Office

Office européen des brevets

(19)

(11) Publication number: **0 468 415 A2**

EUROPEAN PATENT APPLICATION

(21) Application number: 91112259.6

(22) Date of filing: 22.07.91

(51) Int. Cl.⁵: **G01R 33/38**, G01R 33/42, H01F 7/22

(30) Priority: 24.07.90 GB 9016183

(43) Date of publication of application: 29.01.92 Bulletin 92/05

(84) Designated Contracting States: FR GB IT

(71) Applicant: OXFORD MAGNET TECHNOLOGY LIMITED
Sunbury House, Windmill Road
Sunbury-on Thames, Middlesex TW 16 7 HS(GB)

(72) Inventor: Davies, Francis
c/o Oxford Magnet Technology
Oxford OX8 1BP(GB)

(74) Representative: Fuchs, Franz-Josef, Dr.-Ing.
Postfach 22 13 17
W-8000 München 22(DE)

(54) Magnet assembly.

(57) A magnet assembly suitable for use in magnetic resonance imaging has a first superconducting coil assembly (A - D') for generating a first magnetic field and a second superconducting coil assembly (E - F') for generating a second magnetic field. The first and second superconducting coil assemblies are connected in series and are arranged so that a resultant, uniform magnetic field is generated in a working volume, and the second magnetic field substantially opposes the first magnetic field externally of the magnet assembly. The assembly further includes at least one electrically conductive ring (10, 10' 11, 11') disposed outside of the working volume, which is sized and positioned so that current induced in the ring during a transient condition, in which the current in the superconducting coils is changing, such as in the event of a quench, creates a magnetic field which acts to oppose the magnetic fields created by the currents induced in conductive parts of the magnet assembly. This causes the resultant magnetic field prevailing externally of the magnet assembly to remain substantially within or close to the preset limits for normal operation.

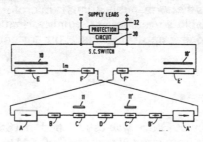

Figure 20.1: A European patent application

British or American specifications would give at (56) the search report, the list of earlier patents which the examiner considers may be relevant in determining the validity of the invention. EPC give search reports at the end of the application, although in this case the report was not compiled in time to be included, so that it was published as an 'A2' document rather than a complete A1. The title and abstract on the front page are provided by the applicant and are, if necessary, altered by the patent office. The lower part of the page is reserved for the most significant drawing.

The structure of the main body of the specification is similar in most countries. First is a statement of the problem which the invention is intended to resolve and the advantages it will provide. This will include a summary of earlier work in the same technical field, the 'prior art', often with references to other patents and to articles in journals. This section is very helpful in identifying specifications of interest, as sometimes the abstract on the front page does not make clear the purpose of the invention. After this comes a concise description of the invention, followed by a very detailed explanation, which may summarize the most significant experimental work and will be closely linked with the drawings, which do not generally carry any text, only reference numbers of components.

Finally, there are the claims, the strictly legal part of the specification which define the limits of the invention. The essential features of the invention will generally be covered in the first claim, though many more may be needed to cover all possible interpretations of the text. For example, a specification for a solar-powered refrigerator must be so worded that the principle cannot readily be adapted for other purposes without infringing the patent.

The length of the specification will depend on the complexity of the invention. Ten pages is typical but it may have only two or three pages, or several hundred. Each patent office charges a flat price for its own specifications, regardless of length. While publishing in paper format is the norm, publications in microform or compact disc are often available, sometimes as the only format. Many libraries, in order to economize on storage space, opt for non-paper formats if available.

British patents

The British Patent Office was established in 1852 although about 14 000 patents were granted prior to that date. From 1916 patents were numbered from 100 001 onwards.

The Patents Act 1977 came into force on 1 June 1978. It provided for an increased threshold of novelty; for a two-stage publication procedure (once as an application, and once as a grant); and patent specifications were numbered from 2 000 001 onwards.

A total of 12 268 British patent applications were published in 1991, of which about a third were from foreigners. Most foreign, and some British companies, prefer to use the European Patent Convention route to secure protection in Britain.

European Patent Convention

Before the European Patent Convention (EPC) came into force in 1978 it was necessary, in order to patent an invention in several countries, to select a patent agent in each country and have the specification translated into the language of the country. Then the application was processed in each country according to the laws of that country. The EPC has been very successful and in 1991 58 204 applications were published (although this does include some under the Patent Cooperation Treaty, see below).

Under the EPC, a single application is made at the European Patent Office in Munich. This can be in English, French or German. Granted European patents are, in effect, a 'bundle' of national patents, one for each of the countries designated in the application. As of 1 August 1992, there were 17 contracting states in the EPC, including all the Member States of the European Community.

The procedure for obtaining a European patent is very similar to that for a British patent. The granted patent has the claims in all three languages. From 1 September 1987, translations of the complete text of German or French granted EPC patents designating Britain must be filed at the British Patent Office.

The EPC has to a considerable extent had the effect of harmonizing the patent systems in contracting countries. This is because the Member States pass legislation to make their own patent systems compatible.

Community Patent Convention

The Community Patent Convention (CPC) has been proposed with the objective of providing a single patent valid in all European Community Member States, but the proposal has not yet been implemented. It will be based at the European Patent Office.

Patent Cooperation Treaty

The Patent Cooperation Treaty (PCT) is administered by the World Intellectual Property Organization (WIPO) in Geneva. It provides facilities for international searching and publication of patent applications and has been operating since 1978. In 1991, 20 179 applications were published.

A single application may be filed designating the countries of interest. Most designate Britain through its membership of the EPC. An international search is made, followed by the publication of the application together with the search report 18 months after filing date. These publications are mainly in English, but they can also be published in French, German, Japanese, Russian or Spanish.

It is important to remember that the PCT does not cover the granting of patents. It provides standardized search facilities, and the applicant may then proceed with national or European patent applications as required. Translations are required if the application was published in a language not handled by that patent office. The European Patent Office, for example, requires a translation if the original is not in English, French or German but otherwise does not reprint the 'Euro-PCT' application. Membership is worldwide and as of 1 August 1992 there were 50 contracting states, including all major countries except China.

Searching

Numerous patent offices or authorities publish patent specifications. Many countries like Britain and the European Patent Office publish virtually the same specification twice, and the same invention may be patented in a number of countries. It has been calculated that this vast quantity of publications represents around 500 000 new inventions a year worldwide and a million documents in all.

It is normal to search for information on online systems, particularly those which have international coverage. Otherwise, compact disc databases, microform indexes and manual sources are available. These are mostly confined to indexing one country each.

The searcher will be guided to those specifications of interest by the subject classification systems and name indexes which also provide useful information on the current activities of individuals and organizations. Redundant specifications can be eliminated by selecting for study only the one in the most convenient language from each patent family. There are two principal methods of searching patent specifications:

1 *Current awareness.* Newly-published specifications or their abstracts are scanned regularly by those wishing to keep up to date with new developments in a particular field, or to monitor the activities of competing firms. The bibliographic details can be obtained from the official gazettes or specifications, which provide up to date information including drawings, but is tedious and

not always arranged as required; or from SDIs run regularly on one of the patent online databases, which are more flexible, but are not so up to date. Alternatively, a professional patent-searching organization may be employed.

2 *Retrospective searching.* This is needed when it is required to look for a solution to a problem or to ascertain whether an idea is new. Since earlier patents are mentioned in the prior art section of specifications, it is not difficult to trace an invention back through its various developments and improvements to the original patent. The list of documents cited in the search report can provide useful leads to other sources of information. Many patents can also be searched for online to see if they were subsequently cited in the search reports of several major patent authorities' patent specifications.

Searching by subject

The classification systems developed for patents are primarily for use by patent office examiners, and only in recent years has it been appreciated that they are extensively used by others. This has meant that the classification is designed to cover the inventive aspects of an invention rather than the context in which it might be used. For example, a scanning device for use with lumber will only be classified under scanning. Determining the correct class in which to search can be quite difficult for the inexperienced. The International Patent Classification (IPC) is now used by all the major industrial countries, and hence in online databases, and Britain and the USA additionally classify their own specifications to their respective national systems. Normally, keywords are used as well as the IPC to improve the precision of the retrieval.

All the systems have a catchword index and a classification schedule running to several volumes. New users should study the introduction explaining the philosophy of the system and pay attention to the many helpful notes scattered throughout the schedules.

INTERNATIONAL PATENT CLASSIFICATION

The International Patent Classification (IPC) was devised for the use of patent offices across the world. The first edition was published in 1968, followed at approximately five year intervals by further editions. The current edition is the fifth, published in 1989. Most patent offices have been using this edition since 1 January 1990. It is now available on a CD-ROM published by the World Intellectual Property Organization, IPC: CLASS.

The whole body of patentable knowledge is divided into eight sections as follows:

A Human necessities
B Performing operations
C Chemistry and metallurgy
D Textiles and paper
E Fixed constructions (i.e. civil engineering)
F Mechanical engineering, lighting, heating, weapons, blasting
G Physics
H Electricity

Each section is divided into classes having a two-digit-number:

G01 Measuring; testing

Each class has a number of subclasses:

G01R Measuring electric variables; measuring magnetic variables

Each subclass is subdivided into groups:

G01R 33/00 Arrangements or instruments for measuring magnetic variables

Figure 20.2 shows a portion of this classification. It includes two classifications allocated to the patent application illustrated in Figure 20.1: G01R 33/38 and 33/42. 33/42 was regarded as subordinate to the idea in 33/38 and this is indicated by the use of four dots rather than three after the symbol. Italics show that the classification was revised for this edition, while the bold numbers in square brackets show in which edition that concept was reclassified.

Frequently, more than one symbol is assigned to an invention, the most relevant being placed first. For example, an invention concerning optical methods for controlling layer thickness, as used in the manufacture of semiconductors, will be classified first under optical means for measuring thickness, but may also be allocated the symbol corresponding to measurement during manufacturing processes. It will not normally be classified under semiconductors, as this feature is incidental and not related to the inventive concept.

Section G, though entitled 'Physics', is essentially about instruments and nucleonics and the remainder of the field of physics is found in other sections.

This concept that the IPC only classifies 'novel' information for the benefit of patent offices is beginning to change with the introduction of 'hybrid' classification. This is where either the functional aspects are catered for by dedicated portions of the classification, or where the ordinary classifications are used to classify the non-inventive aspects of an invention. Their use is indicated on patent documents by a // symbol preceding them and a colon in the classification, e.g. 7:22 rather than 7/22.

G 01 R

31/02	.	Testing of electric apparatus, lines, or components for short-circuits, discontinuities, leakage, or incorrect line connection
31/04	. .	Testing connections, e.g. of plugs, of non-disconnectable joints
31/06	. .	Testing of electric windings, e.g. for polarity (measuring number of turns, transformation ratio, or coupling factor 29/20)
31/08	.	Locating faults in cables, transmission lines, or networks (emergency protective circuit arrangements H 02 H)
31/10	. .	by increasing destruction at fault, e.g. burning-in by using a pulse generator operating a special programme
31/11	. .	using pulse-reflection methods
31/12	.	Testing of articles or specimens of solids or fluids for dielectric strength or breakdown voltage
31/14	. .	Circuits therefor
31/16	. .	Construction of testing vessels; Electrodes therefor
31/18	. .	Subjecting similar articles in turn to test, e.g. go/no-go tests in mass production
31/20	. .	Preparation of articles or specimens to facilitate testing
31/24	.	Testing of discharge tubes (during manufacture H 01 J 9/42) [2]
31/25	. .	Testing of vacuum tubes [2]
31/26	.	Testing of individual semiconductor devices (measurement of impurity content of materials G 01 N) [2]
31/28	.	Testing of electronic circuits, e.g. by signal tracer (checking computers G 06 F 11/00; checking static stores for correct operation G 11 C 29/00)
31/30	. .	Marginal testing, e.g. by varying supply voltage (marginal testing of computers G 06) [2]
31/302	. .	*Contactless testing* [5]
31/305	. . .	*using electron beams* [5]
31/308	. . .	*using non-ionising electromagnetic radiation, e.g. optical radiation* [5]
31/312	. . .	*by capacitive methods* [5]
31/315	. . .	*by inductive methods* [5]
31/318	. .	*Testing logical circuits, e.g. logic analysers (31/302 takes precedence)* [5]
31/32	.	Testing of the switching capacity of high-voltage circuit-breakers (means for detecting the presence of an arc or discharge in switching devices H 01 H 9/50, 33/26) [2]
31/34	.	Testing dynamo-electric machines (testing electric windings 31/06; methods or apparatus specially adapted for manufacturing, assembling, maintaining or repairing dynamo-electric machines H 02 K 15/00) [3]
31/36	.	Apparatus for testing electrical condition of accumulators or electric batteries, e.g. capacity or charge condition (accumulators combined with arrangements for measuring, testing or indicating condition H 01 M 10/48; circuit arrangements for charging, or depolarising batteries or for supplying loads from batteries H 02 J 7/00) [3]
33/00		**Arrangements or instruments for measuring magnetic variables**

33/02	.	Measuring direction or magnitude of magnetic fields or magnetic flux (33/20 takes precedence; measuring direction or magnitude of the earth's field for navigation or surveying G 01 C; for prospecting, for measuring the magnetic field of the earth G 01 V 3/00) [4]

Note

Groups 33/022 or 33/10 take precedence over groups 33/025 to 33/09.

33/022	. .	Measuring gradient [3]
33/025	. .	Compensating stray fields [3]
33/028	. .	Electrodynamic magnetometers [3]
33/032	. .	using magneto-optic devices, e.g. Faraday [3]
33/035	. .	using superconductive devices [3]
33/038	. .	using permanent magnets, e.g. balances, torsion devices [3]
33/04	. .	using the flux-gate principle
33/05	. . .	in thin-film element [3]
33/06	. .	using galvano-magnetic devices, e.g. Hall-effect devices; using magneto-resistive devices
33/10	. .	Plotting field distribution
33/12	.	Measuring magnetic properties of articles or specimens of solids or fluids (involving magnetic resonance 33/20) [4]
33/14	. .	Measuring or plotting hysteresis curves
33/16	. .	Measuring susceptibility
33/18	. .	Measuring magnetostrictive properties
33/20	.	*involving magnetic resonance (magnetic resonance gyrometers G 01 C 19/60)* [4,5]
33/22		*(covered by 33/20)*
33/24	.	for measuring direction or magnitude of magnetic fields or magnetic flux [4]
33/26	. .	using optical pumping [4]
33/28	.	*Details of apparatus provided for in groups 33/44 to 33/64* [5]
33/30	. .	*Probes, e.g. resonators, sample cells, spinning mechanisms* [5]
33/32	. . .	*Radiofrequency excitation or detection systems* [5]
33/34	. . . .	*Constructional details, e.g. surface coils* [5]
33/36	. . . .	*Electrical details, e.g. matching or coupling of the coil to the receiver* [5]
33/38	. . .	*Systems for generation, homogenisation or stabilisation of the main or gradient magnetic field* [5]
33/40	. . . .	*Correction coil assemblies for the magnets* [5]
33/42	. . . .	*Screening* [5]
33/44	. .	*using nuclear magnetic resonance (33/24, 33/62 take precedence)* [5]
33/46	. . .	*NMR spectroscopy* [5]
33/48	. . .	*NMR imaging systems* [5]
33/50	. . . .	*based on the determination of relaxation times* [5]
33/52	. . . .	*based on chemical shift information* [5]
33/54	. . . .	*Signal processing systems* [5]
33/56	. . . . .	*Image enhancement, e.g. image distortion* [5]

Figure 20.2: Extract from the International Patent Classification

ECLA is a more detailed version of the IPC which is used to arrange the search files of the European Patent Office. It is available online, and can be used on the EDOC database (see p. 444) to produce file lists of patents from many countries, often going back several decades or more.

UK PATENT CLASSIFICATION

The UK system as revised in 1963–1964 has a certain resemblance to the IPC, though there are many differences in terminology. Section G, for example, is called 'Instrumentation', not Physics. Besides the classification terms there are, as with the IPC, some indexing terms to identify individual non-inventive features.

British classification is in decline and is not used much now. Manual searching is possible but it is not possible to search online using this classification except through the Patent Office, who can supply 'file lists' of classified patents.

UNITED STATES PATENT CLASSIFICATION

Although the United States classifies by IPC, it does so with a concordance after classifying by its own classification. This often means that the IPCs are inaccurate, since they tend to be broader headings than are truly applicable. Hence, a separate search using American classification is advisable if carrying out a comprehensive search.

This system is probably the most difficult for the newcomer to use. It has about 500 classes, with subclasses further divided into sub-divisions. The latter total nearly 100000, which is more than double the British or the IPC. In addition, there is the problem of selecting the correct keyword or spelling in the American language. A compact disc, ASIST (US Patent and Trademark Office), includes this classification.

Searching by name

Searching by name is comparatively straightforward, as most countries provide a name index to individual issues of their official gazette and this is cumulated quarterly or annually. In principle, the names are listed alphabetically by the significant element of the name. There are, as yet, no internationally accepted guidelines to this procedure and, as all indexes inevitably contain many foreign names, there are important differences in interpretation between countries. One may find Eli Lilly under 'E', or under 'L' as Lilly, Eli; similarly, L'Oreal may be indexed as Oreal L'. 'The Secretary of State for Defence' will be under 'Defence, Secretary of State for' in the UK, but in other countries may well be indexed under 'The'. Acronyms may be placed at the beginning of a sequence or interfiled alphabetically. Sometimes, particularly in indexes originating in Scandinavian countries, letters with umlauts or other diacritical marks are placed after Z. Both personal and corporate names can be searched for online.

Patent online databases

The level of standardization achieved internationally by the use of the International Patent Classification and the identification of the heading data by INID code have greatly facilitated the introduction of online databases. Online searching allows great flexibility in working out search strategies. This is because of the wide variety of databases; the frequent ability to transfer results from one database to another; and the numerous types of searches that can be carried out, more so than with compact discs and especially manual searching.

Several hosts (especially DIALOG, Orbit and Questel) offer a variety of databases for the user. The databases themselves are listed in Sibley, J. *Online patents, trade marks and service marks databases* (Aslib, 1991). None caters specifically for those interested in physics patents, nor does *Physics Abstracts* cover patents. Some important databases are listed in Table 20.1. In some databases the coverage is not complete from the earliest date given.

Two organizations are particularly important in the field of patent information provision. These are Derwent Publications and EPIDOS. Derwent Publications is based in London. It covers patents from around 30 countries with its WORLD PATENTS INDEX (WPI) database. It provides its own English-language abstracts for many of these countries, which facilitates keyword searching. Much of this information is repackaged as microfiche indexes or as published abstracts. These abstracts are either in subject volumes or in country volumes or both. Non-chemical information is held from 1974 onwards.

EPIDOS is a directorate of the European Patent Office, and is based in Vienna. It holds bibliographical data from over 50 countries on its INPADOC database, the most international of all the patents databases. Microfiche indexes are produced from this data. Information is held from 1968 onwards but coverage varies by country and type of data.

COMPACT DISCS

Compact discs (CD-ROMs) are rapidly growing in importance as a means of either storing (and printing) complete patent specifications, or of providing bibliographical or other searching tools. New CD-ROM products are constantly being produced in the field of patents.

CD-ROMs do not offer an international approach to searching at present, although Derwent Publications is planning a series of subject-based CD-ROMs. They also do not offer as much flexibility

Table 20.1 Online patents indexes

Name	Country coverage	Online from	Typical online supplier(s)
INPADOC	50+	1968	(1), (2), (3)
WORLD PATENTS INDEX	30+	1963	(1), (2), (4)
EDOC	18	1877	(4)
CHINAPATS	CN	1985	(1), (2)
CLAIMS	US	1950	(1), (2), (3)
EPAT	EP	1978	(4)
FPAT	FR	1966	(4)
ITALPAT	IT	1983	(5)
JAPIO	JP	1976	(1)
PATDPA	DE	1968	(3)
PATOLIS	JP	1955	(6)
USPA	US	1971	(1)

Online hosts:	Countries:
(1) Orbit	CN = China
(2) DIALOG	DE = Germany
(3) STN	EP = European Patent Convention
(4) Questel	FR = France
(5) ESA-IRS	IT = Italy
(6) JAPIO	JP = Japan
	US = United States

as online approaches. They are, however, cheaper and generally easier to use than online services.

The European Patent Office produces several series of digitized CD-ROMs which contain the full text and drawings of patents, using software which is now accepted as standard. Up to about a thousand documents can be stored on each disc. The series comprise European Patent Convention applications and grants, and Patent Cooperation Treaty applications, and further series for the national patents of the constituent members of the EPC are planned or being produced. Plans are also going ahead to produce discs for the older material. American patents are also available on CD-ROM.

There are a number of CD-ROM databases containing patent information. These include, for American data, APS and CASSIS, and for European Patent Convention data, BULLETIN and ACCESS.

In addition, there is a disc containing several editions of the International Patent Classification (IPC: CLASS) and one that includes American patent classification (ASIST).

Availability of patent documentation and information

The Science Reference and Information Service (SRIS), based in London, is the national patents library. Now part of the British Library, it was established in 1852, as the Patent Office Library. It contains a very extensive collection of industrial property literature, covering most languages and countries. This includes 32 million patents from 30 patent authorities, and many of the Derwent Publications abstracts and the Inpadoc microfiche indexes. It also has a comprehensive collection of books and periodicals in the inventive sciences as well as the modern computer-based searching aids.

SRIS, jointly with the Patent Office, sponsors the United Kingdom Patents Information Network, through which 13 provincial patent libraries are equipped with either patent specifications or abridgments and indexes. They are mostly based in public libraries.

Similar arrangements for facilitating access to patent specifications exist in all the principal patenting countries of the world.

Bibliography

Reference books

Blanco White, T. A. *et al.* (1977) *The encyclopedia of United Kingdom and European patent law*. Sweet & Maxwell.
Chartered Institute of Patent Agents (1990) *CIPA guide to the patents acts*. Sweet & Maxwell.
Chartered Institute of Patent Agents (1988) *European patents handbook*. Oyez/ Longman.
Chisum, D. S. (1978–) *Patents: a treatise on the law of patentability, validity and infringement*. Bender.

Patent documentation

Rimmer, B. (1992) *International guide to official industrial property publications*. London: British Library.
Sibley, J. (1991) *Online patents, trade marks and service marks databases*. London: Aslib.
van Dulken, S. (ed.) (1992) *Introduction to patents information*. London: British Library.

Guides to patenting procedures

Greene, Anne Marie (1981) *Patents throughout the world*. Trade Activities.
Katzarov, K. (1981–) *Manual on industrial property*. Katzarov.

Manual for the handling of applications for patents, designs and trade marks through-
   out the world. (The Dutch manual) Amsterdam: Octrooibureau Los en Stigter.
Patent protection (undated) (Patent Office) [free guide to obtaining patent protection]

Historical works

Baker, R. (1976) *New and improved.* London: British Library.
Davenport, N. (1979) *The United Kingdom patent system.* Havant, UK: Mason.

Note: Most of the reference works and guides are in loose-leaf format and regularly
updated. The publication date above is that on which the book was first published.

Appendix A

The following list of physics journals, abstracted by INSPEC during the period 1 November 1991 to 31 October 1992, is ranked by selection total.

Rank	Title (Country)	Number of items		Percentage of total
		Individual	Cumulative	
1	Physical Review B, Condensed Matter	3 567	3 567	2.91
2	Journal of Applied Physics (USA)	2 268	5 835	4.76
3	Applied Physics Letters (USA)	2 151	7 986	6.52
4	Physica C (NL)	2 088	10 074	8.23
5	Journal of Chemical Physics (USA)	2 085	12 159	9.93
6	Physical Review Letters (USA)	2 042	14 201	11.60
7	Physics Letters B (NL)	1 719	15 920	13.01
8	Journal of Magnetism and Magnetic Materials (NL)	1 552	17 472	14.28
9	Journal of Geophysical Research (USA)	1 421	18 893	15.44
10	Nuclear Instruments and Methods in Physics Research (NL)	1 298	20 191	16.50
11	Surface Science (NL)	1 273	21 464	17.54
12	Chemical Physics Letters (NL)	1 198	22 662	18.52
13	Astrophysical Journal (USA)	1 172	23 834	19.47
14	Physical Review A, Atomic Molecular and Optical Physics (USA)	1 150	24 984	20.41
15	Astronomy and Astrophysics (Germany)	1 127	26 111	21.34
16	Journal of Physics: Condensed Matter (UK)	1 039	27 150	22.19
17	Journal of Materials Science (UK)	994	28 144	23.00

448   *Appendix A*

Rank	Title (Country)	Number of items Individual	Cumulative	Percentage of total
18	Physics Letters A (NL)	992	29 136	23.81
19	Physical Review A, Statistical Physics, Plasmas, Fluids (USA)	970	30 106	24.60
20	Applied Optics (USA)	964	31 070	25.39
21	Japanese Journal of Applied Physics 1 (Japan)	961	32 031	26.17
22	Journal of Crystal Growth (NL)	911	32 942	26.92
23	Nuclear Instruments and Methods in Physics Research (NL)	889	33 831	27.65
24	Journal of Physical Chemistry (USA)	865	34 696	28.35
25	Transactions of the American Nuclear Society (USA)	857	35 553	29.05
26	Soviet Physics – Solid State (USA)	816	36 369	29.72
27	Solid State Communications (USA)	812	37 181	30.38
28	Physica B (NL)	781	37 962	31.02
29	Journal of the Physical Society of Japan (Japan)	758	38 720	31.64
30	Journal of Physics A, Mathematics and General (UK)	713	39 33	32.22
31	Nuclear Physics A (NL)	86	40 19	32.78
32	Physical Review C, Nuclear Physics (NL)	652	40 771	33.32
33	Optics and Spectroscopy (USA)	632	41 403	33.83
34	Journal of the Acoustical Society of America (USA)	630	42 033	34.35
35	Journal of Non-Crystalline Solids (NL)	615	42 648	34.85
36	Optics Letters (USA)	614	43 262	35.35
37	Optics Communications (NL)	613	43 875	35.85
38	Journal of Materials Science Letters (UK)	596	44 471	36.34
39	Journal of Vacuum Science and Technology A (USA)	591	45 062	36.82
40	International Astronomical Union Circular (USA)	571	45 633	37.29
41	Scripta Metallica et Materialia (USA)	562	46 195	37.75
42	Japanese Journal of Applied Physics 2 (Japan)	555	46 750	38.20
43	Soviet Technical Physics Letters (USA)	550	47 300	38.65
44	Journal of Physique IV, Colloque (France)	546	47 846	39.10

45	Physical Review D, Particles, Fields, Gravitation and Cosmology (USA)	545	48 391	39.55
46	Physical Review D, Particles and Fields (USA)	540	48 931	39.99
47	Thin Solid Films (Switzerland)	538	49 469	40.43
48	Semiconductor Science and Technology (UK)	529	49 998	40.86
49	Journal of the American Ceramic Society (USA)	523	50 521	41.29
50	Journal of Physics B, Atomic Molecular and Optical Physics (UK)	505	51 026	41.70
51	Materials Science and Engineering A, Structural Materials (Switzerland)	502	51 528	42.11
52	Geophysical Research Letters (USA)	501	52 029	42.52
53	Ferroelectrics (UK)	488	52 517	42.92
54	Journal of the Electrochemical Society (USA)	477	52 994	43.31
55	Journal of Mathematical Physics (USA)	475	53 469	43.70
56	Monthly Notices of the Royal Astronomical Society (UK)	475	53 944	44.08
57	Soviet Physics – Technical Physics (USA)	469	54 413	44.47
58	Nature (UK)	466	54 879	44.85
59	Soviet Journal of Nuclear Physics (USA)	460	55 339	45.22
60	IEEE Photonics Technology Letters (USA)	448	55 787	45.59
61	Journal of Alloys and Compounds (Switzerland)	448	56 235	45.96
62	Soviet Physics – Semiconductors (USA)	447	56 682	46.32
63	Europhysics Letters (Switzerland)	441	57 123	46.68
64	Physics of Fluids B, Plasma Physics (UK)	440	57 563	47.04
65	Physica A (NL)	437	58 000	47.40
66	Nuclear Physics B (NL)	410	58 410	47.73
67	Physica Status Solidi A (Germany)	409	58 819	48.07
68	Soviet Journal of Quantum Electronics (USA)	409	59 228	48.40
69	Metallurgical Transactions A, Physical Metallurgy and Materials Science (USA)	402	59 630	48.73
70	Journal of Materials Research (USA)	398	60 028	49.06
71	Soviet Physics – JETP (USA)	390	60 418	49.38
72	THEOCHEM (NL)	380	60 798	49.69
73	Applied Surface Science (NL)	379	61 177	50.00

Rank	Title (Country)	Number of items		Percentage of total
		Individual	Cumulative	
74	Nuclear Physics B, Proceedings Supplement (NL)	377	61 554	50.30
75	Acta Metallurgica et Materialia (USA)	370	61 924	50.61
76	Hyperfine Interactions (Switzerland)	369	62 293	50.91
77	International Journal of Modern Physics A (Singapore)	369	62 662	51.21
78	Physica Status Solidi B (Germany)	368	63 030	51.51
79	Modern Physics Letters A (Singapore)	365	63 395	51.81
80	Astrophysical Journal Letters (USA)	359	63 754	52.10
81	Journal of Magnetic Resonance (USA)	355	64 109	52.39
82	Astrophysics and Space Science (NL)	354	64 463	52.68
83	IEEE Journal of Quantum Electronics (USA)	349	64 812	52.97
84	Ultramicroscopy (NL)	349	65 161	53.25
85	Chemical Physics (NL)	346	65 507	53.53
86	Science (USA)	344	65 851	53.82
87	Journal of Molecular Structure (NL)	341	66 192	54.09
88	Journal of Fluid Mechanics (UK)	327	66 519	54.36
89	JETP Letters (USA)	326	66 845	54.63
90	Journal of the American Chemical Society (USA)	319	67 164	54.89
91	Optical Engineering (USA)	315	67 479	55.15
92	Zeitschrift fur Physik C, Particles and Fields (Germany)	314	67 793	55.40
93	Fusion Technology (USA)	313	68 106	55.66
94	Physics of Fluids A, Fluid Dynamics (USA)	312	68 418	55.91
95	Soviet Physics – Doklady (USA)	312	68 730	56.17
96	Journal of Sound and Vibration (UK)	310	69 040	56.42
97	Superconductor Science and Technology (UK)	307	69 347	56.67
98	Molecular Physics (UK)	307	69 654	56.92
99	Solid State Ionics, Diffusion and Reactions (NL)	302	69 956	57.17
100	Astronomical Journal (USA)	296	70 252	57.41

101	Soviet Journal of Optical Technology (USA)	293	70 545	57.65
102	Journal of Physics D, Applied Physics (USA)	290	70 835	57.89
103	Superconductivity: Physics, Chemistry Technology (USA)	290	71 125	58.13
104	Soviet Atomic Energy (USA)	285	71 410	58.36
105	Engineering Fracture Mechanics (UK)	282	71 692	58.59
106	Journal of Solid State Chemistry (USA)	282	71 974	58.82
107	Surface and Coatings Technology (Switzerland)	282	72 256	59.05
108	Soviet Physics Journal (USA)	282	72 538	59.28
109	Journal of the Optical Society of America B, Optical Physics (USA)	281	72 819	59.51
110	Zeitschrift fur Physik D, Atoms Molecules and Clusters (Germany)	269	73 088	59.73
111	International Journal of Quantum Chemistry (USA)	268	73 356	59.95
112	Journal of Statistical Physics (USA)	268	73 624	60.17
113	Nuclear Tracks and Radiation Measurement (UK)	265	73 889	60.39
114	Biophysical Journal (USA)	263	74 152	60.60
115	Nuclear Engineering and Design (NL)	262	74 414	60.81
116	Journal of Molecular Spectroscopy (USA)	256	74 670	61.02
117	Radiation Effects and Defects in Solids (UK)	252	74 922	61.23
118	Journal of the Chemical Society Faraday Transactions (UK)	250	75 172	61.43
119	Classical and Quantum Gravity (UK)	244	75 416	61.63
120	International Journal of Heat and Mass Transfer (UK)	243	75 659	61.83
121	Soviet Physics – Crystallography (USA)	241	75 900	62.03
122	Progress of Theoretical Physics (Japan)	233	76 133	62.22
123	AIAA Journal (USA)	232	76 365	62.41
124	Biophysics (UK)	230	76 595	62.60
125	Materials Letters (NL)	230	76 825	62.79
126	Journal of Lightwave Technology (USA)	230	77 055	62.97
127	Geophysical Journal International (UK)	230	77 285	63.16
128	Journal of Modern Optics (UK)	229	77 514	63.35
129	Journal of the Optical Society of America A, Optics and Image Science (USA)	227	77 741	63.53

Rank	Title (Country)	Number of items		Percentage of total
		Individual	Cumulative	
130	Journal of the Physics and Chemistry of Solids (UK)	225	77 966	63.72
131	Communications in Mathematical Physics (Germany)	222	78 188	63.90
132	Fizika Nizkikh Temperatur (Ukraine)	222	78 410	64.08
133	Physica Scripta (Sweden)	222	78 632	64.26
134	Materials Science and Engineering B, Solid-State Materials for Advanced Technology (Switzerland)	221	78 853	64.44
135	Zeitschrift für Angewandte Mathematik und Mechanik (Germany)	220	79 073	64.62
136	Zeitschrift für Physik B, Condensed Matter (Germany)	218	79 291	64.80
137	Journal of Nuclear Materials (NL)	217	79 508	64.98
138	Fusion Engineering and Design (NL)	215	79 723	65.15
139	Acta Physica Polonica A (Poland)	212	79 935	65.33
140	Strength of Materials (USA)	212	80 147	65.50
141	American Journal of Physics (USA)	211	80 358	65.67
142	Vision Research (UK)	211	80 569	65.85
143	Spectrochimica Acta A, Molecular Spectroscopy (UK)	210	80 779	66.02
144	Geomagnetism and Aeronomy (USA)	209	80 988	66.19
145	Acta Metallurgica Sinica (China)	207	81 195	66.36
146	Zeitschrift für Physik A, Hadrons and Nuclei (Germany)	206	81 401	66.53
147	Journal of Applied Spectroscopy (USA)	205	81 606	66.69
148	Modern Physics Letters B (Singapore)	203	81 809	66.86
149	Philosophical Magazine B, Physics of Condensed Matter (UK)	202	82 011	67.02
150	Bulletin of Materials Science (India)	199	82 210	67.19
151	International Journal of Fracture (NL)	197	82 407	67.35
152	Applied Physics A, Solids and Surfaces (Germany)	195	82 602	67.51
153	IEEE Transactions on Nuclear Science (USA)	195	82 797	67.67
154	Chinese Journal of Lasers (China)	191	82 988	67.82

155	Soviet Physics – Acoustics (USA)	190	83 178	67.98
156	Metallofizika (Ukraine)	188	83 366	68.13
157	International Journal of Solids and Structures (UK)	187	83 553	68.28
158	Comptes Rendus de l'Academie des Sciences II, Mecanique, Physique, Chimie, Sciences de l'Univers, Sciences de la Terre (France)	186	83 739	68.44
159	Journal of Engineering Physics (USA)	181	83 920	68.58
160	Journal of Physics G, Nuclear Particle Physics (UK)	180	84 100	68.73
161	High Energy Physics and Nuclear Physics (China)	179	84 279	68.88
162	Corrosion (USA)	178	84 457	69.02
163	Journal of Luminescence (NL)	177	84 634	69.17
164	Journal of the Japan Institute of Metals (Japan)	177	4 811	69.31
165	Japanses Journal of Applied Physics Supplement (Japan)	176	84 987	69.46
166	Chinese Physics Letters (China)	176	85 163	69.60
167	Journal of Electronic Materials (USA)	176	85 339	69.74
168	Journal de Physique I, General Physics, Statistical Physics (France)	175	85 514	69.89
169	Journal of Applied Crystallography (Denmark)	174	85 688	70.03
170	Medical Physics (USA)	174	85 862	70.17
171	Journal of the Ceramics Society of Japan (Japan)	174	86 036	70.31
172	Physica Scripta, Volume T (Sweden)	173	86 209	70.45
173	Applied Physics B, Photophysics and Laser Chemistry (Germany)	172	86 381	70.60
174	Geophysics (USA)	170	86 551	70.73
175	Journal of Biomechanics (UK)	170	86 721	70.87
176	Philosophical Magazine A, Physics of Condensed Matter (UK)	170	86 891	71.01
177	Journal of Atmospheric Sciences (USA)	169	87 060	71.15
178	Journal of Atmospheric and Terrestrial Physics (UK)	168	87 228	71.29
179	Radiation Protection Dosimetry (UK)	167	87 395	71.42
180	Monthly Weather Review (USA)	166	87 561	71.56
181	Cryogenics (UK)	164	87 725	71.69
182	Journal of the Society of Materials Science, Japan (Japan)	162	87 887	71.83

Rank	Title (Country)	Number of items Individual	Cumulative	Percentage of total
183	Physica D (NL)	162	88 049	71.96
184	Computer Physics Communications (NL)	161	88 210	72.09
185	Astronomy and Astrophysics Supplement Series (France)	160	88 370	72.22
186	IEEE Transactions on Biomedical Engineering (USA)	159	88 529	72.35
187	Theoretical and Mathematical Physics (USA)	159	88 688	72.48
188	Nuclear Fusion (Austria)	158	88 846	72.61
189	Physics in Medicine and Biology (UK)	156	89 002	72.74
190	Solar Physics (NL)	156	89 158	72.87
191	Soviet Applied Mechanics (USA)	155	89 313	72.99
192	Soviet Journal of Plasma Physics (USA)	153	89 466	73.12
193	Nuovo Cimento A (Italy)	152	89 618	73.24
194	Soviet Astronomy (USA)	152	89 770	73.37
195	Tectonophysics (NL)	152	89 922	73.49
196	International Journal of Infrared and Millimeter Waves (USA)	151	90 073	73.61
197	Materials Transactions JIM (Japan)	151	90 224	73.74
198	Radiation Research (USA)	151	90 375	73.86
199	IEEE Transactions on Plasma Sciences (USA)	150	90 525	73.98
200	International Journal of Theoretical Physics (USA)	149	90 674	74.10
201	Acta Physica Sinica (China)	148	90 822	74.23
202	Fluid Dynamics (USA)	148	90 970	74.35
203	High Temperature (USA)	147	91 117	74.47
204	Icarus (USA)	147	91 264	74.59
205	International Journal of Modern Physics B (Singapore)	147	91 411	74.71
206	Transactions of ASME, Journal of Applied Mechanics (USA)	147	91 558	74.83
207	Crystal Research and Technology (Germany)	146	91 704	74.95
208	Fizika Metallov i Metallovedenie (Russia)	146	91 850	75.07

209	Superlattices Microstructures (UK)	146	91 996	75.18
210	Nuclear Technology (USA)	145	92 141	75.30
211	Chinese Physics (USA)	141	92 282	75.42
212	Journal de Physique II, Atomic Molecular and Cluster Physics (France)	141	92 423	75.53
213	Wear (Switzerland)	141	92 564	75.65
214	International Journal of Engineering Science (UK)	140	92 704	75.76
215	Helvetica Physica Acta (Switzerland)	139	92 843	75.88
216	Synthetic Metals (Switzerland)	139	92 982	75.99
217	Nuclear Engineering International (UK)	138	93 120	76.10
218	Bulletin of the Seismological Society of America (USA)	137	93 257	76.22
219	Corrosion Science (UK)	136	93 393	76.33
220	Magnetic Resonance in Medicine (USA)	136	93 529	76.44
221	Czechoslovakian Journal of Physics (Czechoslovakia)	136	93 665	76.55
222	Planetary and Space Science (UK)	135	93 800	76.66
223	Communications in Theoretical Physics (China)	133	93 933	76.77
224	Journal of Polymer Science B, Polymer Physics (USA)	133	94 066	76.88
225	Publications of the Astronomical Society of the Pacific (USA)	133	94 199	76.99
226	Journal of Computational Physics (USA)	132	94 331	77.09
227	Few-Body Systems Supplementum (Austria)	132	94 463	77.20
228	Chinese Science Bulletin (China)	132	94 595	77.31
229	Plasma Physics and Controlled Fusion (UK)	131	94 726	77.42
230	Zeitschrift für Metallkunde (Germany)	131	94 857	77.52
231	International Journal for Numerical Methods in Fluids (UK)	130	94 987	77.63
232	Canadian Journal of Physics (Canada)	130	95 117	77.74
233	Liquid Crystals (UK)	130	95 247	77.84
234	Journal of Computational Chemistry (USA)	128	95 375	77.95
235	Journal of Applied Mechanical and Technical Physics (USA)	126	95 501	78.05
236	Zeitschrift für Naturforschung A, Physik, Physikalische Chemie (Germany)	125	95 626	78.15
237	Nuovo Cimento B (Italy)	124	95 750	78.25

Rank	Title (Country)	Number of items Individual	Cumulative	Percentage of total
238	Proceedings of the Royal Society of London A, Mathematical and Physical Science (UK)	124	95 874	78.35
239	Journal of Korean Institute of Metals (South Korea)	123	95 997	78.45
240	Earth and Planetary Science Letters (NL)	122	96 119	78.55
241	Journal of Quantitative Spectroscopy and Radiative Transfer (UK)	122	96 241	78.65
242	Journal of Structural Chemistry (USA)	122	96 363	78.75
243	Nuclear Physics B, Particle Physics (NL)	121	96 484	78.85
244	Diffusion and Defect Data, Solid State Data Part A (Liechtenstein)	121	96 605	78.95
245	Geofizicheskii Zhurnal (Ukraine)	120	96 725	79.05
246	Astrophysical Journal Supplement Series (USA)	118	96 843	79.15
247	Journal of Nuclear Science and Technology (Japan)	118	96 961	79.24
248	Geology (USA)	118	97 079	79.34
249	Materials Research Bulletin (USA)	118	97 197	79.44
250	Medical and Biological Engineering and Computing (UK)	118	97 315	79.53
251	Theoretica Chimica Acta (Germany)	117	97 432	79.63
252	Indian Journal of Pure and Applied Physics (India)	116	97 548	79.72
253	Journal of the European Ceramic Society (UK)	116	97 664	79.82
254	Journal of Composite Materials (USA)	115	97 779	79.91
255	Journal de Physique III, Applied Physics Materials Science (France)	115	97 894	80.01
256	Mechanics of Solids (USA)	114	98 008	80.10
257	Nuclear Science and Engineering (USA)	114	98 122	80.19
258	Soviet Meteorology and Hydrology (USA)	114	98 236	80.28
259	Solar Energy and Materials (NL)	114	98 350	80.38
260	Hearing Research (NL)	113	98 463	80.47
261	Materials Science and Technology (UK)	113	98 576	80.56
262	Journal of Physical Oceanography (USA)	113	98 689	80.65

263	Pis'ma v Astronomicheskie Zhurnal (Russia)	113	98 802	80.75
264	Acta Crystallographica A, Foundations of Crystallography (Denmark)	111	98 913	80.84
265	Phase Transitions (UK)	111	99 024	80.93
266	Electroencephalography and Clinical Neurophysiology (Ireland)	108	99 132	81.02
267	Transactions of ASME, Journal of Heat Transfer (USA)	108	99 240	81.11
268	Scientia Atmospherica Sinica (China)	107	99 347	81.19
269	Nuovo Cimento D (Italy)	106	99 453	81.28
270	Applied Radiation and Isotopes (UK)	106	99 559	81.37
271	Philosophical Magazine Letters (UK)	106	99 665	81.45
272	Optometry and Vision Science (USA)	106	99 771	81.54
273	International Journal of Radiation Oncology Biology Physics (UK)	103	99 874	81.62
274	IEEE Transactions on Geoscience and Remote Sensing (USA)	102	99 976	81.71
275	Composites Science and Technology (UK)	102	100 078	81.79
276	International Journal of Mass Spectrometry and Ion Processes (NL)	101	100 179	81.87
277	Ultrasound in Medicine and Biology (UK)	101	100 280	81.96
278	Izvestiya Academy of Sciences USSR Physics of the Solid Earth (USA)	100	100 380	82.04
279	Soviet Journal of Nondestructive Testing (USA)	99	100 479	82.12
280	International Journal for Numerical Methods in Engineering (UK)	98	100 577	82.20
281	Physics of the Earth and Planetary Interiors (NL)	98	100 675	82.28
282	International Journal of Remote Sensing (UK)	97	100 772	82.36
283	International Journal of Radiation Biology (UK)	97	100 869	82.44
284	Optik (Germany)	96	100 965	82.52
285	Annals of Physics (USA)	95	101 060	82.59
286	General Relativity and Gravitation (USA)	95	101 155	82.67
287	Journal of Low Temperature Physics (USA)	95	101 250	82.75
288	Optical and Quantum Electronics (UK)	95	101 345	82.83
289	Molecular Crystals and Liquid Crystals (UK)	93	101 438	82.90
290	Acustica (Germany)	92	101 530	82.98
291	Information Bulletin on Variable Stars (Hungary)	92	101 622	83.05

Rank	Title (Country)	Number of items Individual	Cumulative	Percentage of total
292	Acta Crystallographica C Crystal Structure Communications (Denmark)	92	101 714	83.13
293	Journal of Raman Spectroscopy (UK)	92	101 806	83.20
294	Revue Roumaine de Physique (Romania)	92	101 898	83.28
295	British Journal of Radiology (UK)	91	101 989	83.35
296	Soviet Powder Metallurgy and Metal Ceramics (USA)	91	102 080	83.43
297	Acta Physica Hungarica (Hungary)	90	102 170	83.50
298	Russian Metallurgy (USA)	89	102 259	83.57
299	Nippon Seramikkusu Kyokai Gakujutsu Ronbunshi (Japan)	89	102 348	83.65
300	Pramana (India)	89	102 437	83.72
301	Science in China A Mathematics, Physics, Astronomy (China)	87	102 524	83.79
302	Vacuum (UK)	87	102 611	83.86
303	Acta Physica Polonica B (Poland)	86	102 697	83.93
304	Health Physics (USA)	86	102 783	84.00
305	IEEE Transactions on Medical Imaging (USA)	85	102 868	84.07
306	Izvestiya Akademii Nauk SSSR, Fizika Atmosfery i Okeana (Russia)	85	102 953	84.14
307	Acta Geophysica Sinica (China)	84	103 037	84.21
308	Acta Mechanica Sinica (China)	84	103 121	84.28
309	Transactions of the ASME, Journal of Fluids Engineering (USA)	84	103 205	84.35
310	Journal of Climate (USA)	83	103 288	84.41
311	Acta Mechanica (Austria)	82	103 370	84.48
312	Experimental Fluids (Germany)	82	103 452	84.55
313	Ukrayins'kyi Fizychnyi Zhurnal (Ukraine)	82	103 534	84.61
314	Comptes Rendus de l'Academie des Sciences I, Mathematique (France)	81	103 615	84.68
315	Progress of Theoretical Physics Supplement (Japan)	81	103 696	84.75
316	Advanced Materials (Germany)	80	103 776	84.81
317	Review of Laser Engineering (Japan)	80	103 856	84.88

318	International Journal of Pressure Vessels and Piping (UK)	80	103 936	84.81
319	IEE Proceedings J, Optoelectronics (UK)	79	104 015	85.01
320	Fatigue and Fracture of Engineering Materials and Structures (UK)	79	104 094	85.07
321	Muon Catalyzed Fusion (Switzerland)	79	104 173	85.14
322	Mechanics of Composite Materials (USA)	78	104 251	85.20
323	Boundary-Layer Meteorology (NL)	76	104 327	85.26
324	Geophysical Journal (UK)	76	104 403	85.33
325	International Communications in Heat and Mass Transfer (UK)	76	104 479	85.39
326	Journal of Rheology (USA)	76	104 555	85.45
327	Letters in Mathematical Physics (NL)	75	104 630	85.51
328	Heat Transfer – Soviet Research (USA)	74	104 704	85.57
329	Nuclear Technics (China)	73	104 777	85.63
330	Ophthalmic and Physiological Optics (UK)	73	104 850	85.69
331	Foundations of Physics (USA)	73	104 923	85.75
332	International Journal of Multiphase Flow (UK)	73	104 996	85.81
333	Electroencephalography and Clinical Neurophysicology (Ireland)	73	105 069	85.87
334	Journal of Biomedical Engineering (UK)	73	105 142	85.93
335	Clinical Physics and Physiological Measurement Supplement (UK)	73	105 215	85.99
336	Astrophysics (USA)	72	105 287	86.05
337	Tectonics (USA)	72	105 359	86.11
338	Composite Structures (UK)	72	105 431	86.17
339	Transactions of the ASME, Journal of Biomechanical Engineering (USA)	72	105 503	86.22
340	Acta Astrophysica Sinica (China)	71	105 574	86.28
341	Chaos (USA)	71	105 645	86.34
342	International Journal of Non-Linear Mechanics (UK)	70	105 715	86.40
343	Journal of Plasma Physics (UK)	70	105 785	86.45
344	International Journal of Optoelectronics (UK)	70	105 855	86.51
345	Oceanology (USA)	70	105 925	86.57
346	Kerntechnik (Germany)	69	105 994	86.63

Rank	Title (Country)	Number of items		Percentage of total
		Individual	Cumulative	
347	Physics Essays (Canada)	69	106 063	86.68
348	Geochimica et Cosmochimica Acta (UK)	68	106 131	86.74
349	Celestial Mechanics and Dynamical Astronomy (NL)	68	106 199	86.79
350	Journal of Atmospheric and Oceanic Technology (USA)	67	106 266	86.85
351	Journal of Geomagnetism and Geoelectricity (Japan)	67	106 333	86.90
352	Journal of Speech and Hearing Research (USA)	67	106 400	86.96
353	Canadian Acoustics (Canada)	66	106 466	87.01
354	Fluid Mechanics – Soviet Research (USA)	66	106 532	87.07
355	Documenta Ophthalmologica (NL)	66	106 598	87.12
356	Annales Geophysicae, Atmospheres Hydrospheres (France)	66	106 664	87.17
357	Revista Mexicana Fisica (Mexico)	66	106 730	87.23
358	Solar System Research (USA)	66	106 796	87.28
359	Soviet Physics – Lebedev Institute Reports (USA)	66	106 862	87.33
360	Archives of Mechanics (Poland)	65	106 927	87.39
361	Annals of Nuclear Energy (UK)	65	106 992	87.44
362	European Biophysics Journal (Germany)	65	107 057	87.49
363	Chaos, Solitons and Fractals (UK)	64	107 121	87.55
364	Chinese Journal of Physics (Taiwan)	63	107 184	87.60
365	Journal of Molecular Liquids (NL)	63	107 247	87.65
366	Lithuanian Physics Journal (USSR)	63	107 310	87.70
367	Journal of Non-Newtonian Fluid Mechanics (NL)	62	107 372	87.75
368	Journal of Superconductivity (USA)	62	107 434	87.80
369	Philosophical Transactions of the Royal Society A, Physical Sciences (UK)	62	107 496	87.85
370	Acta Crystallographica B, Structural Science (Denmark)	61	107 557	87.90
371	Infrared Physics (UK)	61	107 618	87.95
372	Inverse Problems (UK)	61	107 679	88.00

373	Journal of Electron Spectroscopy and Related Phenomena (NL)	61	107 740	88.05
374	Moscow University Physics Bulletin (USA)	61	107 801	88.10
375	Soviet Physics – Uspekhi (USA)	61	107 862	88.15
376	Journal of the Mechanics and Physics of Solids (UK)	60	107 922	88.20
377	Kinematics and Physics of Celestial Bodies (USA)	60	107 982	88.25
378	Materials Characterization (USA)	60	108 042	88.30
379	Physics Reports (NL)	60	108 102	88.35
380	Proceedings of the Indian Academy of Sciences, Chemical Sciences (India)	60	108 162	88.40
381	Atomwirtschaft – Atomtechnik (Germany)	59	108 221	88.45
382	Canadian Journal of Earth Sciences (Canada)	59	108 280	88.49
383	Magnetic Resonance Imaging (UK)	59	108 339	88.54
384	Journal of Reinforced Plastics and Composites (USA)	59	108 398	88.59
385	Geotectonics (USA)	59	108 457	88.64
386	Journal of the Acoustical Society of Japan (Japan)	59	108 516	88.69
387	Publications of the Astronomical Society of Japan (Japan)	59	108 575	88.73
388	Radiology (USA)	59	108 634	88.78
389	Applied Spectroscopy (USA)	58	108 692	88.83
390	Cosmic Research (USA)	58	108 750	88.88
391	Health Physics (UK)	58	108 808	88.93
392	Applied Acoustics (UK)	57	108 865	88.97
393	Deep-Sea Research A, Oceanographic Research Papers (UK)	57	108 922	89.02
394	Indian Journal of Radio and Space Physics (India)	57	108 979	89.07
395	Steel Research (Germany)	57	109 036	89.11
396	JSME International Journal I, Solid Mechanics Strength of Materials (Japan)	57	109 093	89.16
397	Zisin, Journal of the Seismological Society of Japan (Japan)	57	109 150	89.20
398	Radiobiologiya (Russia)	56	109 206	89.25
399	International Journal of Fatigue (UK)	56	109 262	89.30
400	Journal of Soviet Laser Research (USA)	56	109 318	89.34
401	Nuovo Cimento C (Italy)	56	109 374	89.39

Rank	Title (Country)	Number of items		Percentage of total
		Individual	Cumulative	
402	Chinese Journal of Nuclear Physics (China)	55	109 429	89.43
403	International Journal of Thermophysics (USA)	55	109 484	89.48
404	Journal of the Japan Institute of Light Metals (Japan)	55	109 539	89.52
405	International Journal of Hyperthermia (UK)	55	109 594	89.57
406	Foundations of Physics Letters (USA)	55	109 649	89.61
407	Surface and Interface Analysis (UK)	55	109 704	89.66
408	Computers and Fluids (UK)	54	109 758	89.70
409	Experimental Mechanics (USA)	54	109 812	89.75
410	Indian Journal of Theoretical Physics (India)	54	109 866	89.79
411	International Journal of Mechanical Sciences (UK)	54	109 920	89.83
412	Fluid Dynamics Research (NL)	54	109 974	89.88
413	Journal of the Less-Common Metals (Switzerland)	53	110 027	89.92
414	Nonlinearity (UK)	53	110 080	89.96
415	Transactions of the ASME, Journal of Engineering Materials and Technology (USA)	53	110 133	90.01

Notes

(1) This list has been prepared from data kindly supplied by the staff of Inspec. Titles and country of publication are taken from the INSPEC *List of Journals 1992/3*.

(2) Where title changes have occurred during the year, the total has been adjusted accordingly.

(3) Plenum have announced a number of title changes for Russian journals in translation with effect from January 1993. A list of these is given below for reference:

Old title	New Title
Soviet Physics – Acoustics	*Acoustical Physics*
Soviet Astronomy	*Astronomy Reports*
Soviet Astronomy Letters	*Astronomy Letters*
Soviet Physics – Crystallography	*Crystallography Reports*
Soviet Physics – Doklady	*Physics – Doklady*
Soviet Journal of Low Temperature Physics	*Low Temperature Physics*
Soviet Journal of Plasma Physics	*Plasma Physics Reports*
Soviet Physics – Semi-conductors	*Semiconductors*
Soviet Physics – Solid State	*Physics of the Solid State*
Soviet Physics – Technical Physics	*Technical Physics*
Soviet Technical Physics Letters	*Technical Physics Letters*

Appendix B

Acronyms and initialisms

AACR2	Anglo-American Cataloguing Rules, 2nd edn
AAM	American Academy of Mechanics (Urbana, IL)
ACM	Association for Computing Machinery
ACS	American Chemical Society
AD	ASTIA Document (Now a DTIC/NTIC report series)
ADA	Automatic Data Acquisition
AEA	formerly UKAEA, Atomic Energy Authority (Harwell and Risley)
AEC	US Atomic Energy Commission
AEG	Association of Engineering Geologists
AERE	Atomic Energy Research Establishment
AGARD	Advisory Group for Aeronautical Research and Development
AGU	American Geophysical Union
AI	Artificial Intelligence
AIAA	American Institute of Aeronautics and Astronautics
AIChE	American Institute of Chemical Engineers
AIEE	American Institute of Electrical Engineers
AI-HENP	Artificial Intelligence in High Energy and Nuclear Physics
AIP	American Institute of Physics
ALA	American Library Association (Chicago, IL)
ALEPH	A detector for LEP Physics
ALGOL	Algebraically Oriented Language
A Math S *or* AMS	American Mathematical Society
A Met S	American Meteorological Society
ANN	Analog Neural Network

ANN	Artificial Neural Network
ANSI	American National Standards Institute
APS	American Physical Society
APS	American Philosophical Society
ASA	Acoustical Society of America
ASI	NATO Applied Science Institute Series
ASLIB	Association of Special Libraries and Information Bureaux
ASME	American Society of Mechanical Engineers
ASTIA	Armed Services Technical Information Agency
ASTM	American Society for Testing Materials
ATI	Atomindex (INIS)
BAM	Bundesanstalt für Materialforschung und -prüfung
BASIC	Beginners All-purpose Symbolic Instruction Code
BCS	British Computer Society
BHRA	British Hydromechanics Research Association
BIDS	Bath Information and Data Services
BIPM	Bureau International des Poids et Mesures
BITNET	Because It's Time NETwork
BL	British Library
BLAISE	British Library Automated Information Service
BLDSC	British Library Document Supply Centre (formerly BLLD)
BLLD	British Library Lending Division (now BLDSC)
BMFT	Bundesministerium für Forschung und Technologie (Bonn, Germany)
BNB	British National Bibliography
BNL	Brookhaven National Laboratory
BRD	Bundesrepublik Deutschland (see also FRG)
BRDR	British Research and Development Reports
BRITS	British Theses Service
BRS	Bibliographic Retrieval Service
BROND	Russian Evaluated Neutron Data Library
BRTT	*British Reports, Translations and Theses*
CA	*Chemical Abstracts*
CAAL	Corporate Author Authority List
CAMAC	Computer Automated Measurement and Control
CARL	Colorado Alliance of Research Libraries
CAS	Chemical Abstracts Service (Online)
CASE	Computer Assisted Software Engineering
CASSI	*Chemical Abstracts Service Source Index*
CATEL	Electronic Catalogue of the Office of Official Publications of the European Community
CD	Compact Disc
CDA	*Comprehensive Dissertation Abstracts*
CDF	Collider Detector Facility
CD-ROM	Compact Disc – Read Only Memory
CDST	Centre de Documentation Scientifique et Technique, Paris (See CNRS)

CEA	Commissariat à l'Energie Atomique (Saclay)
CEC	Commission of the European Communities
CEDOCAR	Centre de Documentation de l'Armament (Paris)
CEI	Commission Électrotechnique Internationale (= eng. IEC)
CENDI	Commerce, Energy, NASA and Defense Information Group (USA)
CERN	Organisation Européenne pour la Recherche Nucléaire (European Organization for Nuclear Research, Geneva)
CHEMTRAN	Chemshare Corporation of Houston (Texas)
CHEP	Computing in High Energy Physics
CHEST	Combined Higher Education Software Team
CINDA	Computer Index of Neutron Data
CINDAS	Centre of Information on Numerical Analysis and Synthesis
CIPA	Chartered Institute of Patent Agents
CIPM	International Committee on Weights and Measures
CISM	Centre International des Sciences Mécaniques
CISTI	Canada Institute for Scientific and Technical Information (Ottawa, Canada)
CITE	*Consolidated Index of Translations into English*
CNRS-INIST	Centre National de la Recherche Scientifique – Institut de l'Information Scientifique et Technique, (Vandoeuvre-lès-Nancy, France)
CODATA	Committee on Data for Science and Technology
COMPENDEX	Computerized *Engineering Index*
COMPUMAG	Conference on the Computation of Magnetic Fields
CONF(DE)	A major report series of the US Department of Energy
CONF(FIZ)	Conference database of FIZ Karlsruhe
CONPAPS	*Conference Papers in Applied Physical Sciences*
CONSER	Cooperative Online Serials Program
CORDIS	Community Research and Development Information Service – Online database (Host: ECHO)
COSATI	Committee on Scientific and Technical Information
COSPAR	Committee on Space Research
COSTED	Committee on Science and Technology in Developing Countries
CP	*Computers in Physics*
CPAA	*Current Physics Advance Abstracts*
CPC	Community Patent Convention
CPG(EPS)	Computer Physics Group (European Physical Society)
CPI	*Conference Papers Index*
CRC	Chemical Rubber Company
CSEWG	Computer Science Education Working Group
CTI	Computers in Teaching Initiative (UK)

CUP	Cambridge University Press
DAI	*Dissertations Abstracts International*
DAQ	Data Acquisition
DATASTAR	Online bibliographic retrieval service
DATAPRO	Data-processing service
DBMS	Database Management System
DDC	Defense Documentation Centre (now DTIC)
DDC	*Dewey Decimal Classification*
DDR	*(formerly)* Deutsche Demokratische Republik
DE	US Department of Energy report identification
DECHEMA	Deutsche Gesellschaft für Chemisches Apparatwesen
DESY	Deutsches Elektronen-Synchrotron (Hamburg, Germany)
DG	Directorate-General (of European Communities)
DIALOG	Online Information Retrieval Service
DIANE	(now) *I'M (Information Market)* Guide
DIMDI	Online Bibliographic Retrieval Service (Cologne, Germany)
DNA	Deoxyribosonucleic Acid
DOE	US Department of Energy
DPG	Deutsche Physikalische Gesellschaft
DTIC	Defense Technical Information Centre (Alexandria, VA)
DTP	Desktop publishing
EABS	*Euro-Abstracts*
EAGLE	European Association for Grey Literature Exploitation (The Hague, NL)
EC	European Communities
ECHO	EC Online Information Retrieval Service (Luxembourg)
EDB	Energy Data Base
EEA	*Electrical and Electronics Abstracts*
EEC	European Economic Community
EI	*Engineering Index*
ENDF	Evaluated Nuclear Data File
ENDL	Evaluated Nuclear Data Library
ENSDF	Evaluated Nuclear Structure Data File
EPA	Environmental Protection Agency
EPC	European Patent Convention
EPO	European Patent Office
EPRI	Electric Power Research Institute (Palo Alto, CA)
EPS	European Physical Society
ERA	*Energy Research Abstracts* (or Electrical Research Association, Leatherhead, UK)
ERDA	Electronics Research and Development Agency (or Energy Research and Development Administration)
ESA	European Space Agency (Paris)

ESA-IRS	European Space Agency – Information Retrieval Service
ESD	Electrostatic Discharge
ESONE	European Studies on Norms for Electronics
ETANN	Electronically Trainable Analog Neural Network
ETC	European Translations Centre (now ITC)
EUR	EC report series identification
EURATOM	(now see EUR)
EURONET	European Online Information Network
EUROSAM	European Seminar on Applicable Mathematics
EUSIDIC	European Society of Information Distribution Centres
EXFOR	Exchange Format database for experimental nuclear reaction data
FID	International Federation for Documentation
FIZ	Fachinformationszentrum (Karlsruhe, Germany)
FORSIM	Fortran Simulation
FRG	Federal Republic of Germany
FSF	Free Software Foundation
FTP	File Transfer Protocol
GAMM	Gesellschaft für Angewandte Mathematik und Mechanik
GKS	Graphical Kernel System
GNU	GNU's Not UNIX
GPSS	General Process Simulation System
GRA&I	*Government Reports Announcements and Index*
HEPI	High Energy Physics Index-Database
HMSO	Her Majesty's Stationery Office
HVRA	Heating and Ventilating Research Association
HUST	Huazhong University of Science and Technology
IAA	*International Aerospace Abstracts*
IAEA	International Atomic Energy Agency (Vienna)
IAMAP	International Association of Meteorology and Atmospheric Physics
IAU	International Astronomical Union
IBM	International Business Machines Corporation
IBS	International Bibliographic System
ICDD	International Center for Diffraction Data (Swarthmore, PA)
ICM	International Conference on Magnetism
ICP	*Index of Conference Proceedings Received*
ICSD	Inorganic Crystal Structure Database
ICSTI	International Council for Scientific and Technical Information
ICSU-AB	*formerly* International Council of Scientific Unions Abstracting Board (*now* ICSTI)
IEE	Institution of Electrical Engineers (London)
IEEE	Institute of Electrical and Electronics Engineers (New York)

IFIP	International Federation for Information Processing
IGY	International Geophysical Year (1957)
IKBS	Intelligent Knowledge Based System
I'M	*Information Market Guide*
IM3	International Conference on Magnetism and Magnetic Materials
IMECO	International Measurement Confederation
IMTC	IEEE Instrumentation and Measurement Technology
INID	Internationally agreed numbers for the Identification of Documents
INIS	International Nuclear Information Service
INIS ATOMINDEX	Printed version of ATOMINDEX Database
INPADOC	International Patent Documentation Centre (Vienna)
INP 1(& 2)	International Patents Information (online information service provided by the French Patent Office 1:French and 2:European)
INSPEC	Information Sources in Physics, Electrotechnology and Control, IEE
INTCL	International Class
INTERMAG	International Conference on Magnetics
IOP	Institute of Physics
IPC	*International Patent Classification*
IPG	*International Patent Gazette* (INPADOC)
IPTS	International Practical Temperature Scale
IRS	Information Retrieval Service (Online)
ISBN	International Standard Book Number
ISHTCP	*Inventory of Sources for History of Twentieth Century Physics*
ISI	Institute for Scientific Information
ISIP	*Information Sources In Physics*
ISO	International Standards Organization
ISSN	International Standard Serial Number
ISTP	*Index of Scientific and Technical Proceedings*
ITC	International Translations Centre (Delft)
IUC	International Union of Crystallography
IUPAC	International Union of Pure and Applied Chemistry
IUPAP	International Union of Pure and Applied Physics
IUTAM	International Union of Theoretical and Applied Mechanics
JANET	Joint Academic Network (UK)
JCPDS	Joint Centre for Powder Diffraction Services (now International Centre for Diffraction Data)
JEF	Joint Evaluation File
JENDL	Japanese Evaluated Nuclear Data Library
JETP	*Journal of Experimental and Theoretical Physics*

JICST	Japan Information Center for Science and Technology
JINR-R	*Joint Institute of Nuclear Research Report* (Dubna, Russia)
JPRS	Joint Publications Research Service
JRS	*Journal of Raman Spectroscopy*
JW	John Wiley publishers
KIBV	Kreisler Import BV publishers (The Hague, NL)
KIST	*Keyword Index to Serial Titles*
KWIC	Keyword in Context
LAN	Local Area Network
LAPP	Laboratoire de Physique des Particules (Annecy)
LASER	LaserMedia publishers (Paris)
LASP	Laboratory for Atmospheric & Space Physics (Boulder, CL)
LBL	Lawrence Berkeley Laboratories
LC	Library of Congress (Washington, DC)
LCSH	*Library of Congress Subject Headings*
LEP	Large Electron Positron Collider
LHC	Large Hadron Collider
LLL	Lawrence Livermore Laboratory
LOC	Longman Cartermill publishers
MAG	*IEEE Transactions on Magnetics*
MATHFILE	Database of *Mathematical Reviews*
METADEX	*Metal Abstracts Index* Database
MHD	Magnetohydrodynamics
MIMD	Multiple Instruction Multiple Data
MIT	Massachusetts Institute of Technology
MKS	Metre-kilogramme-second (system of units)
MMM	Conference on Magnetism and Magnetic Materials
MOS	Metal-Oxide Semiconductor
MPL	Microprocessor Programming Language
MT	Magnet Technology
NACA	National Advisory Committee for Aeronautics (now NASA)
NAS	National Academy of Sciences (Washington, DC)
NASA	National Aeronautics and Space Administration
NATO	North Atlantic Treaty Organization
NBS	*(now)* NIST – National Institute for Standards and Technology
NBS-TN	*(now)* NIST – National Institute for Standards and Technology – TN document series
NCTPR	New Computing Techniques in Physics Research
NDB	Numerical Data Base
NDT	Nondestructive Testing
NEA	Nuclear Energy Agency
NGC	Nebular Galactic Cluster
NIM	Nuclear Instrumentation Modules
NIM	*Nuclear Instruments and Methods*

NISC	National Information Services Corporation (Baltimore, MD)
NIST	National Institute for Standards and Technology, formerly NBS)
NLLST	National Lending Library for Science and Technology (*now* BLDSC)
NN	Neural Network
NOAA	National Oceanographic and Atmospheric Administration
NRCC	National Research Council of Canada
NSA	*Nuclear Science Abstracts* (*now* ERA)
NSR	Nuclear Structure References
NSRDS	National Standard Reference Data Series
NTC	National Translations Center (no longer exists) now part of LC and others
NTIS	National Technical Information Service (Springfield, VA)
NUC	*National Union Catalogue*
NUREG	Nuclear Regulatory Commission (Washington, DC)
OATS	Office of Air Transportation Security
OCLC	Online Computer Library Center
OECD	Organisation for Economic Cooperation and Development, (Paris)
OHST	Office for the History of Science and Technology (Berkeley, CA)
OMS	Organic Mass Spectrometry
OO	Object Oriented
OPAC	Online Public Access Catalogue
OPAL	Omni Purpose Apparatus for LEP
ORBIT	Online Retrieval Service (Santa Monica, CA)
OSTI	Office for Scientific and Technical Information, US Dept of Energy (Oak Ridge, TN)
OTS	Office of Technical Services (now NTIS)
OUP	Oxford University Press
PA	*Physics Abstracts*
PACS	*Physics and Astronomy Classification Scheme*
PAS	Patent Applicant Service
PASCAL	Programme Appliqué à la Selection et à la Compilation Automatique de la Litérature
PATSEARCH	Online searching service on US Patents
PC	Personal Computer
PCS	Patent Classification Service
PCT	Patent Cooperation Treaty
PEX	PHIGS Extensions to X
PFS	Patent Family Service
PHIGS	Programmer's Hierarchical Interactive Graphics Standard
PHYS	Physics database of FIZ Karlsruhe
PINET	Physics Information Network

PIS	Patent Inventor Service
PIXE	Proton-Induced X-ray Emission
PL/1	Programming Language (for computers)
PL/2	Programming Language (for computers)
PLM	*Programming Logic Manual*
PPDS	Physical Property Data Service
PRM	PRM Science and Technology Agency Ltd, (London, UK)
QED	Quantum Electrodynamics
QUESTEL/DARC	Online retrieval service suppliers (Oak Ridge, TE)
R&D	Research and Development
RAL	Rutherford Appleton Laboratory
RAND	RAND Corporation
RAS	Royal Astronomical Society
RECON	Online service suppliers (ESA, Frascati and Oak Ridge, TE)
RMP	*Reviews of Modern Physics*
ROM	Read Only Memory
RTL/2	Programming language
SASD	Structured Analysis Structured Design
SASD-RT	SASD Real Time
SCAD	EC online database
SCI	*Science Citation Index*
SCISEARCH	*Science Citation Index*, Data Base
SCOPE	Standing Committee on Problems of the Environment
SCR	Scientific Computer Network
SDC	Systems Development Corporation
SEMT	Science/Engineering/Medicine/Technology
SERB	*Software Engineer's Reference Book*
SGML	Standard Generalized Mark-up Language
SHE	*Subject Headings for Engineering*
SI units	Système International (of physical units)
SIAM	Society for Industrial and Applied Mathematics
SIG	Special Interest Group
SIGGRAPH	Special Interest Group on Computer Graphics
SIGLE	System for Information on Grey Literature in Europe
SIGMICRO	Special Interest Group on Microprogramming
SIGNUM	Special Interest Group on Numerical Control
SIGSAM	Special Interest Group on Symbolic and Algebraic Manipulation
SIMD	Single Instruction Multiple Data
SLAC	Stanford Linear Accelerator Center
SONAR	Sound Navigation and Ranging
SP	SilverPlatter publishing
SPARC	Space Program Analysis and Review Council
SPIE	Society of Photo-Optical Instrumentation Engineers

SPIN	Searchable Physics Information Notices (Database)
SQL	Structured Query Language
SQUID	Superconducting Quantum Interference Detector
SRIM	Selected Research in Microfiche
SRIS	Science Reference and Information Service, British Library (London) formerly SRL
SRL	(now SRIS)
SRM	Standard Reference Material
STAR	*Scientific and Technical Aerospace Reports*
STN	Online information retrieval service
SU3	Unitary Symmetries
SU6	Unitary Symmetries
SUN	Symbols, Units and Nomenclature (International Committee)
SUNIST	Online information service
SUSSP	*Scottish Universities Summer School Proceedings*
TANSA	Database
TCPIP	Transmission Control Protocol Internet Protocol
TID	Technical Information Document (US Department of Energy)
TPRC	Thermophysical Properties Research Center
TRC	Selected Data on Thermodynamics and Spectra (Texas Research Center)
TRI	*Translations-Register Index*
TT	Technical Translations series
UB&TIB	Universitätsbibliothek und Technisches Informationsburo (Hanover, Germany)
UDC	*Universal Decimal Classification*
UIS	*Universal Indexing Schedule*
UK	United Kingdom
UKAEA	(now AEA Technology, Risley, and Harwell Research, UK)
UKNDL	United Kingdom Neutron Data Listing
UMI	University Microfilms International (Ann Arbor, MI)
Unesco	United Nations Educational, Scientific and Cultural Organization
UNISIST	Universal System for Information in Science and Technology
UNIX	Universal time-sharing computer operating system
UP	University Press
USA	United States of America
USAEC	US Atomic Energy Commission (now US Department of Energy)
USSR	Union of Soviet Socialist Republics (now disintegrated) *present US collective term is* 'Central Eurasia'
VDI	Vertical Display Indicator
VDU	Visual Display Unit

VHS	Video Helical Scan
VHSIC	Very High Speed Integrated Circuits
VINITI	Vsesoyuznyi Institut Nauchnoy i Teknicheskoy Informatsii (still describing itself thus, but INITI RAN or Russkoy Akademiia Nauk are also used)
VLSI	Very Large Scale Integration
WATFIV	Waterloo version of FORTRAN IV
WIPO	World Intellectual Property Organization
WMO	World Meteorological Organization
WWW	World-Wide Web
W3	World-Wide Web
XRS	X-ray Spectrometry
XTAL	Crystal Data Identification File

Appendix C

Index of publishers: abbreviations and addresses

The following list shows in the left-hand column (alphabetically) the abbreviation used throughout the text to denote a publisher. In the right-hand column the full name and address of the publisher is listed (note: CAPITALS file ahead of lower case, thus: ACA appears before Aargauer, etc.). Where the publisher is well known, the address is restricted to town, postcode and country except in large cities, e.g. New York, London and Berlin, where a street location is usually given as well. The list includes addresses of some Database hosts.

The addresses have been derived from four major sources:
Books in Print (Bowker, 1992)
Cumulative Book Index 1991 (H W Wilson, 1992)
List of Journals 1992/93 (INSPEC, 1992) and
Scientific & Technical Books and Serials in Print 1991 (Bowker, 1990).

A change of editorship of a journal or the takeover of one publisher's list by another often leads to a change of address and therefore, in case of doubt, the reader is recommended to check the latest editions of these directories.

AAM	American Academy of Mechanics, 212 Talbot Laboratory, Urbana, IL 61801, USA
AAPT	American Association of Physics Teachers, 335 E45S, New York, NY 10017, USA
ACA	American Crystallographic Association, 335 E45S, New York, NY 10017, USA
ACBlack	A & C Black, 35 Bedford Row, London WC1R 4JH, UK
ACLS	American Council of Learned Societies, Charles Scribner's Sons, 597 Fifth Avenue, New York, NY 10017, USA

ACM	Association of Computing Machinery, 11 W42S, New York, NY 10036, USA
ACS	American Chemical Society, 1155 Sixteenth St NW, Washington, DC 20036, USA
AEG	AEG Telefunken, Elitera Verlag, 1 Berlin 33, Germany
AFIPS	American Federation of Information Processing Services Inc., 1899 Preston White Drive, Reston, VA 22091, USA
AGARD	Advisory Group for Aerospace Research and Development, 7, rue Ancelle, 92200 Neuilly-sur-Seine, France
AGU	American Geophysical Union, 2000 Florida Avenue, Washington, DC 20009, USA
AIAA	American Institute of Aeronautics and Astronautics, 370 L'Enfant Promenade SW, Washington, DC 20024, USA
AIChE	American Institute of Chemical Engineers, 345 E 47 St., New York, NY 10017, USA
AIEE	See IEEE
AIIE	American Institute of Industrial Engineers, 25 Technology Park/Atlanta, Norcross, GA 30092, USA
AIP	American Institute of Physics, 335 E45S, New York, NY 10017–3483, USA
ALA	American Library Assoc., 50 East Huron St, Chicago, IL 60611, USA
AMathS	American Mathematical Society, PO Box 6248, Providence, RI 02940, USA
AMetS	American Meteorological Society, 45 Beacon St, Boston, MA 02108, USA
ANSI	American National Standards Institute, 1430 Broadway, New York, NY 10018, USA
APS	American Philosophical Society, 104 S. Fifth St, Philadelphia, PA 19106, USA
ASJpn	Astronomical Society of Japan, Astronomical Observatory, 2–21–1 Osawa, Mitaka-shi, Tokyo 181, Japan
ASLIB	Association of Special Libraries and Information Bureaux, Information House, 20–24 Old Street, London EC1V 9AP, UK
ASME	American Society of Mechanical Engineering, 345 E47S, New York, NY 10017, USA
Aargauer	Aargauer Tagblatt AG, Buchverlag, CH 5001 Aarau, Bahnhofstrasse 39–43, Switzerland
Abacus	Abacus Press, c/o Gordon & Breach, PO Box 90, Reading RG1 8JL, UK
Abrams	Harry N. Abrams Inc., 110 E59S, New York, NY 10022, USA

Academic	Academic Press Inc., 111 Fifth Ave, New York, NY 10003, USA; 6277 Sea Harbour Drive, Orlando FL 32887, USA; and 24–28 Oval Road, London NW1 7DX, UK
Addison-Wesley	Addison-Wesley Publishing Co., Reading, MA 01867, USA
Advanced Technology	Advanced Technology Publications Inc., 1001 Watertown Street, Newton, MA 02165, USA
Akad.Verl.Leip.	Akademische Verlagsgesellschaft, Leipzig, Germany
Akad.Verl.Wies.	Akademische Verlagsgesellschaft, Postfach 1107, D-6200 Wiesbaden, Germany
Akademie	Akademie Verlag, Leipzigerstrasse 3–4, O-1086 Berlin, Germany
Allen & Unwin	Allen & Unwin, Winchester, MA 01890, USA; and 40 Museum St, London WC1A 1LU, UK
Allerton(NY)	Allerton Press Inc., 150 Fifth Avenue, New York, NY 10011, USA
Allyn	Allyn & Bacon Inc., 470 Atlantic Avenue, Boston, MA 02210, USA; and 1 Bedford Rd, London N2, UK
Amacon	Amacon, 135 W 50 St (15th Floor), New York, NY 10020, USA
American Elsevier	American Elsevier Publishing Co., 52 Vanderbilt Ave, New York, NY 10017, USA
Annual Reviews	Annual Reviews Inc., 4139 El Camino Way, Palo Alto, CA 94306, USA
Applied Science	Applied Science Publishing Ltd., Barking, IG11 8JU, UK
Arno	Arno Press, 3 Park Ave, New York, NY 10016, USA
Arnold	Edward Arnold Publishers Ltd., Mill Road, Sevenoaks, TN13 2YA, UK
Artech	Artech House Inc., 6 Buckingham Gate, London SW1E 6JP, UK
Atheneum	Atheneum Publishers Inc., 866 Third Avenue, New York, NY 10022, USA
Athlone	Athlone Press, 44 Bedford Row, London WC1B 3PY, UK
Atomizdat	Atomizdat, Ul. Zvanova 5–7, 103031 Moskva K-31, Russia
Austin	Austin, LBJ School of Public affairs, Univ. of Texas, Austin, TX 78712, USA
BBC	British Broadcasting Corporation Publications, 80 Wood Lane, London W12 0TT, UK
BIW	Bibliografisches Institut Wissenschaftsverlag, Zurich, Switzerland
BL	see British Library

BLAISE	British Library Automated Information Service, National Bibliographic Service, Boston Spa, Wetherby, West Yorkshire, LS23 7BQ, UK
BLDSC	British Library Document Supply Centre, Boston Spa, Wetherby, West Yorkshire, LS23 7BQ, UK
BNL	Brookhaven National Laboratory, Upton, Long Island, NY 11973, USA
BRS	BRS Information Technologies Inc., 8000 Westpark Dr., McLean, VA 22102, USA
BS	see Bowker-Saur
Barnes & Noble	Barnes & Noble Books (a division of Rowman & Littlefield), 4720 Boston Way, Lanham, MD 20706, USA
Barth	J. A. Barth, Salmonstrasse 18B, Leipzig 701, Germany
Basic	Basic Books, 10 E53S, New York, NY 10022, USA, and c/o Harper & Row, London, UK
Bath University Press	Bath University Press, Library, Claverton Down, Bath BA2 7AY, UK
Bell	George Bell and Sons, now Bell & Hyman Ltd., Denmark House, 37–39 Queen Elizabeth St, London SEI 20B, UK
Bell Labs.	Bell Telephone Laboratories, Murray Hill, NJ 07974, USA
Benjamin	see Benjamin Cummings
Benjamin/Cummings	Benjamin/Cummings Publishing Co. Inc., c/o Addison-Wesley Publishing Co., Reading, MA 01867, USA
Bingley	Clive Bingley Ltd., 7 Ridgmount St, London WC1A 1BT, UK
Birkhauser	Birkhauser Verlag, Klosterberg 23, CH-4010, Basel, Switzerland; and Birkhauser Boston Inc., 675 Massachusetts Avenue, Cambridge, MA 02139, USA
Blackie	Blackie (Publ.) Sales Ltd., 7 Leicester Place, London WC1, UK
Blackwell	Blackwell Scientific Publications Ltd., Oxford OX2 0EL, UK
Blaisdell	(now) Xerox College Publishing, Lexington, MA 02173, USA
Blanchard	Albert Blanchard, Librairie Scientific & Technique, 9 rue de Medicis, F-75006, Paris, France
Bowker-Saur	Bowker-Saur Co., c/o Reed Publishing (USA) Inc., 121 Charlon Road, New Providence, NJ 07974, USA; also 59/60 Grosvenor Street, London, W1X 9DA, UK
British Library	British Library Publications, 2 Sheraton St, London W1V 4BH, UK

Bunko Kenkya	Spectroscopical Society of Japan, 1–13 Kanda-Awaji-cho, Chiyuda-ku, Tokyo 101, Japan
Burndy	Burndy Library, Norwalk, CT 06856, USA
Business Books	Business Books, 20 Vauxhall Bridge Road, London SW1V 2SA, UK
Butterworths	Butterworths Scientific Ltd., Westbury House, Guildford, Surrey GU2 5BH, UK; and 88 Kingsway, London WC2B 6AB, UK
CCDC	Cambridge Crystallographic Data Centre, University Chemical Laboratory, Lensfield Rd, Cambridge CB2 lEW, UK
CCM Information	CCM Information Service, Crowell, Collier and Macmillan, 866 Third Ave, New York, NY 10022, USA
CEA	Commissariat à l'Energie Atomique, 91191 Gif-sur-Yvette, Cedex, France
CEC	Office for Official Publications of the EC, 2 rue Mercier, L-2985 Luxembourg
CERN	Centre Européenne pour la Recherche Nucléaire, CH-1211, Geneva 23, Switzerland
CISM	Centre Internationale des Sciences Mécaniques, c/o World Scientific Publishing Co. Ltd., PO Box 128, Farrer Rd, Singapore 9128, Singapore
CNRS	see CNRS-INIST
CNRS-INIST	INIST Diffusion, 2 Allée de Parc de Brabois, F-54514 Vandoeuvre-lès-Nancy, France
CPR	CPR Press Inc., West Palm Beach, FL 33014, USA
CRC Press	CRC Press Inc., 2000 NW24S, Boca Raton, FL 33431, USA
CSA	see Cambridge SA
CSIR(India)	Council of Scientific & Industrial Research, Rafi Marg, New Delhi, India
CUP	Cambridge University Press, The Edinburgh Building, Shaftesbury Road, Cambridge CB2 2RU, UK
Cahners	Cahners Publishing Co., 221 Columbus Ave, Boston, MA 02116, USA
Cambridge SA	Cambridge Scientific Abstracts Inc., Riverdale, MD 20814, USA
Cavitation	c/o McGraw-Hill Book Co., New York and Maidenhead
Chalmers	Chalmers University of Technology, Sven Hultins gata 2, S-41296 Gøteborg, Sweden
Chambers	W. and R. Chambers, Edinburgh EH2 lDG, Scotland
Chapman	Geoffrey Chapman Publ./Johnson & Bacon Ltd., 35 Red Lion Square, London WC1R 4SG, UK
Chapman & Hall	Chapman and Hall, 2–6 Boundary Row, London SE1 8HN, UK

Chatto and Windus	Chatto and Windus, 40–42 William IV St, London WC2 4DF, UK
Chelsea	Chelsea Publishing Company, 432 Park Avenue S, Room 503, New York, NY 10016, USA
Chemical Publishing	Chemical Publishing Co., 80 Eighth Ave, New York, NY 10014, USA
Chemical Rubber	see CRC Press
Chilton	Chilton Book Co., Radnor, PA 19089–0230, USA
Clarendon	Clarendon Press, Walton Street, Oxford OX2 6DP, UK
Compositori	Compositori S.r.l., Via Stalingrado 97/2, I-40128 Bologna, Italy
Comp.Mech.	Computational Mechanics Inc., 25 Bridge Street, Billericay, MA 01821, USA
Computer Science	Computer Science Press Inc., 1803 Research Blvd., Rockville, MD 20850, USA
Constable	Constable and Co. Ltd., 10 Orange St, London WC2H 7EG, UK
Consultants	Consultants Bureau, 233 Spring St, New York, NY 10013, USA
Cork U.Pr.	Cork University Press, University College, Cork, Ireland
Cornell U.Pr.	Cornell University Press, Ithaca, NY 14853, USA
Corsi	P. Corsi, Via Vespucci 16, 10026 Santena, Torino, Italy
Crane Russak	Crane Russak and Co. Inc., 79 Madison Ave, New York, NY 10016, USA
Crosby	Crosby Lockwood Staples, P.O. Box 9, 29 Frogmore St, St. Albans, Herfordshire AL2 2NF, UK
Cummings	see Benjamin/Cummings
DATA STAR	DATA STAR, Plaza Suite, 114 Jermyn Street, London SW1Y 6HJ, UK; also c/o Radio Suisse AG., Lanperstr. 18A, CH-3008 Berne, Switzerland
DIALOG	DIALOG Information Services Inc., 3460 Hillview Ave, Palo Alto, CA 94304, USA
DMG	Deutsche Meteorologische Gesellschaft, eV, D-5580 Traven-Trarbach, Mount Royal, Germany
DOE	US Department of Energy, PO Box 62, Oak Ridge, TE 37830, USA
DPG	Deutsche Physikalische Gesellschaft, eV, Hauptstrasse 5, D-5340 Bad Honnef, Germany
David and Charles	David and Charles, Brunel House, Newton Abbot, Devon TQ12 4PU, UK
Davies	Peter Davies Ltd., 10 Upper Grosvenor St, London W1X 9PA, UK
Dawson	Wm. Dawson & Sons Ltd., Cannon House, Park Farm Road, Folkestone, Kent, CT19 5EE, UK

Dekker	Marcel Dekker Inc., 270 Madison Avenue, New York, NY 10016, USA
Delft	Delft University, Julianalaan 134, POB 5, Delft, Netherlands
Derwent	Derwent Publications, Derwent House, 14 Great Queen Street, London WC2B 5DF, UK
Deutsch	Andre Deutsch Ltd., 105 Great Russell St, London WC1B 3LJ, UK
Douden	Douden, Hutchinson & Ross, 523 Sarah St, Box 699, Stroudsburg, PA 18360, USA
Dover	Dover Publications Inc., 180 Varick St, New York, NY 10014, USA
Droz	Librairie Droz SA, 11 rue Massot, Geneva, Switzerland
Dublin U.Pr.	Dublin University Press, Dublin 4, Ireland
Dunod	Librairie Dunod, 30 rue Saint-Sulpice, 75006 Paris, France
EAGLE	European Association for Grey Literature Exploitation, PO Box 90407, 2509 LK The Hague, Netherlands
ECHO	ECHO – European Communities Host Organisation, 5 rue Höhenhof, Senningerberg, L-1736 Luxembourg
EPRI	Electrical Power Research Institute, 3412 Hillview Ave, Palo Alto, CA 94304, USA
EPS	European Physical Society, Geneva, Switzerland
ERA	Electrical Research Association Technology Limited, Cleeve Road, Leatherhead, Surrey, KT22 7SA, UK
ESA	European Space Agency, 8–10 rue Mario Nikis, 75738 Paris Cedex 15, France; also, Via Galileo Galilei, I-00044 Frascati, Rome, Italy
ESONE	ESONE Committee: H. Meyer, Secretary, CEC Joint Research Centre, Geel Establishment, B-2440 Geel, Belgium
Edinburgh U.Pr.	Edinburgh University Press, 22 George Square, Edinburgh EH8 9LF, Scotland
Edit.Cult&Civ.	Editions Cultures et Civilisation, 115 Avenue Gabriel Lebon, 1160 Brussels, Belgium
Ellis Horwood	Ellis Horwood Ltd., Market Cross House, Cooper Street, Chichester, West Sussex, PO19 lEB, UK
Elsevier	Elsevier Science Publishers BV, PO Box 211, 1000 AE Amsterdam, Netherlands; and 655 Avenue of the Americas, New York, NY 10010, USA
Elsevier Sequoia	Elsevier Sequoia SA, PO Box 851, 1001–Lausanne, Switzerland
Encyclop.Brit.	Encyclopaedia Britannica Inc., 310 S. Michigan Avenue, Chicago, IL 60604, USA
Engelmann	Engelmann Verlag GmbH, Leipzig, Germany

English Universities	English Universities Press, see Hodder & Stoughton
Eurospan	Eurospan Ltd., 3 Henrietta St, London WC2E 8LU, UK
Eyrolles	Eyrolles, 75240 Paris, Cedex 05, France
FIZ	Fachinformationszentrum, Energie, Physik, Mathematik GmbH, 7514 Eggenstein-Leopoldshafen 2, Karlsruhe, Germany
FSIO	Francaise Société Inform. d'Optique, 10 rue de Buci, 75006 Paris, France
Faber	Faber and Faber, 3 Queen Square, London WC1N 3AU, UK
Fitzmatgiz	Foreign-language Scientific and Technical Dictionaries, Moscow
Focal (UK)	Focal Press Ltd., Westbury House, Guildford GU2 5BH, UK
Focal (USA)	Focal Press Inc., 80 Montvale Ave., Stoneham, MA 02180, USA
Foulis	G. T. Foulis & Co. Ltd., Sparkford, Yeovil, Somerset, BA22 7JJ, UK
Foulsham	G. T. Foulsham & Co. Ltd., Yeovil Road, Slough, SL1 4JH, UK
Franklin	Burt Franklin Pub., c/o Lenox Hill Publishing Co., 235 E 44S, New York, NY 10017, USA
Freeman	W. H. Freeman and Co., 41 Madison Avenue (37th Floor), New York, NY 10010, USA
Friedr. Vieweg Sohn	Friedrich Vieweg & Sohn, Verlagsgesellschaft GmbH, Postfach 5829, D-6200 Wiesbaden 1, Germany
Frontières	Editions Frontières, Gif-sur-Yvette, France
GEC(GB)	General Electric Company, 1 Stanhope Gate, London W1A 1EH, UK
GEC (USA)	General Electric Corporation, 120 Erie Blvd., Schenectady, New York, NY 12305, USA
Gale	Gale Research Co., Detroit, MI 48226–4094, USA
Gauthier-Villars	Gauthier-Villars CDR, 11 rue Gossin, 92543 Montrouge, CEDEX France
GeoAbs	Geo Abstracts Ltd., Regency House, Duke St., Norwich NR3 3AP, UK
Georgi	Georgi Publishing Company, CH-1813 St Saphorin, Switzerland
Giard	Giard, Lille, France
Golden Press	The Golden Press (imprint of Western Publ.), 850 Third Ave, New York, NY 10022, USA
Gordon and Breach	Gordon and Breach Science Publishers Ltd., PO Box 90, Reading RG1 8JL, UK; and PO Box 786, Cooper Sta., NY 10276, USA
Granada	Granada Publishing Co., c/o Grafton Books, 8 Grafton Street, London W1X 3LA, UK

Griffin	Charles Griffin & Co. Ltd., High Wycombe, UK
Gruyter	Walter de Gruyter & Co., Postfach 110240, 1000 Berlin, Germany
Gulf	Gulf Publishing Co., PO Box 2608, Houston, TX 77252, USA
HMSO	Her Majesty's Stationery Office, 49 High Holborn, London WC1V 6HB, UK
Halsted	Halsted Press (a division of John Wiley & Sons, New York) 605 Third Avenue, New York, NY 10158, USA
Hamilton	Hamish Hamilton Ltd., 27 Wright's Lane, London, W8 5TZ, UK
Hanover	Hanover Press Ltd., 13 Shepherd's Hill, London N6 SQH, UK
Harcourt-Brace	Harcourt Brace Jovanovich, 1250 Sixth Ave, San Diego, CA 92101, USA; or 24–28 Oval Road, London NW1 7DX, UK
Harper and Row	Harper and Row Pubs. Inc., 10 E53S, New York, NY 10022–5299, USA; and 34–42 Cleveland St, London W1P 5F8, UK
Harrap	Harrap Ltd., 26 Market Square, Bromley, BR1 1NA, UK
Harvard U.Pr.	Harvard University Press, Cambridge, MA 02138, USA; and 126 Buckingham Palace Road, London SW1W 9SD, UK
Harvester Press	The Harvester Press Ltd., 16 Ship St, Brighton BN1 1AD, UK
Hayden	Hayden Publishing Company Inc., 50 Essex St, Rochelle Park, NJ 07662, USA
Heinemann	William Heinemann Ltd., 10 Upper Grosvenor St, London W1X 9PA, UK
Hemisphere	Hemisphere Publishing Corp., 1010 Vermont Ave NW, Washington, DC 20005, USA; and 126 Buckingham Palace Road, London SW1W 9SD, UK
Hermann	Hermann, Editeurs des Sciences et des Arts, 293 rue Lecourbe, F-75015, Paris, France
Hermosa	Hermosa Publishers, Albuquerque, NM 87198, USA
Heyden	Heyden and Son Ltd., Spectrum House, Hillview Gardens, London NW4 2JQ, UK
Heywoods	see Butterworths
Hilger	Adam Hilger Ltd., Bristol BS1 6NX, UK
Hilger(B)	Hilger Press, Techno House, Redcliffe Way, Bristol BS1 6NX, UK
Hirzel	S. Hirzel Verlag, Postfach 347, W-7000 Stuttgart 1, Germany
Hodder & Stoughton	Hodder & Stoughton Ltd., 47 Bedford Square, London WC1B 3DP, UK

Holt-Rinehart	Holt, Rinehart and Winston Inc., 383 Madison Avenue, New York, NY 10017, USA; and 24–28 Oval Road, London, NW1 7DX, UK
Hutchinson	Hutchinson Publishing Group, 20 Vauxhall Bridge Road, London W1P 6JD, UK
Hutchinson Ross	Hutchinson Ross Publ. Co., 523 Sarah St, Box 699, Stroudsburg, PA 18360, USA
IAEA	International Atomic Energy Agency, PO Box 100, A-1400 Vienna, Austria
IBM	International Business Machines Corp. (Armonk), White Plains, NY 10504, USA
ICDD	International Center for Diffraction Data, 1601 Park Lane, Swarthmore, PA 19081–2389, USA
ICTP	International Centre for Theoretical Physics, PO Box 586, Trieste 34100, Italy
IEE	Institution of Electrical Engineers, Savoy Place, London WC2R 0BL, UK
IEEE	Institute of Electrical and Electronic Engineers, 345 E47S, New York, NY 10017, USA
IES	Illuminating Engineering Society of North America, 345 E 47 S, New York, NY 10017, USA
IFLA	IFLA Publications, c/o British Library
IFI/Plenum	IFI/Plenum, 233 Spring St, New York, NY 10013–1578, USA
IFIP	IFIP, c/o: Department of Computer Science, University of Manchester, M13 9PL, UK
IMC (UK)	Institute of Measurement and Control, 87 Gower Street, London WC1E 6AA, UK
IN2P3	Institut National de Physique Nucléaire et de Physique de Particules, 20 rue Berbier Mets, 75013 Paris, France
INIST	Institut de l'Information Scientifique et Technique, 2 allée du Parc de Brabois, 54154 Vandoeuvre-lès-Nancy Cedex, France; also at 83–85 Blvd., Vincent Auriol, 75013 Paris, France
INITI RAN	see VINITI
INSPEC	INSPEC, The Institution of Electrical Engineers, Michael Faraday House, Six Hills Way, Stevenage SG1 2AY, UK
IOP	IOP Publishing Ltd., Techno House, Bristol BS1 6NX, UK
IPC Science	IPC Science and Technology Press, Guildford GU2 5BH, UK
IPPT	Instytut Podstawowych Problemow Techniki, 00–049 Warsaw, Swietokrzyska 21, Poland
ISI	Institute for Scientific Information, Philadelphia, PA 19104, USA

ITC	International Translations Centre, Schuttersveld 2, 2611 WE Delft, Netherlands
IUC	International Union of Crystallography, c/o 5 Abbey Square, Chester CH1 2HU, UK
Iliffe	Iliffe Books Ltd., see Butterworths
Imperial College	Imperial College of Science, Technology and Medicine, South Kensington, London SW7 2AZ, UK
Infotech	Infotech Ltd., High St, Maidenhead, SL6 1LD, UK
Inst.Acoust.	Institute of Acoustics, PO Box 320, St Albans, AL1 1PZ, UK
Inst.Hyd.Res.	Institute of Hydraulic Research, University of Iowa, Iowa City, Iowa 52242, USA
Inst.Las.Doc.	Institute for Laser Documentation, Box 2070, Rolling Hills, CA 90274, USA
Int.Pubns.Serv.	International Publications Service (a division of Taylor & Francis Inc.), 1900 Frost Rd, Suite 101, Bristol, PA 19007, USA
Interscience	Interscience Publishers, New York (a division of John Wiley)
Intertext	Intertext Publ. Ltd., c/o Blackie & Son Ltd., London, UK
I.Pod.Probl.Techn.	see IPPT
Irish U.Pr.	Irish University Press, c/o Irish Academic Press, Blackrock, Co. Dublin, Ireland
Ist.Mus.Stor.	Istituto e Museo di Storia delle Scienze di Firenze, Piazza dei Guidici, Florence, Italy
JICST	Japan Information Center for Science and Technology, c/o USACO Corpn., TsuTsumi Bldg., 13–12 Shimbashi 1–chome, Minato-ku, Tokyo 105, Japan
JOIS	Japan Online Information Service, see JICST
JW	see Wiley
J.Phys.Colloq.	Journal de Physique Colloque, Zone Industrielle de Courteboeuf, BP 112, F-91944 Les Ulix Cedex A, France
John Murray	John Murray Publishers Ltd., 50 Albermarle St, London W1X 4BD, UK
Johns Hopkins	Johns Hopkins University Press, Baltimore, MD 21211, USA
Johnson Repr.	Johnson Reprint Co. Ltd., 24/28 Oval Road, London NW1 7DX, UK; and 111 Fifth Ave, New York, NY 10003, USA
Jones&Bartlett	Jones & Bartlett Publishers Inc., 20 Park Plaza, Boston, MA 02116–9990, USA
Jpn AE Soc.	Atomic Energy Society of Japan, 1–5–4, Ohte-Machi, Chi-yoda-ku- Tokyo 100, Japan
KIBV	Kreisler Import BV, PO Box 93053, 2509 AB The Hague, Netherlands

Kimber	W. Kimber & Co. Ltd., 22a Queen Anne's Gate, London SW1H 9AE, UK
Kluwer	Kluwer Academic Publishers, PO Box 322, 3300 AH Dordrecht, Netherlands; and c/o Walters Kluwer US Corp., 101 Philip Drive, Assinippi Park, Norwel, MA 02061, USA
Knopf	Alfred A. Knopf Inc. (subsidiary of Random House), 225 Park Avenue, New York, NY 10003, USA
Krieger	Krieger Publishing Co., PO Box 9542, Melbourne FL 32902, USA
LASER	LaserMedia, 38 rue de l'Ouest, F-75016, Paris, France
PASP	Laboratory for Atmospheric & Space Physics, Campus Box 590, University of Colorado, COL 90309, USA
LOC	Longman Cartermill Ltd., Technology Centre, St Andrews, Fife, KY16 8QS Scotland
LUP	University of London Press Ltd., see Hodder & Stoughton
Learned Inf.	Learned Information Publishers Inc., 143 Old Marlton Pike, Medford, NJ 08055, USA
Lewis	H. K. Lewis & Co. Ltd., 136 Gower St, London WC1E 6BS, UK
Libr. Sci.	Librairie Scientifique Albert Blanchard, Paris, France
Library Assoc.	Library Association Publ. Ltd., 7 Ridgmount St, London WC1E 7AE, UK
London Mathematical Society	London Mathematical Society, Burlington House, Piccadilly, London W1V 0NL, UK
Longmans	Longmans Group Ltd., Longman House, Burnt Mill, Harlow, CM20 2JE, UK
M & H	Médécine et Hygiène, Geneva, Switzerland
M & T	Markt und Technik Verlag GbmH, D8013 Haar, Hans-Pinsel Str. 2, Germany
MIT	The MIT Press, 28 Carleton St, Cambridge, MA 02142, USA
MUP	Manchester University Press, Oxford Rd, Manchester M13 9PL, UK
Macdonald	Macdonald & Co. (Publishers) Ltd., 165 Great Dover St, London SE1 4YA, UK
Macdonald&Jane's	see Macdonald
Macmillan	Macmillan Publishers Ltd., 4 Little Essex St, London WC2R 3LF, UK
Macmillan NY	Macmillan Publishing Co., 866 Third Ave, New York, NY 10022, USA
Mansell	Mansell Publishing Ltd., Bloomsbury Place, London WC1A 2OA, UK
Marquis	Marquis Who's Who Inc., 200 East Ohio St, Chicago, IL 60611, USA

Martinus Nijhoff	see Nijhoff
Mason	Kenneth Mason Publishers Ltd., Havant, PO9 1EF, UK
Masson	Masson et Cie SA., 120 blvd. Saint Germain, 75280 Paris, France
Math.Sci.Pr.	Mathematical Sciences Press Inc., 53 Jordan Road, Brookline, MA 02146, USA
McGraw-Hill	McGraw-Hill Book Co., 1221 Avenue of the Americas, New York, NY 10020, USA; and McGraw-Hill House, Maidenhead, SL6 2QL, UK
Merrill	Charles E. Merrill Publishing Co., Columbus, OH 43216, USA
Methuen	Methuen and Co. Ltd., 11 New Fetter Lane, London EC4P 4EE, UK
Michigan St.U.Pr.	Michigan State University Press, c/o TABS, 24 Red Lion Square, London WC1R 4PX, UK
Mills and Boon	Mills and Boon Ltd., 18 Paradise Road, Richmond, TW9 1SR, UK
Milton	Milton Publishing Co., London, UK
Min.Soc.	Mineralogical Society of Great Britain and Ireland, 41 Queen's Gate, London SW7 5HR, UK
Mono	Mono Book Corporation, Baltimore, MD, USA
Moos	Heinz Moos Verlag, München, Germany
Morgan-Grampian	Morgan-Grampian (Publ) Ltd., 30 Calderwood St, London SE18 6QH, UK
Multiscience	Multiscience Publications Ltd., 1253 McGill College, Montreal, Quebec, Canada H3B 2Y5
Munksgaard	Munksgaard International Publishing Co., 35 Norre Sogade, DK 1370, Copenhagen K, Denmark
NAS	National Academy of Sciences (US), 2101 Constitution Avenue, Washington, DC 20418, USA
NASA	National Aeronautics and Space Administration, (documents distributed through NTIS, ESA, BLDSC, etc.)
NATO-ASI	NATO Advanced Study Institute Program, B-1110 Brussels, Belgium
NCC	NCC Ltd., Oxford Road, Manchester M1 7ED, UK
NISC	National Information Sources Corp., Suite 6, Wyman Towers, 3100 St Paul Street, Baltimore, MD 21218, USA
NIST	National Institute for Standards, US Department of Commerce, Washington, DC 20234, USA
NRCC	National Research Council Canada, Ottawa, Ontario K1A OR6, Canada
NRLSI	National Research Library for Science and Invention, see SRIS

NTIS	National Technical Information Service, 5285 Port Royal Road, Springfield, VA 22161, USA
NUREG	Nuclear Regulatory Commission, Washington, DC 20555, USA
NYAS	New York Academy of Sciences, 2 E63S, New York, NY 10021, USA
Natl.Bur.Stand.	see NIST
Nauka	Izdatel'stvo 'Nauka', 117864 GSP-7, Moskva B-485, Profsoyuznaya ul. D.90, Russia
Naval Res.Lab.	Naval Research Laboratory, Shock and Vibration Information Centre, Washington, DC 20375, USA
Nelson	Thomas Nelson & Sons Ltd., Sunbury on Thames, Middlesex, TW16 7HP, UK
New Am.Lib.	New American Library Inc., 1301 Ave of the Americas, New York, NY 10019, USA
Newnes	see Butterworths
Newnes-Butterworths	see Butterworths
Nijhoff	Martinus Nijhoff BV, PO Box 566, 2501 CN-The Hague, Netherlands
Noordhoff	see Sijt.Noord.
North-Holland	The North-Holland Publishing Co., PO Box 211, 1000 AC, Amsterdam, Netherlands
ONR	Office of Naval Research, Department of Navy, Washington, DC 20375, USA
OSA	Optical Society of America, Washington, DC 20036, USA
OSI	The Optical Society of India, Applied Physics Dept., Calcutta University, 92 Acharya Prafulla Chandra Rd, Calcutta 700009, India
OUP	Oxford University Press, Walton Street, Oxford OX2 6DP, UK; and Ely House, 37 Dover St, London W1X 4AH, UK
Oak Ridge	DOE Technical Inf. Centre, Oak Ridge Associated Universities, PO Box 117, Oak Ridge, TN 37830, USA
Odhams	Odhams Books, c/o Hamlyn Publishing Group Ltd., 81 Fulham Rd, London SW3 6RB, UK
Ohio State Univ.	Ohio State University, Columbus, OH 43210, USA
Oldbourne	Oldbourne Books, London, UK
Oldenbourg	R. Oldenbourg Verlag GmbH, W-8000 München 80, Rosenheimer Strasse 145, Germany
Oliver and Boyd	Oliver and Boyd, Edinburgh EH1 3BB, Scotland
Olms	Georg Olms Verlag GmbH, W-3200 Hildesheim, Hagentorwall 6-7, Germany
Open	Open Court Publishing Co., PO Box 599, 315 5th St., Peru, IL 61354, USA
Optical Pub.Co.	Optical Publishing Co., PO Box 1146, Pittsfield, MA 01201, USA

Optosonic	Optosonic Press, P.O. Box 883, Ansonia Sta., New York, NY 10023, USA
Orell Fussli	Orell Fussli Verlag, CH-8022 Zurich, Nuschelertrasse 22, Switzerland
Oriel	Oriel Press Ltd., Stocksfield, NE43 7NA, UK; and PO Box 12373, Portland, OR 97212, USA
Oryx	Oryx Press, Phoenix, AZ 85012, USA
Oyez	Oyez Publ. Ltd., 11/13 Norwich St, London EC4A 1AB, UK; and 212 Colgate Ave, Kensington, CA 94707, USA
Oxbow	Ox Bow Press, PO Box 4045, Woodbridge, CT 06525, USA
PRM	PRM Science and Technology Agency Ltd., 261a Finchley Rd, Hampstead, London NW3 6LU, UK
PASCAL	see INIST
Pare	J. R. J. Pare, Establishment for Chemistry Ltd., 1245 Walkley Rd, Suite 1103, Ottawa, ON KIV 9S5, Canada
Parjon	Parjon Information Services, PO Box 144, Haywards Heath, RH16 2YX, UK
Parker	Parker Press Ltd., c/o Prentice-Hall Inc.
Patent Office	Patent Office, 25 Southampton Buildings, Chancery Lane, London WC2A 1AW, UK
Penguin	Penguin Books, Harmondsworth, Middlesex, UB7 0DA, UK
Penn.St.U.Pr.	Pennsylvania State University Press, University Park, PA 16802, USA
Pentech	Pentech Press, 4 Graham Lodge, Graham Rd, London NW4 3DG, UK
Peregrinus	Peter Peregrinus Ltd., c/o INSPEC
Pergamon	Pergamon Press Ltd., Headington Hill Hall, Oxford OX3 0BW, UK; also see Macmillan NY
Perkin-Elmer	Perkin-Elmer Co., Norwalk, CT 06856, USA
Peter Davies	Peter Davies Ltd., 10 Upper Grosvenor St, London W1X 9PA, UK
Petrocelli/Charter	see Van Nostrand
Phil.Libr.	Philosophical Library Inc., 31 W21 St, New York, NY 10010, USA
Philips	Philips Research Laboratories, Eindhoven, Netherlands
Philos.Sci.Assn.	Philosophy of Science Association, Philosophy Dept., Michigan State University, 18 Morrill Hall, East Lansing, MI 48824, USA
Phys.Soc.Jpn.	Physical Society of Japan, Room 211, Kikai Shinko Building, 3–5–8 Shiba Koen, Minato-ku, Tokyo 105, Japan
Physik Verlag	Physik Verlag GmbH, W-6940 Weinheim, Pappelallee 3, Germany

Pineridge	Pineridge Press Ltd., Swansea, West Glamorgan, SA3 SPD, UK
Pion	Pion Ltd., 207 Brondesbury Park, London NW2 5JN, UK
Pitman	Pitman Books Ltd., 128 Long Acre, London WC2E 9AN, UK
Plenum	Plenum Press, 233 Spring St, New York, NY 10013–1578, USA
Pol.Akad.Nauk.	Polska Akademia Nauk, Reymonta 4, 30–059 Krakow, Poland
Pol.Wroclawska	Politechnika Wroclawska, 50–370 Wroclaw, Wybrzeze Wyspianskiego 27, Poland
Prentice-Hall	Prentice-Hall Press, Englewood Cliffs, NJ 07632, USA; also 15 Columbus Circus, New York, NY 10023, USA
Preston	Preston Publications Inc., P.O. Box 48312, Niles, IL 60648, USA
Princeton U.Pr.	Princeton University Press, Princeton, NJ 08540, USA
RAS	Royal Astronomical Society, Burlington House, Piccadilly, London W1V 0NL, UK
READEX	Readex Books, Division of Readex Microprint Corpn., 58 Pine Street, New Canaan, CT 06840–5408, USA
RSC	Royal Society of Chemistry, Burlington House, Piccadilly, London W1V 0BN, UK
Rahara	Rahara Enterprises, El Cerrito, CA, USA
Reidel	D. Reidel Publishing Co., PO Box 17, 3300 AA Dordrecht, Netherlands
Reinhold	See Van Nostrand
Res.Inst.Fund.Phys.	Research Institute for Fundamental Physics, Yukawa Hall, Kyoto University, 606 Kyoto, Japan
Res.Stud.Pr.	Research Studies Press Inc., c/o John Wiley & Sons, 605 Third Ave, New York, NY 10158, USA; and 8 William Way, Letchworth, Herts SG6 2HG, UK
Reston	Reston Publishing Co. Inc., Reston, VA 22090, USA; see also Prentice-Hall
Richelieu	The Richelieu Press, 30 Saint Mark's Crescent, London NW1 7TU, UK
Ronald	The Ronald Press Co., 79 Madison Ave, New York, NY 10016, USA
Routledge	Routledge & Kegan Paul, 11 New Fetter Lane, London EC4P 4EE, UK; and c/o Routledge, Chapman & Hall, 29 W35 St, New York, NY 10001–2291, USA
Rowman	Rowman & Littlefield Inc., 4720 Boston Way, Lanham, MD 20706, USA
Roy.Swed.Acad.Sc.	Royal Swedish Academy of Sciences, PO Box 50005, S-104 05 Stockholm, Sweden

Royal Society	The Royal Society, 6 Carlton House Terrace, London SW1Y 5AG, UK
Rutherford	Rutherford-Appleton Laboratory (SERC) Chilton, Didcot, Oxon OX11 0QX, UK
SIAM	Society for Industrial and Applied Mathematics, 3600 City Science Center, Philadelphia, PA 19104–2688, USA
SLAC	Stanford Linear Accelerator Center, PO Box 4349, Stanford, CA 94305, USA
SML	Science Museum Library, South Kensington, London SW7 2DD, UK
SPIE	Society of Photo-optical Instrumentation Engineers, 1000 20th St, Bellingham, WA 98225, USA
SRIS	Science Reference and Information Service, c/o BL
STN	STN International, PO Box 2465, W-7500 Karlsruhe, Germany; also c/o Chemical Abstracts Service, 2540 Olentengy River Road, Box 3012, Columbus, OH 43210, USA
SUNIST	SUNIST, 950 rue Saint-Priest, BP 1229, F-34184 Montpellier Cedex 4, France
SUSSP	Scottish Universities Summer School Publications, Edinburgh University Physics Department, Kings Buildings, Mayfield Road, Edinburgh EH8 9YL, Scotland
Saunders	W. B. Saunders Co., Philadelphia, PA 19106–3399, USA; and c/o Holt Saunders, Eastbourne, BN21 3UN, UK
Saur	K. G. Saur Ltd., 59–60 Grosvenor St, London W1X 9DA, UK
Sci.Res.Ass.	Science Research Association, Commonwealth House, New Oxford St, London WC1A 1NE, UK
Science History	Science History Publications Ltd., PO Box 493, Canton, MA 02021, USA
Science Press	Science Press Inc., 300 W Chestnut St, Ephrata, PA 17522–0497, USA
Scottish Acad.Pr.	Scottish Academic Press, 56 Hanover St, Edinburgh EH2 2DX, Scotland
Scribner	Charles Scribner's Sons, c/o Macmillan NY
Scripta	Scripta Publishing Corporation, 1511 K St NW, Washington, DC 20005, USA
Shiva	Shiva Publishing Ltd., Nantwich, Cheshire, CW5 SRQ, UK
Sijt.Noord.	Sijthoff and Noordhoff Int. Publ. B.V., PO Box 4, 2400 Alphen-aan-denn-Rijn, Netherlands
Soc.Bibl.	Societas Bibliographica, 7 rue de Genève, Lausanne, Switzerland
Soc.Esp.Opt.	Sociedad Espanola de Optica, Serrano 121, Madrid 6, Spain

Soc.Inform. d'Optique	Française Société Informe D'Optique, 10 rue de Buci, 75006 Paris, France
Soc.Ital.Fis.	Societa Italiana di Fisica, Vialle XII Guigno 1, 40124 Bologna, Italy
Soc.Sci.Fenn.	Societas Scientiarum Fennica, Snellmansgatan 9–11, 00170 Helsinki 17, Finland
Soc.Tel.Eng.	Society of Telegraph Engineers, London, UK
Sotheby	Sotheby Parke Bernet Publishers Ltd., c/o Philip Wilson Publishers, 26 Litchfield St, London WC2H 9NJ, UK
Sourcebook	The Sourcebook Project, PO Box 107, Glen Arm, MD 21057, USA
Spartan	Spartan Books, Hayden Book Co., Rochelle Park, NJ 07662, USA
Spex	Spex International Ltd., 51 E42S, Suite 517, New York, NY 10017, USA
Spon	E. and F.N. Spon Ltd., 11 New Fetter Lane, London EC4P 4EE, UK
Springer	Springer Verlag, Heidelberger Platz 3, W-1000 Berlin, Germany; also W-6900 Heidelberg, Germany and 175 Fifth Avenue, New York, NY 10010, USA
Steinkopff	D. Steinkopff, Darmstadt, Germany
Sussex U.Pr.	Sussex University Press, Refectory Terrapin, Falmer, Brighton BN1 9QZ, UK
Sweet&Maxwell	Sweet and Maxwell Ltd., 11 New Fetter Lane, London EC4P 4EE, UK
TFPL	TFPL Publishing, 22 Peter's Lane, London EC1M 6DS, UK; and 1301, Twentieth Street NW, Washington, DC 20036, USA
Taylor&Francis	Taylor and Francis, 4 John St, London WC1N 2ET, UK
Technical	Technical Publishing Co., 1301 South Grove Ave, Barrington, IL 60010, USA
Technomic	Technomic Publ. Co. Inc., 851 New Holland Ave, Lancaster, PA 17604, USA
Teubner	Teubner, Leipzig, Germany
Tex.A&M.U.Pr.	Texas A&M University Press, College Station, Texas, TX 77843, USA
Texas Insts.	Texas Instruments Inc., PO Box 655303–8338, Dallas, TX 75265, USA
Therm.Res.Cen.	Thermodynamics Research Centre, Texas A&M University, College Station, Texas, TX 77843, USA
Trade Activities	Trade Activities Inc., New York, NY 10001, USA
Trade&Technical	Trade and Technical Press, Morden, Surrey, UK
UB&TIB	University Library & Technical Information Bureau, Am Welfengarten lb, D-3000, Hannover 1, Germany

UMI	University Microfilms International, 300 N Zeeb Rd, Ann Arbor, Michigan, MI 48106, USA
U.Nat.Aut.Mexico	Universidad National Autonoma de Mexico, Ciudad Universitaria, Alvaro Obregon, 04510 Mexico, DF
U.New England	University of New England Publ., Armidale, NSW 2351, Australia
U.Pr.New Eng.	University Press of New England, PO Box 979, Hanover, NH 03755, USA
U.of Arizona Pr.	University of Arizona Press, PO Box 3398, College Station, Tucson, AZ 85719, USA
U.of Cal. Press	University of California Press, Berkeley, CA 94720, USA
U.of Chicago Pr.	University of Chicago Press, 5801 Ellis Ave, Chicago, IL 60637, USA
U.of London	University of London, Senate House, Malet St, London WC1E 7HU, UK
U.of Minn.Pr.	University of Minnesota Press, Minneapolis, MN 55414, USA
U.of Notre Dame Pr.	University of Notre Dame Press, Notre Dame, IN 46556, USA
U.of Pittsburgh Pr.	University of Pittsburgh Press, Pittsburgh, PA 15260, USA
U.of Texas Pr.	University of Texas Press, PO Box 7819, University Station, Austin, TX 78713–7819, USA
U.of Toronto Pr.	University of Toronto Press, Toronto, Canada M4Y 2W8
USAEC	United States Atomic Energy Agency, Technical Information Center, Oak Ridge, TE 37830, USA
U.S.Gov.Pr.Office	United States Government Printing Office, Washington, DC, USA
United Technical	United Technical Publications, 645 Stewart Ave, New York, NY 11530, USA
Univ. Science	University Science Books, 20 Edgehill Rd, Mill Valley, CA 94941 USA
University	University Tutorial Press Ltd., 842 Yeovil Rd, Slough SL1 4JQ, UK
VCH	VCH Publishers, Postfach 101161, W-6940 Weinheim, Germany
VDE	VDE Verlag GmbH, Bismarkstrasse 33, D-1000 Berlin 12, Germany
VDI Verlag	VDI Verlag GmbH, 4 Dusseldorf 1, Postfach 1139, Germany
VINE	VINE, c/o Information Technology Centre, University of Westminster, 309 Regent St, London W1R 8AL, UK
VINITI	Vsesoyuznyy Institut Nauchnoy i Teknicheskoy Informatsii, Baltiiskaia ul. 14, Moscow A-219, Russia

VSP	VSP BV, Postbus 346, 3700 Zeist, Germany
Van Nostrand	Van Nostrand Reinhold Co., c/o Thompson Publishing Corp., 115 Fifth Avenue, New York, NY 10003, USA
Verlag Chemie	Verlag Chemie GmbH, D-6940 Weinheim, Pappelallee 3, Germany
Verlag Dokument.	Verlag Dokumentation (now) K. G. Saur Verlag, 8023 Pullach bei München, Kaiserstrasse 13, Germany
Vieweg	Friedrich Vieweg and Sohn, Braunschweig am Wiesbaden, Germany
Vrin	Librairie Philosophique J. Vrin, 6 Place de la Sorbonne, 75005 Paris, France
Weidenfeld	Weidenfeld and Nicolson, 91 Clapham High St, London SW4 9TA, UK
Where to Buy	FMJ International Publications, Queensway House, 2 Queensway, Redhill RH1 1QS, UK
Whitaker	J. Whitaker & Sons, 12 Dyott St, London WC1A 1DF, UK
Who's Who	See Marquis
Wiley	John Wiley and Sons Inc., 605 Third Ave, New York, NY 10158–0012, USA; and Baffins Lane, Chichester, PO19 lUD, UK
Wiley-Heyden	Wiley-Heyden Journals, Chichester, West Sussex, PO19 lUD, UK
Wiley-Interscience	A division of John Wiley
Wilson	The H. W. Wilson Co., 950 University Ave, Bronx, New York, NY 10452, USA
Wisconsin	Wisconsin University Press, 114 North Murray St, Madison, WI 53715–1199, USA
Wiss.Verlag.	Wissenschaftliche Verlagsgesellschaft, W-7000 Stuttgart 1, Postfach 40, Germany
Wolters-Noordhoff	Wolters-Noordhoff B. K., PO Box 58, Grøningen, Netherlands
World Scientific	World Scientific Publishing, 1022 Hougang Avenue 1, NO 05–3520c, Tai Seng Industrial Estate, Singapore 1953
Wyd.Nauk-Tech.	Wydawnictwa Naukowo-Techicze, Ul. Nazowiecka 2–4, PO Box 359, Warsaw, Poland
Wykeham	Wykeham Books Ltd., c/o Taylor and Francis
Yale Univ.Obs.	Yale University Observatory, Newhaven, CT 06520, USA
Yourdon	Yourdon Press, Prentice-Hall Building, Englewood Cliffs, NJ 07632, USA
ZPS	Zhurnal Prikladnoi Spectroskopii, Institut Fiziki AN BSSR, Leninskii Prospekt 70, Minsk GSP, Byelorussia.

Subject Index

Note: In the alphabetization of entries conjunctions and articles are ignored.